The Amateur Astronomer's Handbook

THE
Amateur Astronomer's
HANDBOOK

Third Edition

James Muirden

PERENNIAL LIBRARY

HARPER & ROW, PUBLISHERS, New York
Cambridge, Philadelphia, San Francisco, Washington
London, Mexico City, São Paulo, Singapore, Sydney

For J. L.,
watcher of the skies

NOTE: This book is written primarily for readers in the northern hemisphere. In the southern hemisphere some of the compass directions given here may have to be reversed, as indicated in the text.

ACKNOWLEDGMENTS

A great number of people have assisted me, in various ways, in the writing of this book, but I am particularly indebted to Dr. W. H. Steavenson and R. M. Baum, two British amateur observers of distinction, who made valuable comments on the first draft of the manuscript. In more general terms I must thank Patrick Moore, the well-known writer and broadcaster, for his help and encouragement over the years.

British, American, and Continental observers have been of material assistance in supplying notes and photographs, and acknowledgment, where due, is made in the text.

I must finally thank my many Greek friends, particularly Giorgos Markoulakis, for their over-flowing hospitality during the time I spent amongst them, observing the sky and writing this book.

Library of Congress Cataloging-in-Publication Data

Muirden, James.
  The amateur astronomer's handbook.

  "Perennial Library."
  Bibliography: p.
  Includes index.
  I. Astronomy—Observers' manuals. I. Title.
QB64.M85 1987   520   81-48044
ISBN 0-06-181622-1   87 88 89 90 91 MPC 10 9 8 7 6 5 4 3 2 1
ISBN 0-06-091426-2 (pbk.)   89 90 91 MPC 10 9 8 7 6 5 4 3

Contents

Foreword to the Third Edition

Progress in astronomy, particularly planetary astronomy, has been so great since the previous edition of this work appeared that extensive revision has been necessary. Space vehicles, both *Pioneer* and *Voyager,* have brought us unbelievable views of Jupiter and Saturn and their satellites, while Mercury, Venus, and Mars have also been visited.

If the planets were static bodies, it would be easy to argue that further amateur observation (interesting and fun though it might be) was futile. But, with the exception of Mercury—which is an elusive world anyway—this is not so. Continuous observation to monitor change is vital; the brilliant glimpse is sensational, but the steady gaze brings its results too. Planetary observation, therefore, can be as useful as ever.

Another field that has been transformed since the first edition of this book, astronomical photography, is now given a section of its own. Modern photographic emulsions are so sensitive that an ordinary camera can record all the naked-eye stars in the blink of an eye. Amateurs with only moderate equipment are now able to embark on serious planetary and stellar programmes. In this new section, I have tried to cover the field in sufficient depth for it to be of value both to the beginner and to the more experienced photographer.

Apart from other revisions and amendments necessitated by the passage of time, the opportunity has been taken of correcting errors noted by readers and reviewers, and of altering some of the illustrations.

In writing this book, I have drawn very largely on the work of other observers. Acknowledgement is made elsewhere to those amateur astronomers who have kindly offered much practical assistance; but it would be a mistake—and is, indeed, an error often made by modern writers on the subject—to overlook the wealth of experience accumulated by those observers who lived before this modern age of research. Times have changed, of course, but not everything has changed with the times. The technique of telescopic observation always has been and always will be based on unending practice and training of the eye. In this connection, the great observers of the past were certainly no less competent than the leading ones of today; especially when we consider the difficulties with which they coped. One cannot become a successful amateur astronomer without the inspiration that is kindled from a love of the night sky and an unshakable determination to succeed.

This book should not be considered as more than a guide for the would-be observer. I have, however, tried to bridge the gap between works that deal with the spectacular "wonders of the sky" and the formal handbooks full of facts. The latter are of great value once a start has been made, but in themselves they are almost meaningless. In short, this is intended to be a survey of the *technique* of amateur astronomy, from the selection of an instrument to the conduct of actual observation. A certain amount of basic astronomical knowledge has been assumed, since no one book can cover everything, and many have failed from trying to be too ambitious!

We must always remember, however, that astronomy has much more to offer us than the delight of scientific investigation. The element of wonder and frank incomprehension is, in many ways, of far more personal account than an explanatory theory. The unknown being more fascinating than the known, I hope my readers will feel inclined to pay as much attention to those branches of observation in which they have no hope of pronouncing a "verdict" as they do to the more orthodox lines of research. By all means let us continue the amateur's tradition of observing Venus, Jupiter, and variable stars, and making valuable reports, but let us also never forget that astronomy loses half its meaning for the observer who never lets his telescope range across the remote glories of the sky "with an uncovered head and humble heart."

I should like to express my thanks to Ian Ridpath and Robin Scagell for drawing my attention to some errors in the previous edition.

JAMES MUIRDEN

PART

I

EQUIPMENT

1

The Telescope and Its Development

We live in the Age of Leisure; and, in this age, recreation has become part of man's estate. There are a thousand pastimes open to us, most of which are not technically "useful," but simply hobbies with which we occupy our spare hours. Yet it is curious that in a world revolutionized by science and technology, so few people consider turning to science in their leisure time.

Perhaps this is not so curious when we remember that the golden age of the amateur scientist has long since passed into history. Universities are now turning out technologists by the thousand, and it is clear that no amateur scientist would have the basic knowledge, let alone the facilities, to perform original work in most of the modern sciences. But there are still a few openings for research, in which the amateur can play a useful, even vital, part. Astronomy, geology, and archeology come to mind at once—and of these, astronomy is perhaps the most demanding and rewarding. Who has never wished, even subconsciously, that he might probe the secrets of a starry night? For there is art as well as science here, and the amateur— humble though his telescope may be—is, in a sense, as well equipped as the professional.

It is mainly the art of astronomy that appeals to the spare-time observer with a small telescope, an art that usually reveals itself in terms of perseverance. Indeed, this is the demarcation between the two classes of astronomer, and it explains why the amateur still has a part to play. For the night sky, which appears so placid and unchanging to the casual glance, is really a turmoil of activity: planets spin on their axes, meteors fly, stars slowly brighten or fade, and sometimes a comet creeps into view from the blackness of outer space. A professional astronomer, working on some particular

3

problem, may not even glance at the sky for weeks on end. But amateurs are tied by no such schedules; they can let their attention wander according to the state of things. Scavenging in this way, they sometimes emerge with quite remarkable prizes, prizes that might otherwise have passed through our hands without notice and without profit.

The fact is that observations requiring endless persistence, and not necessarily needing complex equipment, are highly suitable for the amateur. It must be remembered that a professional astronomer is employed by his university or observatory to get through a certain amount of work in a given time, and this time is obviously severely rationed by the need of other workers to use the same telescope. This means that he is mainly confined to those lines of research likely to show a fairly swift return. No such necessity dogs the amateur. His time is his own, his telescope is his own, and he can observe the same object for years on end if he so wishes, waiting for something unusual to happen. If it does, the chances are that he will be the first to notice it.

A good example of this is Earth's neighbor planet, Jupiter. A small astronomical telescope will show a great amount of detail on its surface, but consistent watching over many months or years is required before valuable results are likely to emerge. This means that the observation of Jupiter's markings is almost entirely in the hands of amateurs scattered all over the world. Even Galileo was technically an amateur when, on that memorable night in January 1610, he pointed his primitive "optick tube" at Jupiter and discovered four bright satellites. Nowadays we can see these bodies through a pair of binoculars, but the wonder of that original observation remains.

Progress in astronomy has gone hand in hand with instrumental developments. It may therefore be enlightening to look back across the 350 years that have passed since Galileo's first peek at the stars. One wonders with what expectations he awaited that first nightfall, his tiny leaden-tubed telescope lying at hand. Did he realize that he was about to plunge into an uncharted universe a hundred times larger than the one that had absorbed men's minds for the past five thousand years? What were his feelings when he began to sweep the sky and found that for every star visible with the naked eye, his telescope showed a dozen? Surely, even at that early stage, he must have started thinking in terms of bigger and bigger telescopes, a yearning shared by every astronomer since. The greater the aperture of the telescope, the fainter and more distant are the stars it will reveal—and the larger the observable universe!

It is not hard to see why this should be. Under perfectly dark conditions, the iris of the human eye has an opening about $\frac{1}{3}$-inch across, with which it can detect stars down to a certain degree of faintness. The luster of the stars is calibrated in terms of *magnitude*, a term that refers to *brightness*, not to size; broadly speaking, the brightest stars in the sky are said to be of zero magnitude, while the faintest visible with the naked eye are of the 6th magnitude, the magnitude number rising with increasing faintness. If we now apply the eye to a telescope whose main lens, or *object glass*, has an aperture of one inch, we shall see a star as nine times brighter, because the light forming the image has been collected from an area nine times greater than that of the human iris. Correspondingly, the faintest star visible through the telescope will be nine times fainter than a 6th-magnitude star. Since a magnitude division represents a step in brightness of about $2\frac{1}{2}$ times, a 1-inch aperture telescope should reveal stars as faint as the 9th magnitude.

Early refractors

Galileo's telescope, which used a lens to gather and focus the light from a star, is known as a *refracting* telescope, or *refractor*. The word refraction refers to the bending of a ray of light when it passes from one transparent medium (in this case, air) through another of different density (in this case, glass). The principle on which the refractor works is shown in figure 1. The light from the star passes through the object glass, is refracted into a cone, and comes to a focus at the bottom of the tube. A star is rather faint for the purpose, but it is easy enough to catch the focal image of the sun or moon on a sheet of paper. However, when the telescope is being used for visual observation, an *eyepiece*, or *ocular*, is employed to magnify the image. This is a small lens placed close to the focal point and adjusted in position until a sharp view is obtained.

This eyepiece acts like a microscope. Just as a magnifying glass amplifies whatever is placed beneath it, so the eyepiece magnifies the image formed by the object glass, and gives a close-up view of the object concerned. Galileo's largest telescope had an aperture of about 2 inches and magnified some 30 times (written × 30). With this instrument, which is still on view in the Museum of Physics in Florence, he not only discovered the satellites of Jupiter but also saw the lunar plains and craters, spots on the sun, and the phases of Mars and Venus. He even made out some peculiar appendages to the planet Saturn. These were later found to be the rings, but he himself died in 1643 without realizing just what they were.

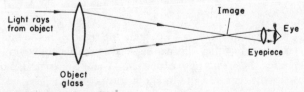

Figure 1. *Principle of the refractor.*

It is often claimed that Galileo was the first telescopic observer, but this is certainly incorrect. Simon Marius in Germany, as well as Thomas Harriot and Sir William Lower in England, was experimenting successfully during the same period. Harriot, indeed, unquestionably preceded Galileo in lunar observation, since he made a sketch as early as July 1609. Yet the Italian's work was of greater consequence, for all his observations were brought into the public eye by the publication of his *Sidereal Messenger* (1610) and *Dialogue* (1632), which discussed his then radical conclusions about the nature of the solar system, which led to his tragic struggle against the disapproval of the Holy Office in Rome.

Chromatic aberration

It soon became apparent, however, that the simple refracting telescope suffered from a grave defect. The image of a star or the edge of a planet was suffused with a colored halo that destroyed sharpness and made it quite impossible to use high magnifying powers. A tiny 1- or 2-inch-aperture telescope gave reasonable results, but astronomers were already impatient for more light and more power.

The explanation of this unhappy state of affairs, not known at that time, is given in figure 2. White light is actually an amalgam of all the colors of the rainbow, which is produced when sunlight is refracted by drops of water in the atmosphere. In addition to focusing light rays, a lens also splits them up into spectral colors. The light of shortest wavelength (blue) is refracted to a greater degree than the long-wave red light; consequently, the image of a star is actually a number of images, each in a different color, at different distances from the object glass. Since it is impossible to focus all these colors sharply at the same time, the image is inevitably faulty. This defect is known as *chromatic aberration*.

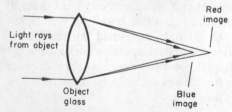

Figure 2. *Chromatic aberration. The effect is considerably exaggerated.*

Aerial telescopes

The early telescope-makers, not knowing the reason for the fault, could provide no real cure, but they did arrive at a workable compromise. This involved increasing the *focal length* of the lens. The focal length is simply the distance between the lens and the image it forms, and the simplest way of finding it is to form a sharp image of the sun on a sheet of paper and measure the distance between the lens and the paper. Galileo's 2-inch object glass had a focal length of about 3 feet; opticians now started grinding lenses of the same aperture, but with focal lengths of 30 feet, or even more. This had the beneficial result of separating out the images of different color, and so reducing the chromatic blur. The first systematic lunar observer, Johannes Hevelius (originally Hewelcke), of Danzig, used one of these *aerial telescopes* to produce the first reasonably accurate map of the moon, published in 1647. The Dutch observer Christian Huygens, well known as the constructor of the first practical pendulum clock, made an instrument giving such good definition that, in 1656, he discovered the true form of Saturn's rings.

Aerial telescopes, like all devices exploiting some form of distortion, enjoyed spectacular, if absurd, progress. Hevelius' main instrument was 150 feet long; Huygens constructed one more than 200 feet long, and a French optician is said to have constructed a 600-foot monster! But, however great their optical advantages, mechanical problems made them almost useless. They had to be used in the open air, since no observatory large enough to accommodate them could be built; only the calmest weather was suitable for their employment, and when they were not in use they had to be dismantled and stored. It is significant that the most memorable discoveries were achieved with relatively modest instruments. Huygens discovered Saturn's rings with a 90-foot telescope, but the Italian observer J. D. Cassini discovered the famous division in the rings, as well as two satellites, with a 17-foot instrument. It is a measure of the modern observer's good fortune that he can see all these features, and others more delicate, with great clarity in a telescope only three or four feet long!

The astronomers of the mid-seventeenth century must have looked at their unwieldy arrangements of masts and rigging and sighed for something more manageable, capable of better performance. But was there any alternative? How could a telescope be made that did not use an object glass? Failing that, how could an object glass be made that did not produce false color?

Early reflectors

It was a Scottish optician, James Gregory, who first suggested a plan for a *reflecting* telescope. As he pointed out in a small book published in 1663, a concave mirror will form an image in the same way as a lens, though by reflection rather than refraction; better still, it cannot produce any chromatic effect. Gregory created a design for what is known as a *Gregorian* reflector but, since his talents were confined to theory, did not

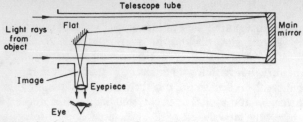

Figure 3. *Principle of the Newtonian reflector. The mirror's curve is greatly exaggerated.*

make such an instrument himself. In any case, the Gregorian reflector has rarely proved satisfactory for astronomical work. The world's first reflecting telescope, constructed in 1668 by none other than Isaac Newton, is still preserved in the apartments of the Royal Society in London.

The principle of the *Newtonian* reflector is shown in figure 3. The light from the star falls on the curved front surface* of the main mirror, or *speculum*, which reflects it back up the tube. Before it forms an image, however, it is again reflected, through a right angle, by a second mirror (this time plane), known as the *flat*. The light then passes through a hole cut in the side of the tube and comes to a focus where the eyepiece is positioned. Thus, instead of looking "up" the tube, as with a refractor, the observer peers more comfortably into the side, thus avoiding backbreaking contortions while observing an object high in the sky. There can be no doubt that of the two types, the reflector is the more comfortable instrument to use.

Newton's telescope was tiny, with a mirror only 1 inch across; astronomically it was useless, but he had shown the way. Opticians gradually turned their attention from the cumbersome refractor to the splendidly compact reflector, whose mirror could be ground to have a relatively short focal length. But the problems they faced were formidable. In the first place, the two mirrors had to be ground much more accurately than the surfaces of the object glass; secondly, there was the difficulty of finding a suitable material. Chemists had not yet discovered a way of precipitating silver out of a solution onto a glass surface, so the early mirrors had to be made of a substance known as *speculum metal* (an alloy of copper, tin, and other elements), which took a good polish but tarnished rapidly; it was also very fragile.

The first really workable speculum-metal reflector was made by an Englishman, John Hadley, in 1720. With the unheard-of aperture of 6 inches, and a focal length of only 6 feet, it caused a sensation; for, despite its handy convenience, it was equal in performance to any of Huygens' or

---

*All mirrors used in optical instruments reflect from the front surface, not from the second surface—as in the case of an ordinary looking glass, which is silvered on the back.

Cassini's unwieldy instruments. Soon after this, James Short, of Edinburgh, developed a technique for the production of fine optical surfaces and set himself up as a manufacturer of reflecting telescopes. By 1740, Short's telescopes were much in demand; but the finished instruments were in no sense permanent, despite their good performance. When their mirrors became tarnished they had to be repolished, and this process was quite enough to affect the delicate "figuring" of the surface, so that the optician had to be called in again to restore the original curve. There was no cure for this. The observer of those days had to choose between the unwieldy refractor or the efficient but impermanent reflector, finding no real satisfaction in either.

Achromatic refractors

However, the second half of the eighteenth century saw spectacular progress. The resurgence of the refractor began when Chester Moor Hall, an English "gentleman scientist," began looking into the root causes of chromatic aberration. After some experimenting, he actually managed to construct an object glass, $2\frac{1}{2}$ inches across and only 20 inches focal length, that produced a relatively colorless image. This he achieved by making the lens from two different kinds of glass: one of low density (crown), the other of high density (flint). The chromatic light produced by the crown lens in focusing the light rays was then recombined into white light by the flint lens, as is shown in figure 4.

This compensating effect occurs because the two lenses are curved in the opposite sense. When a beam of white light passes through a prism, it is dispersed into a spectrum; correspondingly, a spectrum can be recondensed into white light if passed through a suitable prism. It is not hard to see that a beam of white light will emerge as white light if it passes through two correctly arranged prisms; this, in effect, is what happens when it is refracted by the two components of an achromatic object glass.

This was a sensational discovery. The object glass of a refractor could now be worked to almost as convenient a focal length as the mirror of a reflecting telescope. Yet, for some unknown reason, Hall chose to keep this

Figure 4. *Achromatic object glass. The distance between the crown and flint components is increased for the sake of clarity.*

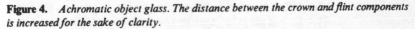
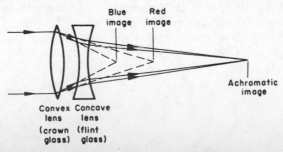

discovery to himself, and it was not until 1758 that the English optician John Dollond grasped the principle and began manufacturing *achromatic* refractors on a commercial scale. We can imagine with what relief astronomers consigned their ungainly aerial telescopes to the rubbish pile, and seized on Dollond's compact and efficient refractors! The new telescopes presented no tarnishing problem, and the definition was even better than that of Short's reflectors. Yet there was one big drawback: glassmakers could not yet cast large disks of glass of sufficient perfection to make big lenses, and a refractor with an object glass of more than 3-inch aperture was a rarity indeed. Therefore, reflectors, having no such drawback, remained on an equal footing. Shortly they were to receive an unprecedented boost when a Hanoverian musician, settled in England, decided to try his hand at telescope making and concentrated, for reasons of light-grasp, on the reflector. His name was William Herschel.

William Herschel

There are few fields of human endeavor in which one cannot point to one figure, whether past or present, as a prototype. The giant of observational astronomy is Herschel; in him, the separate but dependent arts of instrument making, observation, and analysis were combined in the highest degree. Absorbed by the unfathomed challenge of the sky, and faced, for want of money, with the necessity of making his own instruments, he brought the reflecting telescope to so high a pitch of perfection that its performance, in the 1780's, was comparable to that of the best modern instruments. In 1789, Herschel erected in his garden at Slough, England, a monster telescope with a mirror *48 inches across*. This remained the most powerful telescope in the world until, in 1845, Lord Rosse built his 72-inch reflector in Ireland. It is a sobering thought that Herschel's telescope was bigger than any in use in the British Isles today, since the removal of the 98-inch reflector from the Royal Greenwich Observatory.

Herschel, who built other instruments of smaller aperture, was for some time unaware of the excellence of his telescopes. Not until 1782, when visiting Greenwich Observatory in London, did he have a chance to compare one of his 6-inch reflectors with a $9\frac{1}{2}$-inch reflector made by Short. Its mirror was markedly inferior to Herschel's, despite its extra aperture, so Herschel decided to make telescopes for sale, in order to subsidize his more ambitious projects. Since dozens of these first-rate instruments were distributed throughout the world, one might imagine that this would have been of wide benefit to observational astronomy. But nothing could be further from the truth. Only two observers—Pond in England and Johann Schröter in Germany—put them to worthwhile use; the others fell into the hands of dilettantes who had neither talent nor enthusiasm, both of which are part of the equipment of the observational astronomer. Indeed, Herschel might have exhausted himself by working on the means to the end had not King George III financed his further astronomical labors with a royal grant.

The lesson to be drawn from this instance is still applicable. During the last ten or fifteen years, hundreds of firms have been marketing astronomical equipment, some of it worthless, but much of very high quality. Yet this easy availability is not in itself of the slightest benefit to amateur astronomy. If a person has enough enthusiasm, he will find a way of obtaining a telescope; if he is deterred by immediate difficulties, then he will never make a first-class observer!

Resurgence of the refractor

Yet at the same time that Herschel was extending the limits of both his telescopes and his observations, Pierre Guinand, a Swiss optician, was laying the foundation for the resurgence of the refractor, whose limiting factor had proved to be the glass. If a lens is to refract light truly, its material must be perfectly homogeneous, without any flaws or strains that will deflect the tiny star-beams from their correct path; and in Guinand's day, pieces of optical glass more than 2 or 3 inches across were a great rarity. There were several reasons for this: the poor quality of the raw materials, the technique of heating and stirring the molten glass, and the final and immensely important *annealing* process. This is the gradual cooling of the red-hot mass, a process that must take place very slowly and regularly. Uneven cooling will produce disastrous flaws, and possibly cracks; and the larger the pot, the longer the annealing takes. Guinand was the first to study these various aspects of glassmaking; after many years of patient work he saw produced optical disks 4, 6, and even 8 inches across. Now the stage was set for someone with the skill and patience to grind such disks into fine lenses that would justify the caster's efforts.

It was a once poor orphan of Munich, Joseph von Fraunhöfer, who ushered in the age of the modern refracting telescope. Supported in his research by money compensating him for an almost fatal accident in his youth, he soon designed lenses on a slightly different pattern from Dollond's, although the basic crown-flint principle was the same. Joining forces with Guinand, he turned out a series of the finest object glasses the world had yet seen. This labor culminated in the completion of one $9\frac{1}{2}$ inches across, which was mounted in the Dorpat Observatory in eastern Estonia in 1824 and remained for several years the largest refractor in the world.

So far, the story of the telescope's development has been an exclusively European one. But it was not destined to remain so. Soon after the completion of Fraunhöfer's Dorpat refractor, Alvan Clark, a portrait painter living in Cambridge, Massachusetts, became intrigued by his son's efforts to grind an astronomical mirror. The two of them finally succeeded in completing it, only to find that tarnish immediately set in. Discouraged by the caprices of speculum metal, but absorbed by the principles involved, Alvan Clark began making object glasses. By 1853, news of their excellence had reached astronomers throughout the world.

Now it so happened that the founding of Alvan Clark & Son coincided with a sudden awakening of American interest in astronomy. Observatories were springing up in various universities—one of the first was at Harvard College, in Clark's home town—and, since the reflector was at its nadir of disfavor, Clark soon found himself overwhelmed with national demands for instruments. His first really big telescope, an 18½-inch refractor, was completed in 1862, and it is wonderful to record that by the end of the century the Clarks' firm had made what are still the two biggest refracting telescopes in the world: the 36-inch at Lick Observatory in California, and the 40-inch at Yerkes Observatory, Williams Bay, Wisconsin. With more than half a century of use behind them, these magnificent instruments are still in constant employment, eloquent testimony to their makers' genius.

The giant reflectors

It was now time for the refractor's rival to summon itself for the final and conclusive effort. Even in 1845, the great success of Lord Rosse's 6-foot-aperture colossus in detecting extremely faint nebulae proved that the mirror would one day achieve what the lens could not. It was already obvious that object glasses could not grow in size indefinitely. Quite apart from the extreme difficulty of casting huge disks of optically perfect glass, there was the serious drawback of flexure; a lens could be supported only around the rim, whereas a mirror could be held rigidly across the back. So there was no doubt, if only the tarnish problem could be overcome, that reflecting telescopes could be made of a size to shame the greatest refractor.

It was chemistry's turn to aid astronomy; the breakthrough came in 1856, when the German astronomer Carl von Steinheil suggested that a mirror made of glass—an absolutely permanent foundation—could be made highly reflective by the precipitation of silver on the curved surface. If and when the silver tarnished, it could be dissolved off and a fresh coat laid. Moreover, since the glass merely acted as a base for the reflecting layer, it did not need to be as perfect as that required for a lens. The idea was an immediate success, and with the construction in 1879 of a 36-inch reflector by the English amateur A. A. Common, the modern era of giant instruments was truly born. Common himself went further: In 1891, he completed a 60-inch reflector, which must stand as a final tribute to the great age of amateur opticians. For the writing was on the wall. Only the wealthiest of the wealthy could afford to subsidize the building of such tremendous instruments, and such individuals were less and less inclined toward astronomical interests.

Astronomers were now realizing in full measure the importance of more telescopic power. Astounding new facts were coming to light; theories about the stars demanded more research, which itself prompted fresh speculations; and with every increase of aperture, new discoveries were made. On the night of the unveiling of the 40-inch refractor at the Yerkes Observatory by Wisconsin's Lake Geneva, one of the little group of astronomers noticed a faint new star shining beside Vega, the brilliant, blue-white star that passes overhead in late summer in north temperate latitudes. This discovery greatly impressed another in the group, George Ellery Hale, an American solar astronomer.

Indeed, it was Hale who had brought the Yerkes Observatory into being. Having heard that two 40-inch disks, suitable for a colossal new lens, were on the market, he had pestered the Chicago millionaire Charles Yerkes into, first of all, buying them; then into paying the Clarks to grind them; and, finally, into building the entire observatory. Now, realizing that the future hopes of astronomy lay in larger and larger telescopes—and also recognizing that the 40-inch refractor would probably remain king of its kind—he determined to search out funds for new and bigger reflectors. His success is proved by the construction, through his own efforts, of the 60- and 100-inch reflectors at Mount Wilson Observatory in California and the gigantic 200-inch at Mount Palomar. Completed in 1948, it remained the largest telescope in the world until the Soviet 6-meter (236-inch) went into operation at Zelenchukskaya, Crimea, in 1976.

The 200-inch not only began an era of huge telescopes (ten new ones of 100 inches [2.5 meters] aperture or larger have been built since 1959), it also marked the end of an old one. It is a monument to Hale's obsession with "more light." Its principal duty was to concentrate as much light as possible on to a photographic plate smaller than the size of a matchbox. At that time, there did not seem to be any other way of recording fainter objects. The generation of postwar instruments, however, have been conceived on a different principle, which is essentially to refine the image-recording aspect. Some notable gains in emulsion sensitivity have been made, but more significant is the introduction of purely electronic recording. A photographic plate contains a certain inertia: The first light photons reaching it are repelled by the energy of its own atoms. Image-intensifier tubes can respond to a single photon and amplify it by the million.

A second aspect of telescope design, and one which has had more bearing on amateur approaches, is the development of *catadioptric* instruments. The word indicates a telescope that uses both reflection and refraction in the formation of its primary image. As far as image quality on the optical axis is concerned (i.e., that resulting when the telescope is pointed directly toward the object being observed), a straightforward reflector or refractor can approach practical perfection. In the days when telescopes were used entirely visually, this, together with aperture, were the main criteria, since the eye cannot accommodate a wide field of view. The introduction of photography, however, changed the requirements considerably, since a photographic plate can be made of any size. The mirror of the classical reflecting telescope has a paraboloidal concave surface, and the paraboloid gives increasingly poor definition as the object is moved away from the optical axis. This is why the original Palomar optical system could use only tiny plates. The aim of catadioptric systems is partly to widen the field of good definition. Modern large reflectors incorporate a complicated system of lenses just in front of the image, which improves the field of good definition from a few minutes of arc to something approaching the moon's apparent diameter of half a degree. The Schmidt camera, which uses a full-aperture thin glass correcting plate in front of the mirror, permits crisp photography across fields of 10° or more. Two of the most useful instruments in the world today are the similar 48-inch Schmidt cameras at Palomar and Siding Spring, Australia, which between them can cover the whole sky. Research into the possibilities afforded by lens-and-mirror systems has led to interesting new instruments for the amateur, mentioned in Chapter 3.

Sophistication rather than size has been the keynote of postwar professional instrumental development. The same is, to some extent, true of amateur work, which has benefited in particular from improvements in photographic emulsions. Amateur astrophotography, particularly of deep-sky objects, has improved out of all recognition in twenty years. Photoelectric photometry of variable stars, and prominence observations of the sun through special filters, are other fields of work for the advanced enthusiast. But some things have not changed, nor are they likely to. Visual observation of the sky, whether for scientific research or personal enjoyment, remains the art it always has been, an art that must be largely self-taught. No matter how humble his equipment, the amateur is always learning something new, and every clear night conceals new secrets to be uncovered.

2

Simple Telescopes

The construction of a simple refracting telescope should form part of the initiation of every amateur astronomer. For those beginners who cannot yet afford a proper instrument, it will in any case be a matter of necessity; and, no matter how poor the definition, the constructor will certainly learn more about how a telescope works than he can by merely reading a book. It is all very well to have a fine instrument, but if it is to be used with profit it must also be used with insight and wisdom.

This simple telescope will naturally have a single lens for the object glass, and, to combat the chromatic aberration with which the early observers were grimly familiar, it is necessary to choose one with as long a focal length as possible. A very weak, positive spectacle lens is one answer (the sort used to correct farsightedness), but if the local optician does not possess one of the right sort, a camera supplementary lens will do very well. These are designed to fit in front of a camera lens so as to take extremely close-up photographs; they are rated in strengths of 1, 2, or 3 diopters. A diopter is a measure of the focal length of a lens, expressed in terms of the reciprocal of a meter, so that the higher the rating in diopters, the shorter the focal length. A 1-diopter lens has a focal length of one meter, and will serve the purpose admirably; if, in addition, a 2-diopter lens (half the focal length) can be purchased, some interesting experiments can be performed. Both lenses should have the same aperture, and of course be as large as possible—probably about $1\frac{1}{2}$ inches across.

Focal length and image scale

The first task is to measure the focal lengths of the two lenses. This is done by casting an image of the sun on a sheet of paper and measuring the distance from the screen to the lens. If the lenses have been accurately made, their focal lengths should be about 39 inches and 20 inches.

While in this case the focal length is the distance between the lens and its image, this is true only *when the object is at infinity*. If we cast an image of, say, a distant street lamp, the screen will have to be moved a little farther away; the closer the object, the farther away the image is formed. This is why a camera has a focusing screw that varies the distance between the lens and the film. This, however, is of no importance to an astronomer, for all the objects he observes are effectively at infinity.

In the process of measuring the focal lengths, it will become apparent that the sun's image differs in size with each lens; accurate measurement of its diameter will show that the image cast by the 39-inch lens is twice as big as the other. This is because the image size is directly related to the focal length: the longer the focal length, the bigger the image. The relationship can be expressed by a very simple formula:

$$I = \frac{F}{57}$$

Here, I is the scale of 1° and F is the focal length. For example, if the lens has a focal length of 57 inches, an object in the sky 1° across will be exactly 1 inch across on the screen. Since the sun appears as roughly $\frac{1}{2}$° in diameter, it will make an image only $\frac{1}{2}$-inch across. If we want a 1-inch focal image of the sun, the object glass must have twice the focal length: about 9 feet.

Many people begin with the impression that what affects the size of the image is the diameter of the lens. This is quite wrong; the focal length is the only quantity that matters. This can be proved easily enough by placing small diaphragms over the lenses. The smaller the aperture, the fainter the image—but its size remains the same.

Tube and eyepiece

The next task is to mount the two lenses in telescopic form so that their performances can be compared. A cardboard tube will be ideal if it is of the right diameter, with the lens held in at one end by two retaining rings glued to the inside of the tube (figure 5). It is most important to get the lens accurately square-on to the tube; otherwise the image will be distorted. Both refractors and reflectors are equipped with adjusting screws to insure the correct inclination of object glass or mirror. The inside of the tube must be painted mat black to avoid stray reflections when the telescope is pointed to a bright object, and it should be cut off about three inches shorter than the focal length, so that a second, smaller tube can slide stiffly inside for focusing purposes. This smaller tube carries the eyepiece.

Figure 5. *A simple refractor.*

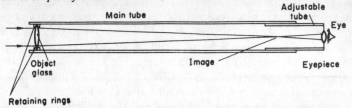

The job of the eyepiece is to magnify the image, which it does by allowing the observer's eye to get very close to the image formed by the object glass. It follows from this that the shorter the focal length of the eyepiece, the greater the magnification. For example, most people cannot focus their eyes on anything less than about 6 inches away; if, however, a 1-inch-focus lens is placed before the eye, an object can be seen sharp at a distance of only 1 inch, which gives a magnification of × 6 over the ordinary view. There is no essential difference between examining an insect on a microscope slide and the image of a planet formed by a telescope's object glass. In each case, if we want a high magnification the lens used must have a short focal length. In this particular case, an eyepiece with a focal length of 1 inch will be ideal; only one is needed, for it can easily be interchanged.

(It should be pointed out here that whereas mirrors and object glasses are rated by their aperture—e.g., 3-inch refractor, 6-inch reflector—eyepieces are rated by their focal length: 1 inch, $\frac{1}{2}$ inch, $\frac{1}{4}$ inch, etc.)

It may well be that a simple, short-focus lens is lying around in a drawer somewhere. It will not be in spruce telescopic condition, but it may serve. Or, you can take the eyepiece out of a pair of binoculars; this will be properly corrected by consisting of several lenses and will give a better image than a single-lens eyepiece. If an astronomical eyepiece is available, so much the better. On the whole, it will be beneficial to use a proper achromatic eyepiece, since the performance of the object glass must be tested and eyepiece flaws will only confuse matters. (See Chapter 4 for types of eyepieces.)

The telescope's performance

Once the eyepiece is fitted into the tube and is found to focus the image properly, it is time to point it at something. A distant building will be a useful beginning. The first drawback of the telescope will appear at once: It is quite impossible to hold it steady by hand, and the tube must be rested on a firm support. With a telescope of this size and weight, the mounting problem is relatively minor; later on, when a full-scale astronomical telescope is being used, its importance will become paramount.

The second drawback is often an unpleasant surprise, for the object viewed appears upside down. A terrestrial telescope, or a pair of binoculars, contains an extra lens or prism that reverses the light rays and gives an erect image; but this extra component does its work at the cost of a slight loss of light. Every time a light ray strikes a glass surface, about 4 per cent of its strength is reflected back, and a further small amount is absorbed by the glass. While this loss is imperceptible by ordinary standards, astronomical telescopes must focus every scrap of available light into the observer's eye; the erecting component is therefore always left out. Thus, *every astronomical telescope gives an inverted and reversed view*. Though confusing to the beginner, this soon becomes an accepted feature of observation.

Once the telescope has been found to focus satisfactorily, it is time to try it out at night. The moon is the best object with which to begin, for it is bright enough to be found easily, and its surface is a mass of detail. *A telescope must never be pointed at the sun without efficient protection.* Even a 1-inch-aperture telescope can focus enough heat to result in permanent blindness, and it is senseless to take such a risk.

The first view of the moon will certainly come as a surprise to anyone who has never before viewed it telescopically. No matter how imperfect the image, a great deal of detail can be made out. The immense dark areas can be seen, and along the line separating lunar day and night the craters cast black shadows and are seemingly thrown up into relief. This is probably a better view than old Galileo ever enjoyed with his tiny spyglass. To realize that these features are brought out by a telescope made with one's own hands doubles the pleasure; it is a great pity that so many amateurs miss this simple delight by immediately buying a far more sophisticated instrument that somehow lacks the romance!

If the telescope is now turned to a star (preferably a white one, for an orange tint will bias the experiment), the effects of chromatic aberration can be examined. When the eyepiece is pushed inside the position of best focus, the star image turns blue, since the eye is now focused on the point where the blue rays come to a focus. If the eyepiece is pulled out slightly beyond the best focus, the image turns red, since the red rays have a longer focus than the blue. At the position of best focus, the star's image is yellowish, surrounded by a purple haze. This is because yellow lies in the middle of the visible color spectrum; also, it is the tint to which the eye is most sensitive. The surrounding haze is composed of the out-of-focus red and blue elements.

Magnification and resolving power

Comparison of the views afforded by the two telescopes will bring home three main differences:

1. The long telescope gives better definition.
2. It shows the moon larger.
3. It shows a smaller area of sky.

Point 1 is accounted for by the fact that the relatively long focal length reduces the chromatic aberration. Points 2 and 3 are explained by the higher magnification.

The *magnification* of a telescope, which is measured in diameters, is found very simply: divide the focal length of the object glass or mirror by the focal length of the eyepiece. Thus, the 1-diopter lens used in conjunction with the 1-inch eyepiece gives a magnification of about × 40; the other lens, whose focal length is only 20 inches, gives × 20. The same eyepiece will therefore give different magnifications, depending on the focal length of the telescope with which it is used. Similarly, if we wish to achieve different magnifications with the same telescope, eyepieces of various focal lengths are required.

This explains why the moon appears twice as large in the long telescope, and it is not hard to see why the actual field of view should be reduced when a higher magnification is used. If the eyepiece is removed and the bright sky viewed through it, a circular disk of light is seen whose angular diameter may be anything from 30° to 60°. This is called the *apparent field of view*. Dividing this value by the magnification afforded by the eyepiece gives the *real field of view*. If the eyepiece in question has an apparent field of 40°, it will show just 1° of sky when used with the long-focus lens (× 40), but 2° when used with the other lens at a magnification of only × 20.

Another way of ascertaining the field of view is by direct measurement of the field lens of the eyepiece. The larger this is, the greater the area of the image formed by the mirror or object glass that can be seen at one time. If the objective has a focal length of 57 inches, which means that 1° of sky is represented by 1 inch at the focal plane, then an eyepiece with a field lens $\frac{3}{4}$ inch in diameter will show $\frac{3}{4}$° of sky at one view.

It is natural to suppose that the amount of detail that can be seen through a telescope depends solely on the magnification used. For example, if we have a glossy black-and-white photograph and wish to see some fine detail in it, we use a magnifying glass. But a glossy photograph is not at all a correct analogy of a telescopic image, which may be compared to a newspaper photograph (a coarse-screen halftone). When viewed from a distance (analogous to a low magnification) its outlines seem perfectly sharp, but from close range (high magnification) the detail is lost in a grid of dots. In the case of a 1-inch telescope of the best quality, the image begins to break down with a magnification of about × 100, and nothing is gained by using more powerful eyepieces; the area of the image will expand, but no additional details will be brought out.

For ordinary terrestrial use, where the magnification is relatively small, this sophistication is of little account. But astronomical telescopes have much greater demands made upon them, and when we start using high magnifying powers it soon becomes apparent that the amount of detail visible—for example, on the moon—varies greatly from telescope to telescope, even though the magnification used in all cases is exactly the same! The best way to prove this is to cut a half-inch hole in a piece of cardboard and fit it in front of the object glass. Take note of the smallest visible lunar feature; then repeat the observation with the diaphragm removed. It will be immediately obvious that much finer detail is now visible. The image is also much brighter, but this is a matter of illumination and has nothing to do with the question of *resolution*, or the ability of the mirror or object glass to distinguish objects that are close together.

Continuing our halftone analogy, we might say that the smaller the aperture, the coarser the screen. Putting it more elegantly, *the resolving power of a telescope is proportional to its aperture*. For instance, a 6-inch telescope will reveal a crater on the moon that is only $1\frac{1}{2}$ miles across, but it requires a 12-inch instrument to make out one only $\frac{3}{4}$ of a mile across. No matter what magnification is used on the 6-inch, it cannot exceed this $1\frac{1}{2}$-mile limit. So, if an observer wants to see fine detail on the moon or a planet, or if he wants to make out individually two stars that are very close together in the sky, he must use a large telescope. The astronomer's hunger for bigger and bigger apertures is as much a desire for improved resolution as for increased illumination.

The stars themselves form the best illustration of this effect, for without exception they are so far away that they appear as only points of light, despite the fact that many are far larger than our sun. No telescope can show even the closest one as a real disk, and it is unlikely that such an instrument can ever be built, for it would need to be several times larger than the 200-inch reflector—itself a masterpiece of both optical and mechanical technique. Instead, a telescope shows a star as an artificial disk, named after Sir George Airy, the nineteenth-century mathematician and British Astronomer Royal. This *Airy disk* is produced as a result of diffraction, the converging light beams interfering with each other and spreading out into a tiny disk rather than forming a true point. Its size depends on the aperture of the telescope: The bigger the telescope, the smaller the disk; hence, the better the resolution. A star viewed with a 3-inch telescope appears to have a diameter of about $1\frac{1}{2}$ seconds of arc (written $1''\cdot5$); the same star, if observed with the 40-inch Yerkes refractor, would appear to be only $0''\cdot11$ across. However, when we remember that $1''$ is equal to the diameter of a silver dollar seen from a distance of 6 miles, it is clear that a star disk appears very small even with a 3-inch telescope!

It so happens that many stars are twins. The two components may be separated by several seconds of arc, or they may be very close indeed. At all events, it is clear that a given telescope cannot resolve such a pair if they are closer than this critical resolving limit. Two stars $4'' \cdot 5$ apart can just be resolved with a 1-inch telescope; if they are closer than this, the instrument will show them as either a single star or an elongated disk, and a larger aperture must be resorted to. These values necessarily assume fine optics; it must not be expected that the primitive refractor we have described will attain its theoretical limit. At any rate, it can be tested on some of the double stars listed in Chapter 20.

However, it is well within the handyman's power to make an astronomical instrument far superior to the simple refractor described. A 6-inch mirror for a reflecting telescope can be ground in a few weeks, and it is such an absorbing business that every amateur, whether or not he wants a complete telescope, is strongly recommended to try it. The glass, carborundum powder, and other necessary materials can be bought in kit form, and worked according to the instructions given in Section V of this book. See also N. E. Howard's *Standard Handbook for Telescope Making*.

Opera glasses and binoculars

Despite the passing of the old chromatic telescope, there is one form—in fact, the original form used by Galileo—that is still in use today: the opera glass. Many amateurs have found a pair to be an invaluable addition to their kit. The opera glass is merely a pair of simple telescopes fitted side by side, having object glasses of very short relative focus and eyepieces giving a magnification of × 2 or × 3. The principle on which it works is shown in figure 6. Note that each eyepiece consists of a concave lens placed inside the focus of the object glass. These eyepieces are never used with astronomical telescopes, since the apparent field of view is very small,

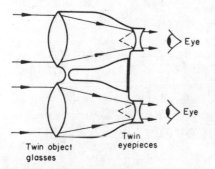

Eye

Eye

Twin
eyepieces

Twin object
glasses

Figure 6. *Opera glass. Note that each eyepiece, which is negative, is placed inside the focal point.*

but in this case they tend to correct some of the color faults of the objective; and since the magnification of an opera glass is extremely low, the field of view is still sufficiently large to make it an invaluable instrument for quickly scanning large areas of the sky. It will presently be seen that magnification is not always desirable; sometimes it is positively objectionable, and it is on these occasions that the opera glass comes into its own.

The big brother of the opera glass, the field glass, has a somewhat larger aperture—about 2 inches—and a magnification of between × 3 and × 6, which somewhat restricts the field of view. If a higher-powered instrument is required, and it will certainly prove to have many uses, the best answer is to buy a proper pair of binoculars.

Binoculars—which are, or should be, truly achromatic—work on the folded-beam principle. Generally speaking, an achromatic object glass gives the best performance when its focal length is between 10 and 15 times its aperture. If the *focal ratio*, as it is called, is less than 10 (written f/10), it becomes extremely difficult to correct the lens adequately for chromatic effects. If it is longer than f/15, on the other hand, the image scale becomes large and it is impossible to get a wide field of view using normal eyepieces. Since object glasses smaller than about 1¼ inches (30mm) admit too little light to give a really bright view, the only way of making an achromatic instrument really portable is somehow to fold up the long focus into a more compact unit. By using two prisms between each object glass and eyepiece, this is exactly what binoculars do (figure 7). For instance, if the instrument has a 2-inch aperture, the distance to the eyepiece is reduced from the nominal 10 or 15 inches to about 6.

Obviously, a pair of binoculars is anything but simple. Both the object glass and eyepiece in each optical system consist of at least two lenses (often three), and the three functioning surfaces of each prism have to be worked to optical accuracy. They therefore contain at least fourteen

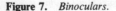

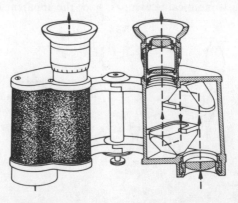

Figure 7. *Binoculars.*

separate surfaces, all scrupulously polished and gauged—so that the wonder is not that binoculars are expensive, but that they are so cheap.

Every amateur astronomer should own a pair of binoculars. They are invaluable for learning one's way around the fainter stars in a constellation, for glancing at the moon to see what formations are well placed for observation with the main instrument, for picking up Venus in daylight or other planets in twilight, for observing bright variable stars, and for examining any really bright comet that may come along. For holding them steady, all that is needed is a broom-handle; sharpen one end to a point and fit a cradle to the other in which they can be rested. The observer can then sit comfortably in a chair, sweeping his glass across the sky and altering his attentions with a freedom and speed that is just not possible at the eye end of a large and ponderous telescope.

When choosing binoculars for astronomical work, the aperture is the principal bargaining point. Magnification is relatively unimportant—except to the extent that the widest field of view goes with the lowest power, so that it is really of negative account.

Binoculars are always catalogued in the form A × B, A referring to the magnifying power and B to the aperture in millimeters. Thus, 8 × 30 means that the magnification is × 8 and the object glasses are 30mm across. This is the most popular formula for terrestrial work, where there is usually plenty of light available; but for stellar observation a 50mm aperture will be far more effective. The two standard magnifications for this aperture are × 7 and × 10; 10 × 50 glasses will show difficult objects, such as the inner satellite of Jupiter, more easily, but the field of view is restricted. For general use, the 7 × 50 pattern is probably the best. These have a field of view of about 9°, whereas the more powerful types include only about 6½° and hence show only half the area of sky.

Binoculars can, of course, be had with apertures much larger than 2 inches; 12 × 60 glasses, widely available, will reveal correspondingly fainter stars, and just occasionally one comes across real freaks, often designed for night work during the war. The binoculars used by George Alcock, the English amateur astronomer who has so far discovered four comets and three novae, are rated 25 × 105; naturally, they require a massive stand of their own. However, such an instrument could hardly be used for casual sky-sweeping, and, all in all, the 7 × 50 model will best suit the requirements of the ordinary amateur.

Finding a usable secondhand opera glass is largely a matter of luck. New 8 × 30 binoculars cost from about $50 (£25); 7 × 50 from about $80 (£40); but real bargains can often be found in government and industrial surplus stores specializing in scientific instruments, which often prove to be an inexpensive source of much optical gear. When buying binoculars, it is most important to insure that each eyepiece can be focused independently to allow for individual anomalies; if possible, a star test should be made before final purchase, since the stellar points of light often reveal errors of definition that might go unnoticed on terrestrial objects. Minor scratches on the lenses do not matter, for they make no noticeable difference to the view, but a network of tiny scores, which can be produced by careless cleaning with a dirty cloth, will diffuse faint light into the field and obscure the dimmest stars.

3

Telescopes and Mountings

There are three ways of acquiring a telescope: to buy it new; to buy it secondhand; or to make it either wholly or partly. There are enough telescopes on the market now for it to be *unnecessary* for the amateur to grind a mirror and knock a tube and stand together, but this does not mean that buying a complete telescope is automatically the best way of starting astronomy. It depends upon one's own inclination. Telescope-making is a superb hobby in its own right, and the observer who has made his own instrument— assuming that it works well—will have the edge over someone else who does not understand the first thing about it if something goes wrong. But, and this is an important point, telescope-making and real observing enthusiasm make a poor mixture. Telescope-makers are rarely satisfied with their product, and want to improve it; this means that the instrument is either stripped down and unusable, or else set aside while another and better telescope is being made. Either way, observing suffers!

This may sound like an amusing exaggeration, but I am writing from experience, and some of the very best observers I know understand practically nothing about how a telescope works, and care even less. There does exist a middle-of-the-road amateur, but he is uncommon. The supreme example of telescope-maker and observer combined was William Herschel; but he made his own telescopes simply because he could not have bought such good ones no matter how much money he had spent. If he could have bought a telescope ready-made, this would have been a much more sensible course than spending months and years grinding and sawing: Every hour lost from the eyepiece was a potential discovery lost to astronomy. So the enthusiastic observer's homemade telescope will reflect his predilection: It will work, and work well, but it will not be much to look at.

25

A story is told of the great Jupiter observer and professional optician James Hargreaves. Pressure of work had made him cease his nightly planetary vigils for some years, until he found that he was in a position to start work again. He took his old telescope mirror from a drawer, went out and bought a bundle of timber and a bag of nails, and after a weekend's hammering and sawing was back in action. It is in this spirit that the telescopemaker and the observer can be most effectively reconciled.

A new 6-inch (150mm) mirror and flat of average quality costs between about $60 and $80. Three different eyepieces, also of average quality, will cost about as much again. Therefore the optical components for an adequate reflecting telescope come to about $150. A new 6-inch reflector complete, with motor drive to follow the stars, and three different eyepieces, will cost upward of $400. If the "timber and a bag of nails" approach is adopted, a lot of money will be saved, but some time will be lost. Most people, fired by enthusiasm to look at the sky, will write a check if they can. There is no right and wrong course. I would only offer this advice: Before buying or making a telescope, join your local society and ask for advice. You will learn a great deal, and you may save a lot of money and time.

Reflector, refractor, or catadioptric?

Anyone planning to *make* a telescope will be well advised to choose the Newtonian type. From the optical point of view, a mirror is less arduous to make than is a lens, although not necessarily less difficult. (Part V of this book is devoted to optical work.) Even if the optics are bought ready-made, however, a reflector is a somewhat easier proposition. Almost anything will serve for a mirror cell in the 6-inch or 8-inch size, whereas an object glass really deserves a metal cell, although mahogany or even plywood can be used; and it is difficult to make a satisfactory object-glass cell without a lathe. A more serious objection to the refractor, however, is the relative length of the tube, which is usually about 15 times the aperture (most object glasses being $f/15$) instead of 6 or 8 times as in the case of a Newtonian. The difficulty of rigidly mounting a tube increases at a rate something like the square of its length. Additionally, the eyepiece of a Newtonian is near the top of the tube, which allows the instrument to be mounted on a low stand, whereas the refractor has to be mounted on a tall tripod or pillar if the eyepiece is to be at a convenient height. Thus, the argument in favor of the reflector is decisive if mounting considerations are the most important ones. The beginner is hardly likely to embark on a catadioptric telescope for his first attempt, particularly since sets of optics for this type of instrument are not widely available.

The considerations to bear in mind when deciding on a ready-made instrument rest largely on performance rather than manufacture, but the point about steadiness must never be forgotten. Mountings will be mentioned later, but one attraction of short telescopes is their relative freedom from vibration. An instrument whose image remains motionless in the field of view when the focusing is adjusted is a joy to use. Too many do not; and a large proportion of them are poorly mounted refractors.

Reflecting telescopes

The Newtonian reflector is the cheapest and simplest type of telescope. Only two optical surfaces are involved, and area for area a mirror and its flat are about one-eighth of the price of an object glass. The images on the optical axis can be as good as those of the refractor and better than practically all the catadioptric telescopes on the market. Light transmission is less good than with a refractor (not more than 70 percent at visual wavelengths, as opposed to 85 percent to 95 percent for a refractor), but it is probably superior to that of a catadioptric when the aluminium coating on the mirror is fresh. The Newtonian is superior to the refractor for most photographic work. However, the quality of the definition some distance from the optical axis deteriorates quite rapidly in Newtonians of $f/8$, uncomfortably so at $f/5$.

In action, the Newtonian is normally much shorter than the equivalent refractor, and the tube of a 6-inch or 8-inch can easily be handled by one person. The two chief practical drawbacks are the need to check and correct, if necessary, the alignment or collimation of the mirrors, and the susceptibility of the aluminized surfaces to tarnish and corrode, with corresponding loss of light-gathering efficiency (though not of resolution).

A great deal has been made of the Newtonian's "tendency" to come out of alignment. On the face of it, there is no reason why a mirror should be harder to mount properly than a lens. No optical component should ever be held tightly (unless it is a small, relatively thick component like an eyepiece lens). There should almost always be a very few thousandths of an inch of clearance, sufficient for a slight rattle to be heard when the unit is gently shaken. It is true that many cheap reflectors mount their optics so sloppily that they cannot hold alignment, or else clamp them down so fiercely that the glass is distorted. But there is a good reason for the reflector being more sensitive than the refractor: The tolerance of a lens or mirror to misalignment at $f/8$ is at least four times less than it is at $f/15$. Therefore, tipping the edge of a normal Newtonian mirror degrades the image a lot more than would the same tip applied to the edge of the corresponding object glass. If all Newtonians worked at $f/15$, or all refractors at $f/8$, the systems would be more closely comparable. But then, object glasses usually have better off-axis performance than do paraboloids, even those of the same focal ratio. This is true because an achromatic lens has four surfaces on which the optical designer can work out ways of widening the field of acceptable definition, while a Newtonian paraboloid has only one surface. All that can be done to change the off-axis performance is to change the focal ratio.

This is an important point. Most 6-inch or 8-inch mirrors are offered at $f/6$ or $f/8$. Small focal ratios give the advantage of a shorter tube and easier mounting problems; they also possess dramatically smaller fields of good definition. For example, consider the definition at the edge of a field measuring ½° across (equivalent to the apparent diameter of the moon). In an $f/4$, 6-inch reflector, a star will appear as a V-shaped blur about $11''$ of arc across, because of the off-axis defect known as coma. At $f/6$ this is down to $4'' \cdot 5$, and at $f/8$ it is about $2'' \cdot 8$. At $f/10$, however, it is only $1'' \cdot 9$, which is about the size a star image often appears, as a result of atmospheric turbulence. If we take $1'' \cdot 0$ as an acceptable size for residual coma, this means that an $f/6$ mirror must be aligned so well that its optical axis passes within 0.017 inch of the center of the eyepiece, whereas the latitude with an $f/10$ mirror is four times as much, or 0.07 inch.

The practical result is that the chance of an $f/6$ mirror being adequately aligned is very small! The frequent statement that "lenses define better than mirrors do" rests in no small degree on the sheer difficulty of collimating optics of small focal ratio. Compared with this, aligning an $f/15$ refractor is child's play. But the chances of an $f/10$ Newtonian running a refractor close, or even surpassing it due to its color-free images, are very good—a lot higher, unfortunately, than the chances of being able to buy an $f/10$ mirror in the first place. If you *can* obtain a good-quality $f/10$, 6-inch or 8-inch mirror, the superb performance on close double stars and planetary detail may well outweigh the extra problems associated with mounting so long a tube. (Remember that Herschel's telescopes were $f/10$, $f/12$, or even longer, and he sometimes used magnifications of several *thousand* times that showed star images "as round as a button.")

The owner of a Newtonian learns to live with the need for scrupulous care of the aluminized surfaces. Both mirror and flat must be covered with a cap or pad when not in use (see page 61). In time, the surfaces will acquire a milky bloom, but this can be at least partly cleaned off, as described on page 61; a coating should not need to be replaced for at least a couple of years if the mirror is washed regularly. Even a very badly corroded mirror is reflecting much more light than might be expected. Bare glass reflects about 4% of the light falling on it, and the worst coating is more reflective than that.

Not much need be said about the Newtonian's close cousin, the Cassegrain (figure 8). It is used widely in professional instruments, but no com-

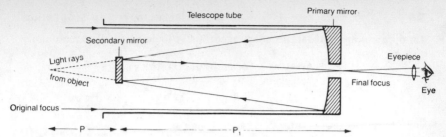

Figure 8. *Principle of the Cassegrain reflector. The focal length of the primary mirror is amplified by a factor P_1/P, the ratio of the distances from the secondary mirror to the final and original foci.*

plete Cassegrain in the 6-inch to 12-inch range is manufactured as a regular line for amateur use. The small convex secondary mirror amplifies the focal length of the primary, usually by a factor of 4 or 5, without necessitating the long tube that would be required had the primary mirror been made to this focal length originally. But the Cassegrain principle has been incorporated in many catadioptric telescopes, as will be discussed below.

Refracting telescopes

The refracting telescope has the advantage, in one sense, of having to work at a long focal length. The reason for this lies in the complex way glass refracts light. Light reflected by a mirror suffers no alteration except change of direction, but the moment it passes into glass it is spread into colors—almost irreversibly. Despite what was said on page 9, the flint lens of an achromatic objective cannot *exactly* recombine the colors produced by the crown. There are two main reasons for this, one being the irrationality of different glasses. If the two components of an object glass were represented by two prisms, then perfect achromatism would result if the spectrum produced by the crown could be exactly condensed back into white light by the flint. In other words, if the crown and flint prisms were each made to produce a spectrum of white light on the same scale, these spectra would be identical, with the colors matching at all points; such glasses would be rational. Unfortunately, rational crown and flint glasses do not exist. As a general rule, the blue end of the spectrum formed by flint is more extended, and the red end more compressed, than that formed by a crown prism. All that the optical designer can do is to fit his two spectra together as best he can, knowing that somewhere along the line the colors will not exactly compensate each other. This residual color is known as secondary spectrum, and its importance varies approximately as the square of the aperture, for a given focal ratio. Taking as a basis the empirical rule that the focal ratio should not be less than three times the aperture expressed in inches, we can make up the following table indicating the shortest permissible focal ratios

for different apertures if secondary spectrum is to be held within reasonable
limits:

APERTURE (INCHES)	FOCAL RATIO	FOCAL LENGTH (INCHES)
3	9	27
4	12	48
5	15	60
6	18	108

In most cases the situation is slightly worse than this, because it is
impossible to make a lens combination equally achromatic from center to
edge. This particular fault is called by the fearsome name of *sphero-chro-matic aberration*. It means that however hard the designer strives to minimize chromatic aberration at some point on his object glass, it will reappear
with concentrated malice elsewhere. He usually aims at getting color as near
right as possible somewhere between the center and the edge, and hoping for
the best. But this fault will be very marked in an $f/9$ object glass, since it
varies roughly as the square of the focal ratio.

As a result of all this, and more, we arrive at the conventional $f/15$
achromatic objective, which in focal-ratio sizes up to 4 inches gives almost
complete suppression of false color. A brilliant object such as Venus, or the
moon's limb, will show a trace of purple. But at $f/15$ the definition on stars
and planets should be superb. It is worth pointing out that no large refractor
can follow the "three times aperture" rule, or its tube could not be mounted.
The Yerkes 40-inch is $f/20$, whereas the rule-of-thumb formula advises
$f/120$! Yet it works very well, despite the excessive false color.

It should also be pointed out that glassmakers have wrestled very successfully with the problem of irrationality, and are now producing crown/
flint pairs that match each other much more closely than any that have gone
before. But these are expensive and have had little impact commercially.
Even an orthodox refractor in the amateur range (which means an aperture
of not more than 6 inches) should give superb results at $f/15$ or, better, $f/20$.

The transmission of an object glass can be very high indeed, and part of
its attraction is the freedom from deterioration compared with a mirror. If
the two components are separated by thin foil spacers, as is usually the case,
about 15 percent of the light falling on the front surface is lost from the
image due to reflections at the air/glass surfaces. Even this, however, is
higher efficiency than that achieved by a Newtonian with freshly-alumi-
nized surfaces. Blooming the four surfaces like a camera lens can reduce the
total light loss to as little as 5 percent, while a film of oil between the two
lenses (assuming that their curves exactly match, which is often the case)
reduces losses still further. At normal focal ratios and apertures, only a little
useful light is lost into the secondary spectrum.

A refractor is undoubtedly difficult to mount. A cheap refractor will probably be an abomination to use, even if the lens is a good one. Apertures of 5 inches and over certainly ought to be mounted on a permanent column. The closed tube has the advantage of sealing out dust and stray light, and since it is relatively narrow the air trapped inside can quickly lose its heat to the outside air. This means that any tube currents, which are caused by air of different temperature and hence refractive index swirling around inside, quickly disappear when the temperature drops at nightfall (see Chapter 5).

Catadioptric telescopes

"Cats" have aroused considerable passion and controversy. Their great merit is extreme compactness and versatility, and successful business ventures have been built up around these undeniable advantages. The question to consider is whether other, possibly more important, aspects have been sacrificed.

The Schmidt camera has already been mentioned. The Schmidt principle, of using a full-aperture optically-worked glass plate to correct the image formed by a mirror, has given rise to a whole family of optical systems: the Schmidt camera, the Schmidt-Newtonian telescope, and the Schmidt-Cassegrain telescope (all illustrated in figure 9). The advantage of the Schmidt system is that the concave primary mirror can be left spherical instead of being figured parabolic. We have already seen that a parabolic mirror suffers from acute coma at short focal ratios. A spherical mirror has no coma, but it has spherical aberration instead, which means that it cannot define a sharp image unless the focal ratio is about 12 or more. The purpose of the corrector plate is to compensate for the spherical mirror's aberration without introducing coma. Hence the huge field of view of the Schmidt camera.

The Schmidt-Newtonian offers a short, wide-field instrument free from coma, but the advantage of the Schmidt-Cassegrain, at least in the form offered commercially, is not really a matter of field, since at $f/10$ or $f/12$ (the usual equivalent focal ratio) the field of view available to the eyepiece of the camera is not very large anyway. What it does do is permit the construction of a telescope with a very short focal-ratio primary; some are as fast as $f/2$. It is easier to mass-produce corrector plates for $f/2$ mirrors than to parabolize them. The plate offers further advantages: It permits the tube to be sealed across the front, and offers a mounting for the secondary mirror, eliminating the need for supporting arms across the light-path. Sealing the tube means better protection for the mirrors' coatings, and some instruments are offered with highly reflective films (efficiency about 99 percent), which are expensive but last much longer than they would do in the ordinary open-tube Newtonian or Cassegrain.

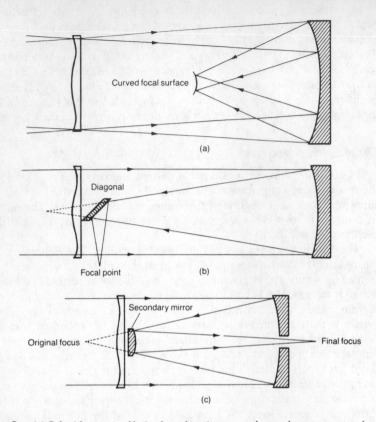

Figure 9.   (a) *Schmidt camera. Notice how the mirror must have a larger aperture than the corrector plate, if a fully illuminated field is required.* (b) *Schmidt-Newtonian.* (c) *Schmidt-Cassegrain, the form of several commercial catadioptric systems. The curve on the corrector plate is enormously exaggerated in these diagrams.*

The close relative of the Schmidt system is the Maksutov. The figured, almost flat plate is replaced by a sharply curved shell (figure 9). By suitable choice of design, the convex secondary mirror of the Cassegrain form can be obtained simply by aluminizing a central spot of the shell's rear surface. When someone talks about catadioptric telescopes, or "cats," nine times out of ten he is referring to one of the Schmidt-Cassegrain or Maksutov-Cassegrain instruments that are now on the market, with primary mirrors of about $f/2$ or $f/2.5$, apertures of 3½ inches and upward, and effective focal ratio about 10.

But the astronomer wants to know how good they are for astronomical work—as opposed to all the other things they are made for: acting as a telephoto lens, for example, or for nature study.

I have used some commercially made "cats," have heard about the performance of others, and have designed and made some myself. The first thing to be said is that the quality of commercial telescopes varies not only from one firm to another but, in some cases, from one instrument to another. Some dealers test each instrument before parting with it, others may not. It is possible that you will find yourself with an instrument whose performance does not match that of the next one on the shelf.

The same could be said of any commercial product that rises above the level of soap powder. One object glass may be better than another; mirrors may differ within a batch. Catadioptric systems are rather special, however, in that two or perhaps three completely separate optical components have to be united together perfectly. Not only must their optical surfaces be sympathetic; they must be in precise mutual alignment. Of course it can be done, and probably it is done in most cases. But if it is not done properly, or things come out of alignment, the image quality suffers far more seriously than it would in an ordinary reflector or refractor.

But assuming that all is well as far as optical quality and alignment are concerned, what then?

A perfect optical system, as we saw on page 20, produces a star image of finite diameter, even though the apparent diameter of the star in the sky is practically infinitely small. In seconds of arc, the diameter of the star image is effectively $4''.56/D$, where D is the aperture in inches. But not all of the light collected by the objective goes into this Airy disk. Even in a perfect system, only 84 percent does. The rest is scattered by diffraction at the edge of the object glass or mirror, and produces a series of very faint rings around the spurious disk of a star. It is possible to eliminate these rings by treating the margin of the objective, with the result that the light transmission or reflection fades gradually to nothing instead of coming to a sudden stop. They can, correspondingly, be exaggerated by putting another "edge" in the light path such as by sticking an opaque disk over the center of the mirror or lens. It is interesting to experiment with fixing larger and larger disks over the front of the telescope tube. When the diameter of the obstruction attains about 90 percent, the star image practically disappears, and a series of brilliant concentric rings will surround the position where it ought to be.

Loss of light into the diffraction rings is particularly serious in planetary work. The telescopic image of a planet is made up of a grid of diffraction images, with the diffraction rings of one image overlying the spurious disk of the next. The effect of this is to reduce the contrast, as though a gray veil had been spread over the planet's surface. At all costs, these diffraction rings must be kept as faint as possible; and this is where the catadioptric design as usually marketed can be most seriously faulted. Whereas the diagonal mirror in a normal Newtonian can and should have a diameter of a quarter or less than that of the primary mirror (a 1/5th diameter diagonal is feasible with an $f/8$ system, but it has to increase as the focal ratio is reduced), the central obstruction of the typical commercial catadioptric system is about ⅓. Under these circumstances, the diffraction rings must be noticeably brighter than with the orthodox reflector, and therefore the contrast of planetary detail is reduced.

The actual obstruction of light by the large central obstruction, although exceeding 10 percent, is less serious than the fact that it is throwing light away from where it is wanted to where it very definitely *isn't* wanted!

Large central obstruction is not an inescapable feature of the Schmidt— or Maksutov-Cassegrain. H. E. Dall long ago designed and constructed systems of this type with a one-fifth-diameter obstruction (as well as an erect and unreversed image). But this feature, when found in a commercial product—together with the almost obligatory right-angled prism or mirror (an extra and undesirable optical component) to turn the image into a suitable observing position—means that, at least on planetary detail, its performance will be inferior to that of the equivalent $f/8$ reflector or long-focus refractor.

It may be worth quoting a passage from a test report on a well-known make of catadioptric telescope, at the point where it was compared with an $f/8$ Newtonian of similar aperture (*Popular Astronomy,* July 1981, page 89):

> *Despite the fact that the mirrors badly need realuminising, [the Newtonian] gives good images. I was pleased, from my point of view, to find that it gave a good, sharp image of Saturn as soon as I had set it up after a car journey with the Newtonian in the car (mirror near the heater) and the [catadioptric telescope] in the [trunk]. . . . The Newtonian consistently gave images of better contrast, with the shadow of the rings on the globe visible as a hard black line, while the [catadioptric] view, while sharp, did not have the same biting quality to it. This again is a consequence of the large central obstruction and optical design of the Schmidt-Cassegrain.*

It must in fairness be pointed out that these compact telescopes lend themselves very well to most branches of astrophotography, where critical definition is less important than rigidity and ease of following.

Choice of instrument

There are three major considerations to be taken into account when deciding what sort of telescope to obtain: (*a*) expense, (*b*) portability, (*c*) type of observing to be done.

Item (a) needs no further explanation here, but the question of portability does. There used to be a rule of thumb which stated that Newtonians larger than about 8 inches, and refractors larger than about 4 inches, should always be mounted on solid columns concreted into the ground. Nowadays some enthusiasts are towing telescopes of 16 inches aperture and even larger on trailers to get them to favored sites in the country. But some general points ought to be made. First, a concrete column is more rigid than a portable stand, no matter how well made the latter. Second, having a telescope already set up, rather than stored in a shed or requiring towing, is better for the instrument and provides more incentive for the observer. Finally, if the telescope is on an equatorial mounting, the polar axis must be satisfactorily aligned with the Earth's axis, and this cannot easily be achieved with most portable instruments, certainly not to the precision possible with a permanent mounting.

The need for portability depends upon the sort of observing that is envisioned. The lunar, solar, or planetary observer does not need to seek country skies; the deep-sky observer probably does. The astronomical photographer requires a well-aligned equatorial mounting; the planetary observer is less fussy. All in all, a permanently mounted instrument should be superior, as a telescope, to a portable one, but there are many factors to consider. Reflectors are definitely more portable than refractors of the same aperture.

Item (c) is fairly straightforward. If the main aim is the observation of faint objects, the Newtonian offers the cheapest way of acquiring light-collecting surface. A new 5-inch object-glass costs roughly as much as a new 16-inch Newtonian mirror and flat.

For critical definition without frequent attention to the alignment, a long-focus Newtonian or a refractor is the best choice.

For general observing—which means being able to cover every likely field reasonably well—a catadioptric telescope, if it is a good one, will be satisfactory. It may be particularly good for celestial photography. But it is not a specialist's instrument: The deep-sky enthusiast would spend the same money on a large mirror, the planetary observer would seek perhaps the same aperture but a more incisively defined image.

To sum up so far: If a foolproof, ready-to-use telescope is required, $500 (£250), or so will buy a new 3-inch refractor. Alternatively, by very careful perusal and selection (an astronomer friend is of great service here), a second-hand refractor can be picked up for less than $200 (£100). It may need slight attention here and there, but so long as the object glass is good, the focusing movement works smoothly, and the mounting is steady, the telescope should give excellent results.

It is not easy to give general advice on the purchase of a reflecting telescope, for there are many different models to choose from; but it can, in general, be said that quality is commensurate with price. A new 6-inch reflector will cost about $400 (£200); but a small extra outlay may well bring better optics and a much sturdier mounting, making all the difference between an instrument which soon reveals its own drawbacks and one which is always serviceable, no matter what other telescopes may be acquired later on. Spindly axes and undersize drive wheels characterize the cheap telescope at first glance, and should discourage any further investigation; tiny finders also betray the maker's parsimony. Undoubtedly the best way of deciding the telescope to buy is to join the local astronomical society and seek direct advice from those active observers who have purchased instruments of their own. It may even happen that a good secondhand reflector may be traced in this way and a bargain unearthed.

The biggest imponderable in a new telescope is the optical quality. Stiffer competition has undoubtedly led to a rise in standards, but it is still surprising how many bad mirrors "slip through," and a star test by an experienced observer is invaluable. When a telescope is bought through mail order, this is hardly practicable, so that there is much to be said for arranging a demonstration through a retailer. This is another advantage of buying a secondhand instrument from a local source.

It is generally accepted that for a reflecting telescope to give reasonable definition, its optical surfaces should be within about ⅛th of the average wavelength of light (or about 0.000002 in.) of perfection. Accordingly, practically every manufacturer feels himself obliged to offer "⅛th wave" or even "¹⁄₁₀th wave" optics, without always checking to ensure that every mirror is really within this limit! Moreover, some makers misleadingly call their mirrors "¹⁄₂₀th wave," meaning ±¹⁄₂₀th, which is equivalent to a total error of ¹⁄₁₀th. The purchaser would be well advised to take all such claims with a grain of salt, and be guided solely by what he sees through the eyepiece.

A simple mounting

Here is a description of a very cheap and simple mounting suitable for a small or medium-sized Newtonian reflector, made almost entirely from wood. It is an altazimuth, which means that the axes operate at right angles to one another, one being vertical and the other horizontal. Perhaps a word should be said about this kind of mounting, because in a market that is dominated by the equatorial (page 43), even mentioning the altazimuth may give cause for amusement.

In order to follow the apparent movement of a celestial object across the sky, the user of an altazimuth has to adjust the telescope tube using both axes simultaneously, and there is no straightforward way of giving it an automatic drive. This is the main burden of complaint against it. Practically no commercial telescope, apart from some low-power sky-sweeping instruments, is offered on this type of mounting, and very few books give it more than passing mention. Yet it has the following features that, in my opinion, make it worthy of consideration for any beginner setting out to make a telescope for himself: (*a*) it is easier to make than an equatorial; (*b*) it is inherently steadier; (*c*) it requires no on-site alignment.

Its main disadvantage is its unsuitability for celestial photography. But the statement, or assumption, found in much literature, that serious *visual* observation with an altazimuth is impossible, is sheer nonsense, and it is time that the altazimuth was regarded in a fresh light. After all, most equatorial mountings are so out of balance that they require counterweights, or massive axes, or both. The moment the overhanging tube is moved inboard, the need for large axes disappears. The altazimuth stand described here uses nothing more sophisticated or expensive than door hinges for the altitude axis and a large bolt for the azimuth axis; yet if properly made it will rival the best commercial equatorials for rigidity. The fact that the latest generation of big reflectors is going back to the altazimuth mounting, with computerized drives on both axes to allow perfect guiding on a star, is a tacit admission of the mechanical defects of the equatorial.

Of course, given a rigid motor-driven equatorial telescope, accurately aligned and with a clear view of the sky from its site, one would have to be obtuse not to prefer it to an altazimuth. But the case is not nearly so clear, given an equatorial only moderately rigid, particularly one not set up properly so that stars drift through the field (as may well be the case if it has to be moved around in order to avoid obstructions). Also, the eyepiece with an equatorial can assume awkward inclinations, and it is a nuisance when trying to find objects near the celestial pole.

This simple mounting (illustrated in figure 10), consists of the following parts:

1. A very rigid three-legged stand about 30 inches high, made of 3- x 2-inch or similar timber, cross-braced, screwed, and glued. At the top is a thick circular disk 8–10 inches across—a pair of ½-inch plywood disks screwed and glued together are suitable.

2. A plywood isosceles triangle (¾-inch is best) of such a size that it just covers the disk. A strong bolt is passed through their centers so that the triangle can rotate. Three pads of linoleum or Formica are glued to the underside of the triangle, giving it a 3-point seating on the disk. Do not paint or varnish the top of the disk, since this may cause the pads to stick or snatch.

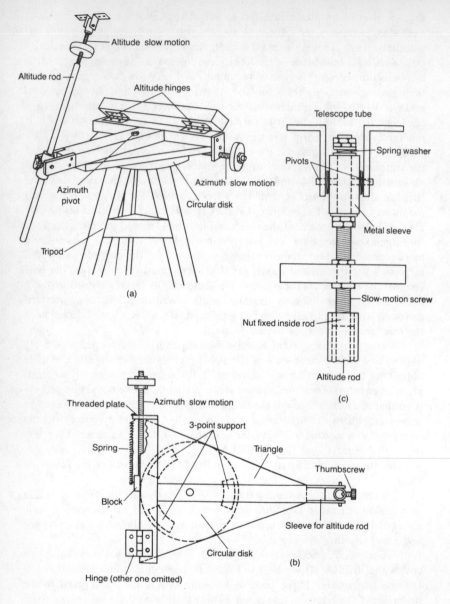

Figure 10. *A simple altazimuth mounting.* (a) *General view.* (b) *Plan view of the triangle and fittings.* (c) *Front view of the slow-motion control in altitude.*

3. A slow-motion in azimuth. A piece of stout screwed rod about 10 inches long has a knob fitted to one end to give the hand something to turn. This can be a plywood disk locked on with a couple of nuts and washers. A nut to take the rod is fitted to the back of the triangle, so that the end of the rod bears horizontally against a block fitted to the edge of the lower disk (a hole in a metal plate can be drilled and tapped for the purpose, but a short piece of tube with a nut hammered tightly into each end eliminates any wobble in the rod). A spring keeps the end of the rod bearing against the block, so that turning the rod causes the triangle to rotate slowly on the disk.

4. Two robust wooden spreaders, at least 2 inches square in cross-section and about 15 inches long. They are linked together at the ends by stout door-hinges. One spreader is fitted to the triangle, above the slow-motion rod, the other takes the telescope tube. These form the altitude axis.

5. A steady-rod and slow-motion in altitude. Another piece of threaded rod is made a captive fit inside a metal sleeve, which is pivoted on the underside of the telescope tube. A wooden disc is locked on to the rod just below the sleeve, allowing it to be turned easily. The rod screws into a nut fitted tightly into the end of a long metal tube. This tube passes easily through another sleeve pivoted on the front corner of the triangle. A thumb-screw allows the rod to slide through this sleeve until the required altitude is obtained. It is then tightened, and a slow-motion is achieved by turning the screwed rod.

In conclusion, it should be emphasized that smooth slow-motions, without backlash, are absolutely essential for any telescope that will be used with a magnification of about × 75 or more. It is frustrating trying to observe with an instrument that can be moved only in a series of jerks: Equatorial or altazimuth, it must be possible to keep the object in the field of view with the minimum of conscious attention.

One often hears would-be purchasers asking whether such-and-such a stand is "portable." In this, they are begging for trouble, for it is virtually impossible to combine real solidity with true portability. To perform well, a stand must not only be rigid; it must also be heavy. Nor should it contain any folding joints; they can never be locked as rigidly as unjointed members. In short, a light, collapsible telescope stand that can be packed away in the car will never perform satisfactorily when high magnifications are used; any commercial mounting advertised as being "portable" should therefore be viewed with the deepest suspicion.

Naturally, this is not meant to imply that the instrument should be totally immovable. One of the great advantages of a small telescope is that it can be moved over short distances to avoid trees and other obstructions that might block the view of some celestial object. Large apertures really require a well-laid concrete base if they are to be truly steady; if such a telescope is subsequently acquired, the smaller instrument will be found very useful when some infuriating blockage occurs. But this is an entirely different matter. Any compromise between steadiness and portability will inevitably be to the telescope's disadvantage.

Tubes and fittings

Before we discuss the various types of mounting, the telescope tube and immediate fittings deserve attention. The layout of a typical refractor is shown in figure 11; this varies little from model to model. The object glass itself is safely enclosed in a cell, which is screwed into a mount at the top of the tube. This mount is secured to the main tube by three sets of adjustable screws, as shown, so that it can be set and locked absolutely square-on. This is a most important adjustment, since if the object glass is even slightly tilted the telescopic image will be faulty. Projecting in front of the lens, for a distance of at least three diameters, is a tube with a blackened interior called a *dew cap;* this, as its name suggests, decreases the tendency for the lens surface to become dewed in damp weather (see p. 63). Inside the tube are one or two stops, which serve to suppress stray reflections; and, of course, the inside of the tube must be painted mat black. At the other end of the tube is a large knurled knob that allows the eyepiece (which itself screws or slides into a narrower tube) to be focused correctly. In the best telescopes, this is worked by a rack-and-pinion drive inside the tube; some cheaper ones use a simpler friction method. All that matters is that the resultant motion be slow and regular.

A reflecting telescope (figure 12) is rather more complicated. The mirror should be held inside a metal cell, which of course has a solid back; this cell is secured to the base of the tube by three adjustable screws which allow it to be lined up correctly. Near the top of the tube, four thin vanes support a metal disk, to which the small flat mirror, in its own cell, is secured. Because of diffraction effects, a three-vane support produces six rays of light around the image of a bright object. A four-vane support gives only four, and is therefore preferable. The flat is elliptical, so that when inclined at the correct angle of 45° and viewed through the eyepiece tube, it appears circular. The flat itself must be adjustable; so must the length of the vanes so that it is located exactly in the center of the tube. It can be seen

Figure 11. *Refractor: tube and fittings.*

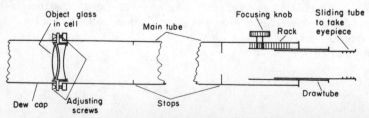

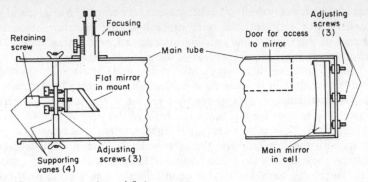

Figure 12. *Reflector: tube and fittings.*

at once that the business of lining up, or *collimating*, a reflector is rather more complicated than with a refractor; there are also more components to adjust.

The last essential is the focusing mount, opposite the flat, which adjusts the position* of the drawtube into which the eyepiece fits; also, both the mirrors need covers to protect them from dirt and damp when the telescope is not in use.

All refractors have solid (i.e., totally enclosed) tubes; so have many reflectors, but many observers have found that reflecting telescopes work better with framework tubes, which reduce the disturbing tube currents. Briefly, the argument is this: During the day, the telescope and all its components warm up; at nightfall, when the temperature drops, they begin to cool. The metal tube cools very rapidly, giving up its warmth to its surroundings; which means that the air in contact with the tube's surface is cooled, gains density as a result, and sinks down the tube. Soon a system of convection currents is set up, and the swirling air inside the tube so disturbs the image that it is impossible to do any observing until the whole system has cooled down.

The "open tube" enthusiasts point out that if a mere framework is used, the air currents can flow freely away from the light-path. This is undoubtedly true; but it also means that other heat currents, such as those from the observer's body or from the rest of the stand, can get in the way. With an open tube there is also less protection from stray light, although if the observer lives in the country this is of little account. Other experimenters have found that heating effects are less marked if the tube or framework is made of material that is a poor conductor, such as wood or plastic, but these substances are not, unfortunately, as rigid as metal, and the

*The drawtube must always be fitted in a horizontal position to the east side of the tube (which means that the mouth of the tube is to the observer's left and the mirror to his right). Any other position will upset the orientation of the image.

mirrors will probably need to be collimated more often. If a solid metal tube is desired, one excellent idea is to make it considerably larger than the aperture of the mirror. The swirling air currents, which tend to spiral up around the wall, are then kept clear of the incoming light rays.

Users of catadioptric telescopes experience particular trouble from thermal effects, particularly since these portable and valuable instruments are usually stored indoors. This means that the air in the tube is appreciably warmer than the outside air. Since the front of the tube is enclosed by the corrector shell or plate, the warm air is trapped inside, and can cool only by contact with the corrector and the tube walls. In a long-tubed refracting telescope, the cool air sinks to the bottom of the tube and is rapidly replaced by the remaining warm air, so that equilibrium is reached relatively quickly—tube currents in the typical amateur refractor never last long. But the short, stubby tube of the typical catadioptric is not sufficiently long to encourage rapid circulation. Instead, the cold air tends to remain clinging to the outer margin of the light path, and since its refractive index is different from that of the warm air nearer the center of the tube, a spurious effect of spherical aberration is introduced. Warm currents can sometimes be detected swirling up from the metal light-baffle tube that passes through the main mirror. It is a common experience for the image quality of a catadioptric to be poor for some time after it has been set up.

A survey of the mass of material available on the subject suggests that the telescope's site has at least as great an influence on image steadiness as has the form and material of the tube. Some solid metal tubes perform quite adequately, especially when inside an observatory, while identical instruments elsewhere prove troublesome; there is no hard-and-fast rule, and every amateur will soon develop his own ideas on the subject.

Equatorial mountings

The mountings that have been considered so far have had motion in both altitude and azimuth, and are for that reason known as *altazimuth* mountings. Provided they are well made, they perform satisfactorily enough; illustrious proof of this is afforded by Herschel, who mounted all his telescopes in an altazimuth style of his own. As a more recent example, the famous English planetary observer William Frederick Denning was awarded the Royal Astronomical Society's Gold Medal for his work on Jupiter, done with a 10¼-inch Browning reflector on an altazimuth stand. Many contemporary observers own similar equipment. But this type of mounting does have the drawback that adjustments must continually be made in both altitude and azimuth in order to compensate for the earth's rotation. This effect, though normally imperceptible to the naked eye over a period of five or ten minutes, can carry an object right across the field of a

Figure 13. *Principle of the equatorial mount.*

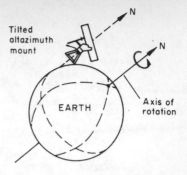

high-power eyepiece in ten or fifteen seconds, so that the observer's hands are kept fully occupied manipulating the slow-motions. When he must make a drawing at the telescope, his problems are multiplied!

It is encouraging, therefore, to find that there is a way in which this objection can be partly or entirely overcome. The stars and planets seem to move across the sky because the earth is rotating on its north-south polar axis. If one of the telescope's axes is tilted so that it is exactly parallel to the earth's axis, it is clear that rotation about this axis—in a sense, opposite to that of the earth—will keep the telescope firmly pointed to any particular celestial object (figure 13). This so-called *equatorial* mounting therefore requires continuous adjustment to one axis only (the *polar* axis), and, since it rotates at a constant speed—once in 23 hours 56 minutes—it can be controlled by a geared-down motor. A synchronous electric motor is commonly used for this purpose, its rate being automatically regulated by the frequency of the AC power supply.

At right angles to the polar axis is the *declination* axis. This needs adjusting only when finding and locking on to the star; after this, it is clamped and should not require readjustment.

The equatorial stand is clearly a great advance on the simple altazimuth, for even without an automatic drive, the observer need do only half the work to keep the object in view. In addition, such a stand is essential if any sort of celestial photography is to be undertaken. For these reasons, it is worth examining the various available forms of equatorial.

The simplest way of making an equatorial mounting is merely to tilt an ordinary altazimuth stand until its azimuth axis points at the celestial pole (the North Pole is marked approximately by Polaris, the pole star). The telescope shown in figure 8 could be converted in this way, but the result would not be very satisfactory, for the tube could not point to the low southern sky. The mounting would also be out of balance, throwing an unnecessary strain on the slow-motions. Clearly, some reorganization is necessary.

The commonest equatorial mounting is the *German* type, invented by

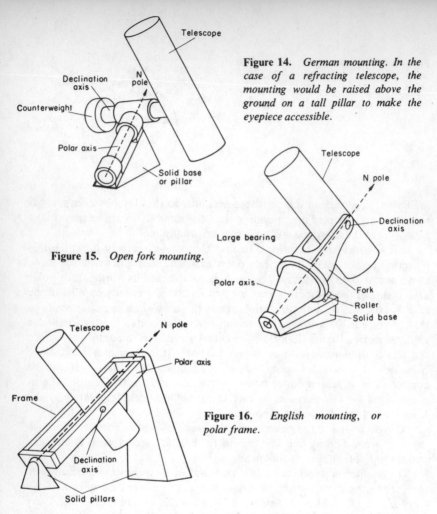

Figure 14. *German mounting. In the case of a refracting telescope, the mounting would be raised above the ground on a tall pillar to make the eyepiece accessible.*

Figure 15. *Open fork mounting.*

Figure 16. *English mounting, or polar frame.*

the Jesuit astronomer Christoph Scheiner in Galileo's time, but which first became popular at the hands of Fraunhöfer. Virtually all refractors are mounted in this way, and so are many small reflectors; the principle is shown in figure 14. The mounting is in the form of a T; the stem forms the polar axis, which rotates once a day, having the crosspiece as the declination axis. This carries the telescope tube at one end; at the other end a counterweight maintains the balance. The German-mounted telescope is extremely compact—an especially valuable feature should any sort of observatory be involved.

Some astronomers, however, object to the counterweight's excess baggage; and, in certain positions, the tube can foul the supporting pillar.

When this happens, the polar and declination axes must both be turned through 180° to bring the telescope to the other side of the pillar, a process known as "reversing." The *open fork* mounting (figure 15) does not suffer from this defect; it also avoids the use of a counterweight, since the telescope tube is slung between the arms of the fork at the end of the polar axis. Unfortunately, it has drawbacks of another sort. In the first place, it cannot be used with a refracting telescope, since the eyepiece is inaccessible when the tube is pointed near the celestial pole; secondly, the weight of the telescope at the end of the fork means that the polar axis must be extremely rigid, with an especially large bearing at the north end.* A well-built mounting of this type is illustrated.

The *English* mounting (figure 16), also known as the *polar frame,*

*This book is intended for readers in the northern hemisphere. The polar axes of telescopes used in the southern hemisphere must point to the south rather than the north.

German mounting.   *Below left: A 12-inch Calver reflector on a rugged stand made by Calver. It originally belonged to T. E. R. Phillips, a famous planetary observer, and is now in the observatory of A. W. Heath. The upper part of the tube rotates to bring the eyepiece into a convenient position.*

Open fork mounting.   *Below right: A 10-inch reflector made and used by G. Turner. The large dimensions of the north bearing on the polar axis insure stability. (G. Turner).*

English mounting. *A. Sanderson's 10-inch reflector, mounted in the English style with the declination axis raised above the frame so that the telescope can point to the north celestial pole. The construction is entirely of wood. (The Staveley Iron & Chemical Co., Ltd.)*

avoids this second objection, though at the cost of taking up much more room, by swinging the tube inside a closed frame that is pivoted at both ends. This design makes for extreme rigidity; structurally, it is the most stable of the three. Once again, it is unsuitable for refractors because of the length of tube involved, and the region around the celestial pole is blocked by the north bearing, but this latter difficulty can be overcome by raising the tube pivots above the level of the frame. The 60-, 100-, and 200-inch telescopes of the Mount Wilson and Palomar observatories are all mounted on this principle; the 120-inch reflector of the Lick Observatory is the open fork type, and the big refractors all have German stands.

There are, naturally enough, many modifications of these mountings. As figure 17 shows, the frame of the English mounting can be reduced to a single beam, with the telescope pivoted at one side and a counterweight at the other; this, the *modified English* mounting, is a halfway stage between the English and German types. There are plenty of other varieties too, but these three patterns provide the basic choice from which the amateur can design his own mounting. It is worth pointing out that virtually all commercially made stands, whether for reflectors or refractors, are of the German form, since this is economical of both materials and space and can take either type of telescope.

Proof that an equatorial mounting need be no more difficult to construct than an altazimuth is given by the photograph of a 6-inch reflector mounted in the English style. The construction is almost entirely of wood, and the bearings are made stiff enough to hold the telescope firmly in any position and yet allow free motion when required. If this is not considered sufficient,

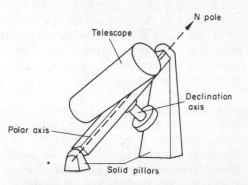

Figure 17. *Modified English mounting.*

English mounting. *A very simple equatorial mounting for a 6-inch reflector, made by L. Pointon. The construction is of wood. There are no slow motions, the pivots being sufficiently stiff to hold the telescope firmly in any desired position. (G. Turner).*

a driving wheel can be added to the polar axis, the main difficulty being in finding a sufficiently large wheel. In order to guide smoothly, especially if a mechanical drive is being used, a diameter of at least 6 or 8 inches is required, and the bigger it is the better. Motion is supplied through a worm wheel.

The English mounting is probably the easiest type for the amateur to construct, since it is well balanced and simple bearings will serve quite well. The principal problem is making the frame stiff enough to prevent flexure. It must be built massively; if of wood, 3 × 2-inch beams are the minimum size suitable for a 6-inch reflector, and 4 × 3 would be better still. It must be remembered that such a telescope will often be used with magnifications of × 200 or × 300, and under these conditions the tiniest tremor is magnified to disastrous proportions. One other very efficient basis for a mounting (the modified English) is a car's complete rear axle. The shaft forms the polar axis, revolving on the hubs at either end, and the telescope is pivoted on the central differential, with a suitable counterweight at the other side. This will hold a 6- or 8-inch reflector very firmly, and a rear axle is not difficult to obtain. Really, anything goes in the world of amateurs' telescopes; there is no need to be shy of using unorthodox materials or ideas! Water-softening units and old geysers, or hot water boilers, are just two of the many surprising items to be found in homemade mountings, and large-diameter water pipe has a multitude of applications.

An entire book could be devoted to this subject; but since we must be brief, here is a short summary. There can be no doubt at all that an equatorial offers more convenient observing conditions than an altazimuth, and for photography it is indispensable. On the other hand, for anyone who has "horizon problems" and needs to move his telescope around to avoid obstructions, an altazimuth stand will be handier, since an equatorial mount must be aligned accurately with the earth's axis, and this adjustment should preferably remain undisturbed. An equatorial can also be used to find very faint stars at night, or bright planets during the day, by a process employing celestial latitude and longitude; the same objects can usually be found with a telescope on an altazimuth stand, but the task is more difficult.

It has frequently been stated that for "serious" observation an equatorial mounting is "essential." This is utter nonsense. Apart from celestial photography, there are hardly any amateur observations that are impossible with an altazimuth stand as against an equatorial. The main difference is that with an altazimuth an amateur has to take more care, since life is that much harder for him; but, by learning to overcome these difficulties, he will gain experience as an observer and a much greater appreciation of the advantages offered by an equatorial telescope.

Two mountings to avoid

Just occasionally one comes across unwise mountings, and it is as well to be warned in advance. In the writer's experience there are two: ancient and modern. The traditional deathtrap of any telescope (usually a small refractor) is the *pillar and claw* mounting. Also known as the *table stand*, it consists of a vertical pillar about a foot high supported on three crablike legs. On the top of the pillar is a universal joint, to which the telescope is attached, and the whole device is supposed to be stood upon a table. It is hardly necessary to add that since tables are not normally made five feet high, it is quite impossible to look through the telescope at a star without grovelling on one's knees; in any case, so lightweight a mount lacks the necessary stability. Three-inch refractors on table stands often come at bargain prices and are well worth purchasing, provided the stand itself is disposed of and replaced by a proper tripod. In this connection, it is amazing how many manufacturers design their tripods for a race of dwarfs; the legs should be quite six feet long, so that the observer can stand comfortably erect when using the instrument.

The second device originated in the United States. It is known as the *Springfield* mounting, and is intended to satisfy the needs of the observer who wishes to remain stationary in a chair. The Springfield is certainly ingenious, for by the use of an extra mirror in the optical train, the eyepiece of a reflecting telescope is made absolutely stationary. However, this advantage is gained at the cost of a slight loss of light and, possibly, definition; in addition, the image is reversed, as compared with the ordinary telescopic view, which makes it almost impossible to refer to charts. The reasoning behind the Springfield may be sound, but the lack of published observations made with it suggests that it is the plaything of the amateur optician and mechanic rather than useful to the practical astronomer.

Drive systems

Most commercial telescopes are equipped with a manual or motor drive on the polar axis. In a 6-inch instrument, this usually takes the form of a worm wheel a few inches across, this wheel having not more than 360 teeth and usually far fewer. Amateurs making their own equatorial mounting may have studied the problem of successfully imparting a smooth, accurate motion to the polar axis without going to the expense of buying a worm and wheel.

They would do well to do so, since the wheels supplied with most standard mountings are totally inadequate. Consider the typical 6-inch f/8 Newtonian. The focal length is 48 inches, and the linear diameter of a high-power eyepiece field ($\times$ 250, say, with a field of view of about 12′) is less than 1/5 inch across. If the drive wheel has a diameter of 6 inches (which is larger than many), the radius is only 1/16 of the length of the tube. Yet this wheel is expected to mesh with a worm and hold the tube so rigidly that a celestial object will remain near the center of this tiny field of view when the focusing knob is turned or a breeze blows.

To hold the polar axis in this way, the worm must mesh so tightly that the bearings or drive must be damaged. It is mechanically impossible to combine an easy running fit with rigidity, unless the radius of the drive wheel bears some reasonable comparison with the length of the tube. One great virtue of the typical compact catadioptric is that the drive wheel is, or can be, a large proportion of the tube length in radius. Catadioptrics are characteristically rigid and pleasant to use because their bearings and wheels can afford to be relatively large.

One way of simulating a large, fine-toothed drive wheel is the tangent arm. Its principal drawback—the fact that it has a limited duration of operation before it has to be reset—is countered by the fact that this duration can be made longer than most amateurs would ever drive their telescope without wishing to pause for at least a couple of minutes.

The tangent drive, shown schematically in figure 18, consists of a length of screwed rod turning inside a nut or threaded aperture attached to the base of the mounting and operated by hand or a motor; the other end bears against a plate carried by the polar axis, at Q. Let the distance from the center of the polar axis to Q be B, and suppose that the angle at Q at the beginning of operation is 90°. Then the screw must advance by $B/57.3$ in order to turn the polar axis through an angle of 1°. Since the Earth's true or sidereal rotation on its axis takes 23 hours 56 minutes (see page 448), it requires 239.3 seconds to turn through 1°. Working out the required screw advance per minute, it comes to $0.004375 \times B$.

It will be found best to calculate the distance B on the basis of the available screw threads and drive rates, and a list of various possible combinations is given on page 351. For example, a screw with 20 teeth per inch, driven at a rate of 1 rpm, would require B to be 11.43 inches: probably a very convenient length for a small telescope. This is simulating a drive wheel measuring almost two feet across, with 1,436 teeth.

Unfortunately, this simple drive begins to lose rotational speed as the minutes pass. When the polar axis turns, the angle Q becomes less than a right angle, and the effective length of B increases. After 20 minutes, for example, the telescope will be pointing 45″ to the east of the star or planet, and after this time the error accelerates seriously; even so, for much visual work, this is adequate. Some compensation can be obtained by increasing the speed of the screw, or, more conveniently, by shortening the length of B.

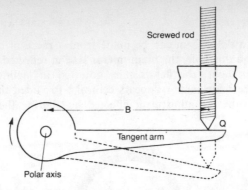

Figure 18. *Principle of the tangent drive as fitted to the polar axis of a telescope.*

If this is done correctly, the telescope drives slightly too fast to begin with, then slows down, so that the image wanders a little to the east before drifting back to the west. A 0.37 percent increase in speed will produce a total image wander of only 30″ over 24 minutes. But the best solution, both observationally and mechanically, is to have the bearing point at Q captive, so that the distance B remains constant. In this case, a drive rate 0.2 percent fast will reduce the image drift to only 15″ over 40 minutes, which in practice is comparable with all but the best standard motor drives.

The beauties of the tangent drive are (*a*) it is cheap; (*b*) it requires no accurate machining facilities; (*c*) its radius and pitch can be chosen from a wide range of combinations.

In manual form, the tangent arm makes the best possible slow-motion adjustment for the declination axis of an equatorial telescope.

Collimation and testing

Any telescope, whether new or secondhand, must be tested before purchase to see how well it defines. Although the sight of a reputable maker's name on the side of the tube can inspire confidence, its excellence should not be assumed on this basis. This is because every mirror or object glass has to be given its finishing touches or "figuring" by a skilled craftsman; and as the skill varies from worker to worker, so will the resultant quality. There is no question of mass-producing fine optics; as a result, every telescope must be examined on its own merits. A star test is the safest guide to quality, but before this can be made, the optical components must be collimated.

Aligning a reflector's mirrors is an operation best done by daylight. First of all, remove the lenses from a high-power eyepiece so that only the diaphragm with the central hole is left, and screw this into the drawtube (alternatively, a diaphragm can be made from a piece of cardboard). On looking through this hole, the main mirror is seen reflected in the flat as a circle of bright light; the flat is then adjusted in inclination until the reflection of the main mirror appears central.* Provided the flat is in the center of the tube (an adjustment effected by altering the length of the

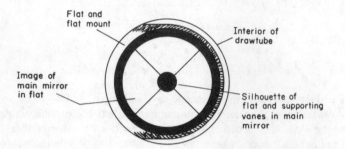

Figure 19.   *Adjustment of a reflector. This shows the correct alignment of the mirrors when viewed through the drawtube with the eyepiece removed.*

vanes) and is directly opposite the drawtube, it can be locked and does not need further attention.

The outline of the flat itself is reflected in the main mirror, appearing as a central black spot, with arms (the vanes) radiating from it. The most tedious part of the operation is to adjust the main mirror, using its three screws, until the reflection of the flat is directly in line with the drawtube and symmetrical in the reflection of the mirror (figure 19). It is useful to remember that if the flat appears to one side of the mirror's reflection, the screw on the *opposite* side should be *advanced*. It is a great help if a friend can be enlisted to do the adjusting while progress is followed through the drawtube.

The collimation of a reflector is critical, since a mirror working at f/8 has a relatively narrow field of good definition, and if it is even slightly off axis the image will be imperfect. With a focal ratio as short as f/6, the adjustment is even more vital, and the reflector's bad record so far as performance goes may be partly due to the user's neglect in insuring that the mirrors are correctly aligned. It is an excellent idea to make a habit of checking the collimation before an observing session.

The adjustment of a refractor must be carried out at night. Select a fairly bright star (the polestar is a good choice for northern observers),

---

*Strictly speaking, the image of the mirror should appear slightly displaced towards the bottom of the tube, since a cone does not pass through an inclined ellipse symmetrically. However, this refinement is negligible with apertures smaller than f/5 or f/6.

and examine its image under a high power. If the alignment is perfect, the star will appear as a tiny spot that expands into a circular disk or system of rings both inside and outside the correct focus. If the expanded disk is elliptical, however, it means that the object glass is not correctly squared-on, and it will be impossible to get a perfectly round, sharp image at the focus. In this case, the adjusting screws must be attended to until the necessary symmetry is obtained.

The optical quality of a telescope can most easily be ascertained by comparing the intrafocal and extrafocal images of a star. At the focus, it should appear as a tiny spot, almost a point, of light, and if the telescope is a refractor this spot should be surrounded by two or three faint and narrow rings called *diffraction rings*. Atmospheric conditions must, however, be steady to show this well. If the eyepiece is then moved slightly inside the point of best focus, the image expands into a disk; as it is moved farther inside, this disk should break up into a number of rings. The same sequence of events, giving precisely similar effects, should occur when the eyepiece is moved outside the focus.

A reflecting telescope, because of its shorter focal ratio, is very unlikely to show diffraction rings either around the focused image or in the ex-panded disk. Instead, these disks should appear evenly illuminated, with a black spot in the center representing the outline of the flat. If the mirror is perfect, the disks should appear identical at equal distances inside and outside the focus.

The advantage of the test just described—which should be performed with the most powerful available eyepiece—is that it can be carried out under almost any atmospheric conditions, even when the air is too unsteady to show the focused star as anything but a tiny, jiggling blur of light with no trace of the fair-weather diffraction rings. The disadvantage is that it is perhaps *too* sensitive. Many telescopes perform splendidly even when the expanded images are somewhat anomalous, and it can safely be said that if the disks are "more or less" alike, the telescope is a good one.

A reflector is perfectly achromatic, but a refractor needs to be examined for color correction as well as definition, and this can be assessed at the same time. A well-corrected lens will show a lilac-green fringe to the disk outside the focus, to be replaced by crimson when the eyepiece is advanced inside; at the point of focus there should be no extraneous color at all, unless the object is very brilliant, when a bluish halo will be apparent. If the lens is overcorrected (i.e., compensated too strongly) the outer fringe will be orange, the inner one bluish, and a very obvious bluish haze will surround the focused image. The characteristics of undercorrection, on the other hand, are turquoise and orange borders to the extrafocal and intrafocal images, respectively, and a red glare at the position of best focus.

The experienced observer can tell at once when a telescope's optics are first-rate. The planet Jupiter is a fine test object, and so is Venus when seen in a twilight sky. If the lens or mirror is a good one, the edge of the planet should come up critically sharp, and the slightest movement of the eyepiece should make an obvious difference to the definition. Atmospheric turbulence may make the image ripple and boil, but there is a clear difference between this effect and the misty outlines associated with poor optics. It is hardly necessary to point out that the eyepiece used in these tests should be of established excellence, for any inherent faults will certainly prejudice the telescope's performance.

If a star expands into an elliptical rather than circular disk, and the axis is crossed at a right angle inside and outside focus, *astigmatism* is present. This may have been caused by faulty alignment, but if this is not the case, it must be caused by one of the components. Astigmatism is almost always the sign of poor optical work or strained glass. Such an astigmatic telescope is worthless.

The villain can be weeded out by a simple process of elimination. An astigmatic component always distorts at a constant angle relative to itself, since it is effectively a good lens or mirror that has subsequently been slightly "bent" across the diagonal. Hence, rotation of the culprit will produce a corresponding rotation of the expanded ellipses. A reflector's flat is a likely source of trouble, or the fault may even lie in the observer's eye, although with an eyepiece of high power this effect is unlikely to manifest itself (see p. 79). Provided the mirror or object glass is good, however, it may be worth going ahead with the purchase and paying a little extra money to replace the faulty component.

4

Maintenance, Accessories, and Observatories

While the fundamentals of the telescope—objective, tube, and mounting—must rightly receive first attention, there are other important items to attend to. Indeed, unless the accessories are of the same quality as the instrument itself, its efficiency will certainly be reduced. In this category are eyepieces, finders, and other immediate accessories; the larger question of whether or not to build a proper observatory; and the all-important matter of keeping the optics of the telescope in good condition.

Types of eyepiece

Eyepieces, or oculars, are probably the chief culprits behind poor telescopic performance. It is amazing how many people spend a considerable amount of money on a first-class telescope, then buy a battery of cheap eyepieces that give inferior results. It is also a fact that many makers have the same bad habit; so it is clearly worth knowing something about the different kinds of eyepiece and their various characteristics.

Almost all eyepieces consist of two or more lenses. In some types, these provide better color-correction, while in others they give a wider field of sharp definition; a single lens gives good results only in the very center of the field, which is why it is rarely used. With a two-lens eyepiece, the component nearest the eye is called the *eye lens;* the other one, facing the objective, is called the *field lens.*

The most common type of eyepiece is the *Huygenian,* named after the seventeenth-century astronomer; it consists of two plano-convex lenses (i.e., one side flat, the other convex), with both flat sides facing the eye. Between the two components, at the focus of the eye lens, is a pierced diaphragm known as the *field stop*, which cuts off the badly defined margins of the field of view. A variation on the Huygenian eyepiece is the *Ramsden* type, in which the convex surfaces face each other. (Both types are shown in figure 20.) The advantage of the Ramsden is that the edge of the field is not so blurred; but neither of these eyepieces is truly achromatic, and for the finest definition a more sophisticated type must be used. This is especially important in the case of a reflecting telescope, since a Huygenian or Ramsden eyepiece will badly upset the perfect achromatism of the mirror.

An *Achromatic Ramsden*, similar though not identical to the *Kellner* (with which it is often confused), is often chosen for low magnifications. This consists of a plano-convex field lens and an achromatic eye lens, consisting of two separate components. An Achromatic Ramsden gives a fairly wide field of view, but it has one most annoying feature: The surface of the field lens is in the focus of the eye lens, so that any dust specks that happen to be present are painfully obvious. Another drawback is its tendency to internal reflections, so that bright stars often appear to have fugitive companions. These "ghosts" can be most misleading, and it is as well to remember which eyepieces in the battery are "haunted."

The *Orthoscopic* is another haunted eyepiece, but it gives a very wide field of view and a colorless image, although the definition falls off at the margins. It makes a very suitable eyepiece for low and medium powers. For high-power work, involving critical definition, the Tolles and Monocentric are very effective. The *Tolles* type, which is especially effective with reflectors, consists of a single glass cylinder, with both ends convex; the *Monocentric* contains three lenses cemented together. Since there can be no internal reflections in either of these types, they are quite ghost-free, and so are exceptionally useful for hunting faint stars in the vicinity of bright ones. They are also ideal for planetary work, in which study their small fields of view matter little. The *Erfle,* another eyepiece that has become popular in recent years, is somewhat similar to the Orthoscopic and is useful for low powers since it has an extremely wide field of view, although it is badly haunted. Another eyepiece that has gained in popularity is the *Plossl,* formerly called the *Dialsight* since it was frequently used in military and sur-

Figure 20. *Various types of eyepiece.*

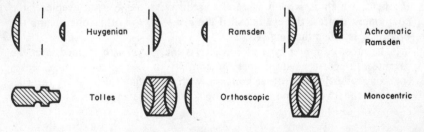

Huygenian   Ramsden   Achromatic Ramsden

Tolles   Orthoscopic   Monocentric

veying instruments. It consists of two similar achromatic lenses almost touching, their flint surfaces outward, and has the important advantage of large *eye relief*, combined with a reasonably wide, flat field. Eye relief is a measure of the distance between the face of the eye lens and the optimum point at which the pupil should be situated if it is to see the entire field of view simultaneously; it is, in fact, the point at which the eyepiece forms a little image of the telescope objective. The Ramsden, classical Kellner, and Tolles have so little eye relief that they are rather uncomfortable to use. Eyepieces with long eye relief, such as the Plossl and Orthoscopic, should have an eyecup incorporated to restrain the observer's tendency to bring his pupil too close to the eye lens.

It is worth noting that, of the types described, both outer surfaces of the Monocentric and Erfle, and the eye lens of the Achromatic Ramsden, are made of flint glass, which is softer than crown and easily marked by careless handling and cleaning.

By no means are all of these eyepieces easily available, certainly not in new condition. Probably 90 percent of all those sold today are Achromatic Ramsdens, Orthoscopics, and Plossls. Eyepiece fittings have been standardized in the United States with a 1¼-inch push fitting, but those coming from Japan, a major source of supply for European amateurs, have only a 24.5mm barrel. Old telescopes often had a 1¼-inch screw threaded fitting.

High-power eyepieces can be obtained with focal lengths as short as 4mm (eyepiece focal lengths are usually given in millimeters), but such tiny lenses are uncomfortable to use. One way of achieving very high powers without the discomfort of using tiny eyepieces is to invest in an *Achromatic Barlow lens*. This is a biconcave "negative" lens, so called because it diverges the light rays passing through it instead of converging them to form an image, as does the usual convex "positive" lens. It is supplied in a short tube that fits inside the telescope's drawtube, some inches in front of the eyepiece. The Barlow effectively increases the focal length of the objective without necessitating a corresponding increase in the length of the tube, so that the magnification is stepped up by an amount depending on the actual position of the Barlow. Some enterprising opticians have designed variable-power eyepieces on this principle, incorporating a negative lens that can be slid to and fro to produce a wide range of magnifications. Barlow lenses (also known as negative amplifiers) are usually designed to double the focal length, although some treble it. A good Barlow costs no more than an ordinary eyepiece, and since it doubles the range of magnifications available with a set of eyepieces, it is a worthwhile investment.

Because of the longer focal ratio of the objective, Barlow lenses work better with refractors than with reflectors; however, if well made and truly achromatic, there is no reason why they should not give excellent definition with either type of telescope. The one slight disadvantage is that the introduction of an extra lens into the system makes the image slightly dimmer, since each glass surface reflects away about 4 percent of the light falling on it; in the case of bright objects, such as the moon and Jupiter, however, this is of minor importance. If the image does appear too faint, though, it may be worth experimenting with a single lens, such as the eye lens taken from a high-power Huygenian. A single lens transmits more light than a compound eyepiece, and the extra brilliance of the image may well compensate for the lack of achromatism and sharpness away from the very center of the field. As a rough guide, the magnifying power of a Huygenian eyepiece is increased by about one third when the eye lens only is used.

Magnifying powers

To use a telescope to its fullest capacity, a number of different magnifications are necessary. A very low power, with a consequently wide field of view, gives the best results on extended objects, such as star clusters and comets; a medium power will show a large area of the moon in considerable detail; and a high power is necessary for the glimpsing of fine planetary markings and the resolution of close double stars. Depending on circumstances, each is essential, and it is no use trying to compromise. On the other hand, there is no point in buying great numbers of eyepieces. Four different magnifications are all that a small or moderate telescope really requires. Thus:

	Low	Medium	High	Very high
3-inch refractor	× 20	× 70	× 150	× 200
6-inch reflector	× 30	× 100	× 180	× 250
10-inch reflector	× 40	× 100	× 200	× 300

On excellent nights, it is possible to use powers as high as × 350 with a 6-inch or × 400 with a 10-inch, but it is doubtful whether such eyepieces will show detail more clearly than a lower power, for with very high magnifications the image of the moon or a planet is spread out over a large area and so appears much dimmer than when viewed with a lower power. The stars, being virtual points of light, are an exception to this rule and take magnification better. As a rough indication, it has been found that planetary observers using telescopes of between 6 and 12 inches aperture rarely employ powers greater than between 30D and 40D, where D is the aperture expressed in inches. A 3-inch refractor can give good results with powers as high as 70D; on the other hand, the great refractors at Lick and Yerkes give splendid performances with much lower relative powers. E. E. Barnard, one of the greatest observers of modern times, found that a power of × 1000 was on the whole too high for planetary work with the 36-inch telescope;

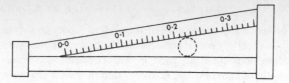

Figure 21. *Dynamometer. The value of the scale divisions depends on the angle between the two edges.*

and that powers of between × 350 and × 600 (roughly, between 10D and 20D) gave the best results.

The reason for this general deprecation of high powers lies not in the quality of the telescope but in the instability of the atmosphere. It is usually a gyrating mass of air currents at various temperatures, which refract the light irregularly and cause the image to swirl and flicker; thus, stars "twinkle" on very unsteady nights. Only when the air is calm and the image placid are very high magnifications feasible; even then, depending on the object under review, they may not be necessary. It will soon be learned from experience that, far from always using high powers, the lowest magnification that shows sufficient detail is much to be preferred. With a powerful eyepiece, not only is the image made dimmer and the field of view much reduced, but also every slight tremor of the telescope is augmented.

If a telescope comes equipped with its own eyepieces, the magnification given by each one may be inscribed somewhere on the barrel or eyecup. Usually, however, the eyepiece simply carries a note of its own focal length, and in the case of some ex-equipment eyepiece that has been stripped down, it may not even have that. Eyepiece focal lengths are not always quoted accurately, particularly on older equipment, but it is not difficult to check the magnification given by any eyepiece, and hence to find out its focal length if this is required.

The way to do this is first of all to focus the telescope on a distant object, and then to measure the diameter of the exit pupil, which is the little image of the objective formed by the eyepiece just beyond the eye lens. To do this properly, take a magnifying glass to examine the image, and use a transparent millimeter scale or a vernier caliper. A home-made *dynamometer* (shown in figure 21) could also be employed. Divide the diameter of the exit pupil into the diameter of the objective to find the magnification, and divide this result into the focal length of the objective to determine the focal length of the eyepiece.

When measuring the diameter of the exit pupil, it is a mistake to turn the telescope toward the bright daylight sky, since irradiation will make the disk appear larger than it really is. A white wall, or a sheet of paper, makes the best background for the objective.

You can pay from about $20 (£10) for a simple Achromatic Ramsden to

$50 (£25) or more for a fine Plossl, depending upon quality and source. They therefore deserve careful treatment; cramming them into the observer's coat pocket does them no good at all! The best container is a small wooden box with a hinged lid and false bottom, in which are cut a number of small holes to take the eyepiece barrels. In this way, they cannot shift around and scratch each other, and if they are always stored in the same order it becomes a simple matter to pick out the right one in the dark. Should the lenses become dirty, as may be the case after several nights' use, the offending matter should be gently whisked off with a soft, damp brush. Rubbing the surfaces with a cloth will certainly produce scratches.

The care of objectives

Second-hand object glasses are often not only dirty on the outside; they may also be stained on the inside surfaces. This often happens when the cold lens is taken indoors; dew forms on the glass, and it cannot evaporate quickly in the confined space. The same "sweating" sometimes occurs outdoors when the air temperature happens to rise rapidly, and ordinary dewing is a common feature of autumnal nights. Since prolonged moisture is bad for the lenses, and may even stain them permanently, it is a good idea to give such an object glass a proper cleaning. After that, if reasonable care is taken, there is no reason why it should ever have to be taken apart again.

Many writers insist that a lens should be dismantled only by a qualified optician; but there is no real reason for this, provided elementary care is taken. First of all, unscrew the cell from the end of the tube and place it face down on a pad of soft cloth. Next, unscrew the retaining ring in the back of the cell. The two components can now be lifted out by placing a handkerchief over the rear (flint) lens, turning the whole assembly face up, and lifting the cell away from the lenses.

Before they are separated, examine the edges for two adjacent pencil marks. These indicate the correct orientation of the components with respect to each other and the lenses must be reassembled in the same way. Few object glasses are quite insensitive to orientation, usually through fractional errors in lens thickness at different points on the circumference; if this precaution is not observed, therefore, the performance will suffer.

Now lift off the convex lens and lay it aside. It may be that three tiny pieces of foil, known as *spacers*, have been placed around the circumference, to keep the two lenses slightly apart. If so, these too must be kept in a safe place. The lenses themselves can now be cleaned, preferably in warm soapy water, washed down afterwards with clean water and stood on edge to dry. Special care must be taken with the flint component, for flint glass scratches more easily than crown. Once they are thoroughly dry they can be reassembled, with great care to match up the edge marks. The retaining ring, incidentally, should be left with a tiny amount of slack, so that the object glass gives a slight rattle when gently shaken; a tight grip may strain the lenses and spoil the performance.

While all large lenses are of this "air-spaced" type, the components of some small ones may be cemented together with Canada balsam, a colorless adhesive. It need hardly be added that such cementing should not be disturbed—in any case, no dewing could possibly occur between the surfaces of such an object glass.

The mirrors of a reflecting telescope are altogether more accessible. If they are aluminized, a six-monthly wash in detergent will clean off the dirt and keep them highly reflective for years on end, provided they are kept covered when not in use, and barring heavy industrial contamination. The old process of silvering has almost entirely vanished, although, since it reflects visual wavelengths better than does aluminium, and since also the sulphur compounds that are its greatest enemy have largely been banished from the atmosphere, there might be a case for reintroducing it. Silver, like aluminium, can be deposited by evaporation, and gives an optically smoother reflecting surface—hence less scattered light; but it reflects the blue end of the spectrum rather poorly.

Aluminium-coated mirrors are particularly affected by salt, for aluminium chloride is a dull white compound. For this reason, the surface must never be touched with the fingers. If a mirror is very badly tarnished, its condition can be improved in two ways that may sound drastic but do in fact work. The first is to wash it thoroughly in detergent, making sure that no hard particles remain on its surface. Lay it face-up. Place a dab of jeweler's rouge or, preferably, optical rouge on a clean plate and rub it with a damp tissue until the paper is well stained. The point of this is to break down any hard clumps that may be present. Then lightly rub all over the mirror's surface, using a circular motion, until the coating has become as bright as it will go without being removed altogether, and wash and dry. Another way is to dissolve a few pellets of domestic caustic soda in an eggcup-full of water and to pour this gently over the cleaned mirror's surface, swabbing it with a tissue so that all the coating is covered. (Caustic soda, as its name implies, is dangerous stuff and must be handled with rubber gloves.) The solution will, if all goes well, dissolve off the surface layer of tarnish and leave a bright film underneath. Wash off and rinse thoroughly immediately the treatment seems to have worked, or the whole film may disappear!

Mirrors must always be kept covered and dry when not in use. A circular disk of cardboard, covered with soft cloth and padded with cotton wool, is the best protection for the main mirror; this can be held down by the cap that fits over the cell. If the tube is of the solid variety, a small door somewhere near the bottom is required for the insertion and removal of the cover. A much lighter cap will serve for the flat. Similarly, in the case of a refractor, it is a good idea to put a pad of dry cloth inside the cap that covers the object glass.

Finders

The finder, a small, wide-field telescope fitted to the main instrument to allow it to be pointed quickly at the required object, is a most important accessory. This is another item that many manufacturers, being only remotely acquainted with the observer's many requirements, tend to dismiss summarily. They are often inconveniently placed, and may not even be adjustable. But the main trouble is that most finders are far too small. The usual type supplied with a 3-inch refractor has an object glass about ¾ inch across; and, while this is large enough to find the moon or Jupiter, it is far from satisfactory where fainter objects are concerned. The finder should be considered as an astronomical telescope in its own right. Nothing less than a 1½-inch aperture will give a good view of the sky, and a 2-inch telescope is better still. Such an instrument will not only make it easy to pick up difficult objects, but it will also show very extended features, such as star clouds and bright comets, even better than the main instrument.

Some really excellent low-power telescopes are available in government and industrial surplus stores specializing in scientific instruments. These make ideal finders for any telescope of from 3- to 12-inch aperture. I once bought a 7 × 50 monocular with a prism system similar to that used in binoculars, and found that it makes an excellent finder for a 3½-inch refractor. It had originally been designed as a gun sight, and the cost was less than $12 (£5); so that two, if mounted side by side, would work out considerably cheaper than binoculars of the same specification and quality. The only drawback in using a relatively large finder is that the telescope tube needs some sort of counterweight to restore the balance. Such a telescope also has another advantage: It gives an erect image and so provides a direct comparison with the naked-eye view.

The finder should contain cross wires to mark the exact center of the view, so that this point can be matched with the field of a high-power eyepiece on the main telescope. If these are not already included, they can be supplied by stretching two pieces of spider web, or even human hair, across the field stop of the eyepiece, so that they appear in focus against the sky. Unfortunately, if the observer is favored with the truly black skies of rural sites, he will find that the cross wires disappear from view! In this case, provision must be made for illuminating the wires so that they stand out against the sky. The way of doing this is to cut a small hole in the side of the eyepiece so that the light from a low-voltage bulb, wrapped in red cellophane, can shine on the wires. The illumination must be sufficient to make the wires clearly visible, but no more, or the faintest stars in the field will be lost. A dodge employed by some observers, particularly those who usually seek bright objects, is to place a scrap of colored filter at the center of the eyepiece field lens; when the object in the finder changes color, it must be at the center of the field. The method works best if a Ramsden-type eyepiece is used, since the field lens is very near the focus of the eye lens.

Dewing troubles

On a night when the air temperature falls rapidly and the air is calm, the active astronomer is apt to have a frustrating time. Such conditions often offer splendid atmospheric conditions, and the steady telescopic images are conducive to observation; but they also provide a heavy fall of dew, and this precipitation must at all costs be kept off the optical surfaces. It is infuriating to have to disrupt the observing session to let a component dry out; it is also harmful to aluminized surfaces, for repeated dewing can undermine the reflective film. Prevention being more satisfactory than cure, it is wise to see what can be done to combat this menace.

Dew falls on cold, exposed surfaces, and, in the case of a refractor, the best way of preventing precipitation on the surface of the object glass is to provide a dew cap. This is a long cylinder extending out over the lens. To be serviceable under all conditions its length should be about three times the aperture of the object glass; those supplied by the manufacturer are invariably too short, and it is best to make another one—preferably of some light plastic material—and line it with black blotting paper. If it is too long, however, the effective aperture of the telescope may be cut down slightly, and it is best to make it in the form of a funnel rather than a perfect cylinder. The object glass of the finder needs similar protection. If, during a night's work, the image begins to turn hazy—a sure sign of dewing—a piece of warm, dry cloth, such as a handkerchief, should be placed loosely inside the dew cap and a cover put over the end. The dew will vanish in a very few minutes. The shell or plate of a catadioptric telescope is particularly troublesome, since an effective dewcap has to be about as long as the tube itself! Many observers have found that a thick brown paper bag forms an effective, if hardly elegant, deterrent, the bottom having been snipped open until it fits over the tube. All dewcaps should be made of a poor heat conductor, so that the material tends to remain slightly above air temperature.

In the case of a reflecting telescope, dewing troubles are dependent on the nature of the tube. It is almost impossible for the main mirror to dew up if protected by a solid tube, but the flat is more troublesome. In this case, it depends on how far the walls of the tube project beyond the position of the flat. The material of the tube can also have an influence, metal offering less protection than some nonconductor of heat such as wood.

There is no way of directly shielding a flat, as there is for an object glass; but some observers have successfully overcome dewing tendencies by heating the glass. This is not as drastic as it sounds; a tiny amount of heat will raise the temperature of the surface fractionally above that of the air, and under these conditions no dewing can possibly occur. The best way of applying the heat is to insert a small flashlight bulb, wrapped in tin foil so that no light can escape, inside the mounting of the flat. The wires can be led away along one of the vanes and thence to the power supply. A transformer working from the domestic supply is convenient, provided that all contacts are insulated against damp.

When wiring up the telescope in this way, it is extremely useful to include a dim red or green lamp for general illumination. This can be fixed to the tube or stand so that it shines down on the notebook; if a variable resistance is included in the circuit, the brightness can be adjusted so that it is just enough for reading or making notes without eyestrain. Too bright a light will dazzle the eye and make it unfit for viewing faint objects, so it is essential to achieve just the right level.

If the reflector has an open tube, the main mirror is far more susceptible to dewing troubles. A certain amount of protection is offered by wrapping some material, such as black cloth, around the lower half of the tube; but if damp persists in forming, it may be removed by placing the cover, previously heated in some warm place, over the cell. A mirror must never be closed up for the night when damp, as the reflective film will suffer.

The most dew-prone item of all is the ocular's eye lens, which quickly mists up in the presence of the observer's body warmth. As soon as the lens itself warms up, the trouble vanishes. The best way of accelerating the process is to wrap the eyepiece in a clean handkerchief and carry it in an inside pocket for some minutes before the observing session begins.

Notebooks and atlases

The organized observer, before going to his telescope, adopts a plan of campaign. He decides just what he is going to observe and what materials he will need. This decision will be dictated, in part at least, by weather conditions. If the sky is very dark and the stars twinkle a great deal, it is the sign of a transparent but unsteady atmosphere. Planetary observation on such a night will probably be a waste of time; on the other hand, it will offer splendid opportunities for hunting faint objects, such as nebulae or any comet that happens to be about. These must be marked on a star chart, and the purchase of a good star atlas is just as important as the acquisition of a telescope.

The best-known and most universal publication, used by astronomers and observatories all over the world, is *Norton's Star Atlas and Reference Handbook*. The atlas covers the whole sky in sixteen separate charts showing all the stars visible with the naked eye and some fainter ones. It also includes a great many telescopic objects of interest. The associated handbook is no less useful, containing a multitude of facts and tables, and the work can rightly be termed essential for every practical observer.

Another widely available atlas is the *Atlas of the Heavens* (A. Becvár), published by Sky Publishing Corp., Cambridge, Massachusetts. This shows stars down to about magnitude 7·75, which is roughly the limit of a pair of 8 × 30 binoculars, and includes a catalogue that gives details of the color and type of all the naked-eye stars. While the atlas is more comprehensive, it is much more bulky than *Norton's* and not so convenient to use at the eyepiece. Becvár has followed this up with three large-scale atlases: *Borealis* (north), *Eclipticalis* (central), and *Australis* (south). Together, these atlases show more than 100,000 stars brighter than the 9th magnitude. These are obviously very useful if a very faint star or minor planet has to be identified, but the newcomer to astronomy is unlikely to require such sophisticated maps. Even more comprehensive is the *Photographic Star Atlas* compiled by a German amateur, Hans Vehrenberg. This shows stars as faint as the 13th magnitude, which is roughly the limit of an 8-inch telescope. Another useful publication is H. B. Webb's *Atlas of the Stars,* which shows stars down to magnitude 9·5 but does not cover the far southern regions. For observers in north temperate latitudes, however, this is no great handicap. N. E. Howard's *The Telescope Handbook and Star Atlas* includes fourteen maps showing stars down to the 6th magnitude, with the telescopic stars printed on transparent overlays. It also includes a listing of the Messier objects by season rather than by number.

Two famous atlases, which the amateur is unlikely to possess but which may be consulted in astronomical libraries, are Argelander's *Bonner Durchmusterung* and the Beyer-Graff *Stern Atlas*. The *B.D.* covers the northern half of the sky down to about 10th magnitude; the Beyer-Graff atlas descends farther south but shows stars only to the 9th magnitude.

Besides having a chart of the permanent features of the sky, the amateur needs to keep in touch with its changing aspects. The standard work on celestial movements throughout the year is the *Astronomical Almanac,* the former *Nautical Almanac,* which is published annually in both the United States and Great Britain. This contains daily positions of the sun, moon, and planets; details of the planets' rotation and the positions of their satellites; eclipse information; and a wealth of detail of far more complexity than the amateur is ever likely to require. The essential matter is included in another valuable publication, *Handbook of the British Astronomical Association,* which also gives details of expected comet appearances for the year in question. Unexpected events, such as new comet discoveries are relayed through the 24-hour I.A.U. Telegram Bureau at Cambridge, Massachusetts, and the *Circulars* of the British Astronomical Association. In addition to these, notes on coming phenomena are included in *Sky & Telescope,* an excellent monthly journal of astronomical affairs.

A scientist's worth is revealed by the state of his notebooks; from the beginning, the amateur astronomer must take the business of documentation seriously. If an observation is worth making, it is worth making well; one never knows whether a few innocent notes may not have an important bearing on some subsequent discovery. To begin with, when one is engaged in "learning the constellations," notes can be written up in a general observing book, preferably with alternate ruled and blank pages to allow drawings to be included. Later, when definite fields of observation are adopted, a separate observation book should be kept for each object; but it is also a good idea, as a sort of index, to make a brief note of such observations in the general book, using this for notes at the telescope, copying them later—a process, incidentally, that should not be delayed until memory has faded. Every observation must be prefaced by the date and time (preferably using Universal Time, which is 5 hours fast on E.S.T.), details of telescope and magnification, and the state of the air. The amount of care taken in documentation can make all the difference between an interesting observation and a definite record of permanent value.

Some luxury items

The observer's physical comfort is of paramount importance. At the instant of making a critical observation, every muscle in the body must be relaxed to allow the retina full play in picking up the faintest sensations. An observer's eye, as we shall see, must be educated to accomplish feats of vision considerably beyond what one might otherwise consider possible, and it is unreasonable to expect superhuman feats in the half-stooped, cramp-riddled posture one has to adopt with so many carelessly designed mountings. It is agonizingly uncomfortable to crouch halfway between standing and sitting in order to see through the eyepiece.

A set of *observing steps* provides the answer to this dilemma. Household steps (kitchen ladders) are serviceable, but are unlikely to be very comfortable to sit on, and a better answer is to make a special set, with ledges about 10 inches wide and rising to a height of about 3 feet. An attached rail or post allows one to keep perfectly still without clinging to the telescope itself, an act that will set the image dancing so violently that nothing can be seen.

All astronomers prefer to observe an object when it is high in the sky, since atmospheric conditions always improve with increased altitude, but a refractor is not altogether comfortable to use when the object under observation is almost overhead. One answer is to restrict one's observations to the period before or after the object crosses the *zenith*, or overhead point, since atmospheric conditions do not improve noticeably above about 60°. If this should prove inconvenient, one way of preventing vertebral contortions is to use a *zenith prism*, which effectively turns the eyepiece through a right angle and allows one to view at a much more comfortable angle. However, a zenith prism inverts the image from north to south, which is a nuisance when comparing the telescopic view with a chart, and it also involves a slight loss of light. Considering these severe drawbacks, it is amazing how many manufacturers apparently include them as more or less standard equipment with refractors—even for observation at low altitudes! Actually, zenith prisms do more harm than good, and they should be avoided wherever possible.

As already mentioned, a reflecting telescope is a most agreeable instrument to use, for when mounted on an altazimuth stand the eyepiece is always horizontal. An equatorial mounting, however, can give the observer some bending and crouching problems, and a rotating tube is a useful refinement. The usual way is to have the tube freely held in a cradle on the declination axis so that it can be twisted to bring the eyepiece into a convenient position, while some models have just the upper portion of the tube rotatable. Unless both mirror and flat are in the very center of the tube, this differential rotation may bring the optics out of alignment; so the former method is, on the whole, preferable.

Painting and upkeep

Only small and medium-sized telescopes can be either wholly or partly carried indoors at the end of each observing session, and, as we have seen, it is preferable for an equatorial mounting to be left permanently in position. Larger telescopes must endure the elements, but there is no reason at all why they should come to harm. The optical parts must be covered with scrupulous care, and all moving parts must be well greased; as for the rest, aluminium paint is an excellent protector of metal, and has the additional advantage of making the instrument conspicuous in the dark. The inside of the tube should be given an occasional sweep to remove any particles of rust or paint that might fall on the mirror and harm the surface when the cap is replaced, but except for these elementary precautions a well-made instrument should need no special attention.

The main war to be waged on behalf of an outdoor instrument is against insects. Beetles and cockroaches may take up residence in the stand, and spiders can spin their webs in the tube—even in the nether regions of a refractor. An open-tube reflector is particularly susceptible to this menace, and little preventive action can be taken. However, a lid can be fitted into the mouth of a solid tube, and an old eyepiece fitted into the drawtube will keep both insects and dust out of the interior. It is strange but fortunate that, even though the observer takes care to collect only good eyepieces, he always acquires a store of old Huygenians that are quite useless optically, but do serve this important function. Inevitably, too, the amateur will become intimately acquainted with an insect marauder when it crawls across the field lens of his eyepiece and appears projected with astonishing relief against the moon!

Observatories

Every amateur who has spent the quiet watches of the night communing with the stars and coping with the various problems of nature—dew, wind, and chill—has sighed for a proper observatory in the remote countryside, with a clear view of the whole sky and no artificial lights to dim out the fainter stars. Alas, few of us have much choice in the matter, and there is certainly little chance of being able to build an impressive domed structure in the back garden. But, even so, there is really no reason why a person who is capable of building a telescope mounting should not be able to construct some sort of observing shelter. The simplest contrivance is better than none at all, and will certainly add much to the comfort of observing. Astronomers are, by tradition, hardy folk, but this is no reason for courting unnecessary agony!

The simplest protective cover for an observatory is a run-off shed, which is simply a box on wheels that rolls away to reveal the telescope. This is very suitable for a reflector, since this type of telescope is compact, especially if on an altazimuth or German stand. A rigid timber or angle-iron framework provides the basis for the covering, which can be asbestos sheeting, wooden slats, or even heavy-gauge plastic. The box rolls on rails that are sunk into the ground. When it is in position with the door closed, the telescope is perfectly protected; moreover, if made sufficiently large it also forms a shelter in which the observer can write up notes during a night's work. Some form of illumination and a shelf for books and eyepieces makes this kind of shelter a very cheap and handy device. But it does not serve one of the implied functions of an observatory, which is to give the observer, as well as the telescope, some protection from wind and dew while at work.

The gravest drawback of this type of observatory is its instability. It is almost certain to be higher than it is broad, and if a strong wind chances to

Simple observatory. *A. W. Heath's rotating hut.*

blow on either beam it may be upset, with possible disaster to the telescope. One way to prevent this possibility is to mount a second rail *over* the wheels, so they cannot lift off the ground. A less satisfactory solution is to fit low projections, either to the shed itself or in the ground, that will hold the observatory steady if it starts to tilt; these, however, may bring possible disaster to the observer!

The next step in sophistication is to have the walls of the shed permanent and merely slide the roof back. This, the sliding-roof observatory, is the best type of all, since it is cheap, protects the telescope and observer from wind, and permits coverage of the complete sky at one time. This is something that cannot be had from inside a dome, where only a segment of the sky is clear; in fact, a dome is a grave disadvantage for anyone who does not know the night sky intimately, since only a few constellations can be seen at any one time.

The main secret of building any sort of observatory that involves having permanent walls around the telescope is to allow plenty of room for the observer. An observatory should never be less than ten feet square, and preferably twelve; cramped conditions will merely hamper the observer's efficiency, and he would do better with a moveable shed. A second point is that the walls must be sufficiently low to allow the telescope to point down to the horizon. Since the tube of a reflector is usually mounted low down, this immediately cancels one of the advantages of this type of observatory, since the low walls will provide little or no protection. All things considered, the sliding-roof observatory works best with a refracting telescope.

Everyone will have his own ideas on the details of construction. The walls can be of slatted wood, or even brick, while the roof, which must slope, can be covered with fiberglass sheeting. Generally speaking, it is best to have the roof sliding off to the north, since this part of the sky is likely to receive the least attention and the obstruction will not be serious; but, whichever direction is adopted, a free-running motion is essential. Nothing is more infuriating than a roof or dome that persistently sticks. Depending on the observer's agility, he can either admit himself by a door or else vault over the wall when the roof is slid back.

The best observatory for a reflector is a totally revolving affair, like that illustrated here. In this type, the entire hut rotates, with a wide slit running across the roof and down one (or both) of the walls, so that the telescope commands an uninterrupted view from horizon to zenith. A revolving observatory requires a properly leveled concrete foundation, with a circular track to take the wheels, but otherwise it is a much less complicated affair than might be imagined; in the case of the observatory in the photograph, the only part that had to be made professionally was the 8-foot-diameter angle-iron ring, $\frac{1}{4}$-inch thick and with 2-inch sides, that is built into the base of the hut. This rotates on eight pulley wheels, which are held in the concrete. The rest of the building, made of wood and asbestos sheeting, involved only simple carpentry.

The disadvantage of a revolving observatory is that, since the slit is relatively narrow, it requires constant rotating if a large area of sky is being covered during the observing session. This particular structure houses a 12-inch reflector that is used almost exclusively for planetary work, and it functions very well since the only shifting required is that needed to follow a planet's drifting across the sky. For anyone with free-range ideas, however, this type may prove rather limiting. It is difficult enough to make a critical observation, without having to attend to the observatory itself every five minutes.

Nevertheless, owning an observatory gives one a tremendous sense of purpose. Everything is at hand; books, atlases, eyepieces, and the miscellany of minor items that the observer is likely to need at a moment's notice, can all be reached immediately. Moreover, there is another point. When observing faint objects, it is vital to get the eye thoroughly dark-adapted. It takes about half an hour for the retina to become fully sensitive, and the interruption of having to return to a brilliantly lighted room ruins the adaptation, which must then start all over again. For this reason, the observatory should be lighted by the bare minimum of deep red light; it is even better if the actual intensity can be varied at will. It is also handy to have an ordinary white light for use before the session begins; but in this case, as in a photographic darkroom, the switch must be easily distinguishable in the dark.

When weighing the pros and cons of the matter, it is essential to keep a sense of proportion. An observatory is a desirable luxury, but no more, and money should certainly not be spent on building an elaborate structure that might better be invested in a larger telescope. Tarpaulins are cheap and will protect a small or medium-sized telescope from the worst ravages of weather; and no observer with the necessary zest will be put off by a frosty wind in his face. Some of the world's greatest amateurs have had nothing between them and the elements. Herschel put his telescope out on the lawn at the back of his house; so did Denning, one of the greatest planetary observers of recent times; and so do many leading observers today. The worth of a telescope is not to be measured in terms of its immediate environment; in the long run, its own quality and, above all, the quality of the man at the eyepiece, are what matter.

5

Atmosphere and Observer

Many beginners in astronomy are discouraged and their enthusiasm is sharply diminished when their first glance through a telescope proves disappointing. And with just one celestial exception—the moon—this will almost certainly be so. Great expectations of a canal-streaked Mars or a cloud-scribbled Jupiter come to nought when the novice's telescope shows a tiny, trembling blob of light on which little or no detail can be seen. His immediate reaction is that the telescope must be at fault, or the state of the atmosphere; everything is blamed except what is really responsible—the inexperience of the observer!

Learning to see

A little thought will show that this initial disappointment is only to be expected, for the discernment of faint detail through an astronomical telescope is perhaps the severest possible test of discipline for the human eye. In just the same way as a violinist's left hand is trained to a quite exceptional degree of suppleness and control, so an observer's eye must refine its reactions far beyond the level necessary for ordinary day-to-day vision. When the slightest glimpse of a wispy marking may herald a world-shaking change on the surface of a planet millions of miles away, it is hardly surprising that perceptual observations can be made only by the experienced eye. Tales are legion of how old hands, using a small telescope—perhaps a 3-inch refractor or 6-inch reflector—have perceived new details long before they became apparent to other amateurs possessing technically superior equipment, most of whose potential was nullified by the observer's inexperience. Sir William Herschel, whose visual acuity has been surpassed by few, summed the matter up as follows:

You must not expect to see *at sight. Seeing is in some respects an art which must be learned. Many a night have I been practising to see, and it would be strange if one did not acquire a certain dexterity by such constant practice.*

The lesson is clear. No opportunity should be lost to train the eye to work with the telescope; to observe the same object with different powers so as to see the effect of magnification; to try to see faint stars; and to draw planetary markings. In the beginning, to be sure, this may all seem to be wasted effort; the observing book will fill up with valueless sketches and brief notes of failure. But this apparently empty labor is absolutely essential; for, as the weeks pass, a steady change will be taking place. Objects considered difficult or impossible to see will now be discerned at first glance, and fainter specters will have taken their place. Indeed, these former features will now be so glaringly obvious that the observer may suppose that some radical improvement has occurred in the observing conditions. But the credit belongs entirely to the eye, which, in Herschel's expression, is "learning to see"—and the more practice it has, the better it will see. Success in astronomical observation comes only with persistence; it cannot be emphasized too strongly that nominal telescopic power counts for relatively little against *the observer's perceptive powers.* Again and again, one hears a beginner sighing for a bigger telescope because the one he possesses "isn't powerful enough"! A 3-inch refractor—assuming, of course, that it is of good quality and well adjusted—will show far more detail than may at first be suspected; and its full potentiality must be mastered before the special advantages of a bigger telescope can be appreciated. The amateur who immediately arms himself with, say, a 12-inch reflector, will be severely handicapped over the observer who patiently educates himself with a small and manageable instrument.

Just occasionally one comes across feats of astonishing vision, far beyond that expected to be acquired from experience alone. The American observer S. W. Burnham discovered with a 6-inch refractor new double stars that had previously been seen as single with telescopes of two or three times the aperture. An Irish amateur, Isaac Ward, glimpsed two of the satellites of Uranus with a 4·3-inch refractor; they are usually fair tests of vision with even a 12-inch telescope! An English clergyman, W. R. Dawes, was also famous for his ability to detect very faint stars. Such achievements are too rare to be of application to the vast majority of amateur and professional observers, but they form interesting instances of exceptional development of one of the natural senses.

The atmosphere and "seeing"

Part of the reason the eye has to play an active rather than a passive part in making an observation lies in the dense atmospheric layer that extends for about ten miles above the earth's surface. Above this, the air is too thin to have any effect on the telescopic image, but the lower level, being hardly ever calm, can ruin it. The heating of the ground below this level and the movement of winds act together to mix the lower atmosphere into a soup of strata at different temperatures, and since the light rays from a star or planet are slightly refracted each time they change strata, the net result is frequently to produce a wobbling, blurred image, an effect that astronomers term "bad seeing."

Even on the best nights, there is still a slight tremor in the air, which is enough to mask the finest detail on a planet's surface; but just occasionally, for periods of a second or two at a time, the image suddenly shrinks and steadies itself into a neat, sharp disk. This is the moment when the eye must play its crucial part, rapidly picking out the faintest shades of texture so that the hand, when the moment of perfect vision has passed, can transfer them to the sketch. The same is true of attempts to see very faint stars, which may be glimpsed only when the image stops boiling and concentrates into a point. On such nights, by far the greatest proportion of the observing period is spent waiting for these moments.

Bad seeing affects different apertures in different ways, being more serious the larger the telescope; but it is certainly not true to say, as do some observers, that a small telescope is equal or even preferable to a large one! This misapprehension has probably arisen because there is greater general steadiness of the image in a small telescope. On most nights air waves of between 6 and 12 inches across, travel quite slowly at a height of several hundred feet above the ground. It is not hard to see that if these waves are larger than the aperture of the telescope, their effect will be to move the image bodily rather than to break it up into fragments, so that while the image of a planet will wobble no matter what telescope is used, it will maintain its general outline rather better with a small aperture. On the other hand, critical definition can be obtained only when a period of true steadiness occurs, and in such an interval a large aperture will show far more detail than a small one; a night on which a small telescope is constant- ly superior to a large one can be written off as quite impossible.

Low seeing, as this atmospheric condition is called, is produced by local agencies. For instance, anyone living on the lee side of a city will experience very bad conditions—at any rate, until the early hours of the morning, by which time the heat from chimneys and factories has died down. On the other hand, an observer actually in the city may enjoy quite favorable seeing, since he is in the center of the "warm spot" and the air above his head is relatively homogeneous; the haze associated with such conditions may also help to steady the lower atmosphere. Hilly surroundings also help produce turbulence, especially when a wind is blowing, and it is for this reason that so many big observatories have been built on plateaus at heights of between 4,000 and 7,000 feet, above the densest and dirtiest layer of the atmosphere.

Of more frustration to the observer, in the sense that nothing can be done about it, is *high seeing*. This effect occurs at heights of between five and ten miles, where winds spring up as a result of large-scale pressure changes associated with meteorological "fronts." Once again we have the disruptive effect of layers of air at different temperatures; but since the waves are much farther away from the telescope, the blurring effect is greater and the wobble less than that produced by low seeing. Being essentially a climatic phenomenon, high seeing varies with the observer's position on the earth's surface, being less serious at low latitudes because of the more uniform pressure.

The third atmospheric effect, *ground seeing*, is at least partly under the observer's control. This is the disturbance produced by heat waves rising from the warm ground in the immediate vicinity of the telescope; naturally, this occurs mostly during the day and in early evening. The trouble can be minimized by surrounding the telescope with grass or scrub, as has been done at such observatories as Mount Wilson and Palomar, since this radiates very little heat; but a town-based telescope will inevitably prove troublesome when used during daylight hours. Also, the observatory itself can radiate a great deal of heat from its walls and roof, and, since the air inside warms up during the day, it should be opened up at sunset to allow free circulation to take place.

Even more locally, ground seeing can affect the air lying within a few feet of the ground, producing a thick stratum of warm air that may extend only as high as the observer's shoulders. When this condition is severe, it may be found that a refractor on a tall tripod outperforms a reflector on a low stand, especially if the reflector has a framework tube that leaves the mirror immersed in the warm layer. In this connection, experiments with thermometers placed at various heights above the ground may prove of interest.

Transparency is also important to the observer. This has nothing to do with the stability of the air, but is instead a measure of how bright the stars appear. In this area of observation, the rural observer is at a permanent advantage; the clear skies of country districts show the stars in all their majesty, a sight denied the town-dweller, who can never hope to see diaphanous objects, such as nebulae and comet tails, in their true form.

Atmospheric steadiness, which frequently occurs in the haze and smoke of a large settlement, is of little help in observing such extended objects; transparency is the deciding factor. When trying to observe very faint stars, however, it is essential for the air to be still as well as clear, for turbulence will expand the tiny pinpoint of light, and may make it invisible.

Sky transparency varies from night to night. Under the best conditions, when the moon is absent and the sky a pure black, stars as faint as magnitude 6·5 can be seen by a keen-eyed person. (It is worth noting, in this connection, that naked-eye acuity gives little or no guide to a person's telescopic vision.) On other nights, the faintest stars are nowhere to be seen and the Milky Way is visible only as a vague stain. This may be due to haze, but it can also be caused by faint auroral activity in the upper atmosphere, making these nights, to the rural observer, noticeably brighter than others.

Scales of seeing

1. Image usually about twice the diameter of the third ring.
2. Image occasionally twice the diameter of the third ring.
3. Image of about the same diameter as the third ring, and brighter at the center.
4. Disk often visible; arcs (of rings) sometimes seen on brighter stars.
5. Disk always visible; arcs frequently seen on brighter stars.
6. Disk always visible; short arcs constantly seen.
7. Disk sometimes sharply defined. (a) Rings seen as long arcs. (b) Rings complete.
8. Disk always sharply defined. (a) Rings seen as long arcs. (b) Rings complete, all in motion.
9. (a) Inner ring stationary. (b) Outer rings momentarily stationary.
10. Rings all stationary. (a) Detail between the rings, sometimes moving. (b) No detail between the rings.

Users of refracting telescopes may care to class their images in accordance with this scale, devised by W. H. Pickering on the basis of observations carried out with a 5-inch refractor. Seeing 1–3 is considered very bad; 4–5 poor; 6-7 good; and 8-10 excellent. Since few reflecting telescopes will show diffraction rings, however, users of reflectors must base their estimates on the smallness and sharpness of the star disk itself. There is usually a difference of some three scale divisions between the seeing at altitudes of 20° and 70°.

Pickering's scale, being based on the appearance of star disks, is not very suitable for planetary work, where the observer is scrutinizing an extended image. The Greek astronomer E. M. Antoniadi, well known for his work on Mercury and Mars, produced the following seeing scale for planetary work:

I. Perfect seeing, without a quiver.
II. Slight undulations, with moments of calm lasting several seconds.
III. Moderate seeing, with large tremors.
IV. Poor seeing, with constant troublesome undulations.
V. Very bad seeing, scarcely allowing the making of a rough sketch.

The scale used in recording observations is best specified by writing Pickering's in Arabic and Antoniadi's in Roman numerals. Note that the two numerical sequences work in the reverse order.

Sky conditions and "changes"

By now, the reader should be aware that making a critical telescopic observation demands reconciliation of a number of independent factors. The three most obvious are the observer's acuity, the telescope's quality, and sky conditions. On top of these are the less definable influences such as dark adaptation, the telescope's collimation, the observer's mood, the particular eyepiece used, and so on. But of all these, atmospheric conditions probably have the most far-reaching effect. No two nights are exactly alike.

Burnham, the famous double-star observer, once said:

> An object glass of 6 in. one night will show the companion to Sirius perfectly: on the next night, just as good in every respect, so far as one can tell with the unaided eye, the largest telescope in the world will show no more trace of the small star than if it had been blotted out of existence.*

Two facts emerge from this: Each night must be treated on its merits; and the most scrupulous care must be taken when comparing observations made under different conditions. It stands to reason that the telescope and magnification should remain unchanged if comparable views are to be had; but unless conditions are similar, "changes" will almost certainly be observed, for which responsibility must be placed not on the object under observation, but on the *atmosphere*.

Another overhasty conclusion indulged in by many people concerns the myth that some regions have "clearer" skies than others, in the sense that the air is more transparent. Thus, exotic regions, such as the Mediterranean, are endowed with star-crammed nights, whereas the much-maligned British climate is considered incapable of producing skies to the same specifications.

*Sirius, the brightest star in the sky, has a faint star close to it which is very difficult to see.

It is true, of course, that the Mediterranean and North African climate, like that of Arizona, favors cloudless nights; but on the question of transparency there is probably nothing to choose among different regions of the world. For vindication of British skies one need look no further than the work of Sir William Herschel; and the comet-hunter G. E. D. Alcock, who observes from a site near the large town of Peterborough, England, has made the following interesting comments:

> Conditions have deteriorated in the Peterborough area during the past 35 years, but this has never been an ideal part of England for astronomical observation, and it is very easy to exaggerate any changes that may have taken place.
>
> Obviously, there are two. First, there is the glow from street lighting and advertisements; second, the polar air which comes with the passage of cold fronts is more filled with smoke from Birmingham, the Potteries, and above all the Sheffield and south Yorkshire industrial zone. Formerly, we could expect transparent dark skies with N.E. winds, but now the visibility falls to less than three miles, and the smoke is lit up by artificial lights.
>
> However, we can still have superb conditions; occasionally the skies of the East Midlands can surpass those of Italy and North Africa. On October 12, 1948, and again at dawn on December 4, 1964, the sky was so brilliant that the Galaxy [the Milky Way] appeared "granulated" with faint stars along the entire band of light. Between 1942 and 1945 I never suspected this effect in the Mediterranean area.

Just occasionally, geophysical effects can interfere with sky conditions. The faint auroral glow is one, but there are sometimes more spectacular agents. The explosion of the Indonesian volcano Krakatau on August 27, 1883, released a tremendous amount of fine dust into the upper atmosphere and produced noticeable obscuration for many months; while the eruption of the Mt. Agung volcano on the Indonesian island of Bali, on March 17, 1963, reduced sky transparency all over the world, the effects lingering until the end of 1964. It is significant that the lunar eclipses of 1964 were unusually dark. Unusual atmospheric phenomena followed the eruption of Mt. Helena in 1978.

Astigmatism

We have already seen that exceptionally sharp naked-eye sight does not necessarily mean that a person will excel in telescopic observation. This is because a telescope can be focused to anyone's individual requirements, so that even chronic shortsightedness or farsightedness is of no account. There

Figure 22. *Test for astigmatism.*

is, however, one inherent eye defect that can be troublesome, and this is astigmatism, caused by the corneal lens being strained in one direction. Even slight astigmatism in any of a telescope's optical components is ruinous to its performance, so it is scarcely surprising that astigmatic eyesight can also cause trouble. The best way to find out if one's eyes are astigmatic is to look at a system of radiating lines, as in figure 22. A normal eye will show all the lines equally sharp and dark, but, should astigmatism be present, one or more lines will be more sharply defined than the rest. Astigmatism can ordinarily be solved by wearing glasses, but it is difficult or impossible to use a telescope in this way.

Luckily, astigmatism makes itself really objectionable only when low-magnifying powers are being used. The reason is not hard to understand. Every eyepiece has what is known as an *exit-pupil,* which effectively is the diameter of the beam of light leaving the eye lens and passing into the observer's eye. In measuring the circle of bright light seen in the eye piece, to determine the magnification, we were measuring the diameter of the exit-pupil, the relationship being

$$E = \frac{D}{M}$$

where E is the diameter of the exit-pupil, D the diameter of the objective, and M the magnification. The lower the magnification, the larger the exit-pupil; and the larger the exit-pupil, the greater the area of the corneal lens that is used to form the image. If a high magnification is used, and the exit beam is narrow, only a small proportion of the corneal lens is called upon, and any inherent astigmatism will have less effect. However, this does not help the observer with moderate or severe astigmatism who wishes to use a low-power instrument. (I can write from personal experience here, having suffered from serious astigmatism throughout my observing life.)

Binoculars such as 7 × 50, 10 × 50, and even 8 × 30, show stars as short lines or crosses, depending upon how they are focused. Only with high magnification on an astronomical telescope do stars appear as round dots. The star images can be corrected by wearing glasses, but now the eye cannot reach the exit pupil of most eyepieces, so that only the center of the field can be seen.

The problem can be solved, however, by anyone with access to a lathe, drill press, some coarse carborundum powder, and a few pieces of scrap. In other words, call in the services of one or more members of the local society! Remove the lenses from an old, unwanted pair of spectacles. Manufacture a "trepan," a piece of thin brass tube about ¾ inch across, mounted axially on a shaft that will fit into the drill chuck. The tube need not be more than an inch long. Stick each lens in turn to a wooden board, using pitch or waterproof adhesive tape. At slow speed, and with a mixture of carborundum powder and water gently dabbed onto the surface of the glass, carefully grind through the lens. With care, and luck, a circular compensating lens will result. My first view through binoculars equipped with correctors made in this way was a revelation after many years of astigmatic star images. The lens must, of course, be oriented correctly, so that the line of correction matches the direction of the eye's astigmatism.

Low and high magnification

But even if the eye is perfectly formed, there is still a practical limit to low magnification; this is decreed by the aperture of the telescope. It is clear that, after struggling to obtain as large a telescope as possible, the observer will not want to waste any of his hard-won light; yet this is what will happen if he uses a magnification of, for example, × 6 with a 3-inch refractor. The exit-pupil will now be ½ inch across—about twice as large as the maximum pupil opening—so that only a quarter of the light focused by the object glass is actually going into the eye.

In perfect darkness, the normal pupil expands to about ⅓ inch; in dim twilight, it is about ¼ inch. This means that to use a magnification of less than 3D or 4D results in a waste of light; theoretically at least, the lowest feasible power with a 3-inch telescope is between × 9 and × 12; with a 6-inch, between × 18 and × 24; while a 12-inch telescope can never profitably use less than × 36 or × 48, depending on the conditions of darkness.

However, such low powers do not show the faintest stars that the telescope is capable of revealing. The reason for this lies in the background luminosity of the night sky. Even in country skies totally free from artificial light, the atoms of the upper air are releasing bursts of radiation that give rise to a permanent airglow, while at times auroral activity can brighten this glow considerably. This means that although the telescope is forming an image of a star in the focal plane, it may be submerged in the sky brightness. Now, the image of a star is practically a point of light, and considerable magnification is needed to make its spurious disk obvious, so that for present purposes its apparent size can be regarded as being independent of the magnification used. The same is not true of the circle of skylight in which the star lies. Let us suppose that the apparent fields of view of the eyepieces in a set are all 50°, and that they give magnifications of × 25, × 50, and × 100. The × 25 eyepiece will therefore show a disk of sky 2° across expanded into an apparent diameter of 50°, while the × 50 and × 100 eyepieces will accept only 1° and ½° of the sky respectively, but will expand them into the same apparent size. Therefore the 50° apparent field of the × 25 eyepiece will contain all the background light from a circle of sky that has four times the area of the × 50 field and 16 times the area of the × 100 field. The relative darkness of the background, when viewed through the eyepiece, must be in these ratios.

This simple theory is slightly misleading. It does not mean that stars 16 times as faint can be seen by using four times the magnification, since the star image itself does in practice lose a slight amount of intensity through magnification. But the sky-darkening effect of a powerful eyepiece undoubtedly does bring fainter stars into view, and the effect is most obvious in bright urban skies.

As we have seen, the highest useful magnification is dictated mainly by atmospheric conditions; but even on nights of superb seeing (which occur mainly in observers' dreams), a power of more than about 60D for 3- to 6-inch telescopes, and about 40D for those of up to 12-inch aperture, will show no detail that is invisible with smaller magnifications. They effectively exceed the telescope's limit of resolution, so that for most purposes such magnification is "empty."

Temperature effects

The temperature drop at nightfall can, as we have seen, cause undesirable tube currents in a reflecting telescope, which can be overcome by using a square or framework tube. It can also, however, affect the actual quality of the image by temporarily distorting the mirror. This thermal effect can be very important indeed in the case of large telescopes, and is well worth investigating.

To give perfect definition, an astronomical mirror must be figured to within a very few millionths of an inch of the correct curve. This curve is not spherical, but *paraboloidal*, which means in effect that the mirror is slightly deeper than it would be were it a sphere. If the mirror were merely spherical, the image would be faulty and the defect would be termed "under-correction." If the mirror's curve were too deep, or *hyperboloidal*, it would be "over-corrected," and once again a star or planet would have a hazy appearance. These errors, however, are minute by all except optical standards. For instance, the curved surface of a 6-inch f/8 mirror is about 1/20-inch deep—hardly noticeable to a casual glance. To deepen such a spherical curve to a paraboloidal one involves the removal of a mere 1/80,000 of an inch of glass; yet the difference bridges the gap between an almost useless mirror and a splendid one!

The heating troubles arise because the glass has to be thick; were it thin, the mirror would sag under its own weight and ruin the definition. This explains why object glasses, where the lenses have to be supported around the edge, have a definite restriction on size. The disk for a 6-inch mirror is normally an inch thick, and the 6:1 ratio for diameter to thickness is a standard one; hence, large mirrors are very heavy indeed. As a result, when the glass is warm it expands and distorts the curve, and it may take an hour or more for a 12-inch mirror to cool down thoroughly by just a few degrees. During the time it is radiating heat, the correction is increased. A paraboloid becomes hyperboloidal, whereas an undercorrected mirror turns temporarily paraboloidal if the error is just right.

Most astronomical work is done during the night; this is why most good opticians intentionally make their mirrors slightly undercorrected, so that they perform well in the cooling air. The method is not, of course, foolproof, for unexpected temperature variations can have a prejudicial effect on the curve. The effect is usually slight, but for critical work it can be obvious enough, and it certainly pays to "know one's mirror" under various conditions of heating and cooling.

The observer

The final and most vital link in the whole chain is the observer himself, and on his competence, more than on any other factor, depends the success of the observation. Telescopic observation is an art, a fact too easily forgotten when large instruments are glamorized on their own merits, as if they themselves recorded the objects to which they were pointed! It is true that the introduction of photography has made many fields of observation a simple technical exercise; but in most amateur investigations the eye reigns supreme, the telescope being simply a tool that will do the job well or badly, depending on the skill and experience of its user. There are perhaps three broad headings under which an observer must discipline himself.

EDUCATION. We have already seen that the eye requires plenty of experience in telescopic work before it can hope to discern all that the telescope is capable of revealing. The interesting thing is that even after experience has been gained, many observers find their eyes particularly gifted in just one field of observation. E. E. Barnard, the famous American comet-hunter who also made stellar observations with the 40-inch Yerkes refractor, could glimpse very minute stars; yet he saw much less planetary detail than other observers using small telescopes. The same was true of the "eagle-eyed" Dawes, who had exceptionally acute stellar vision. Conversely, many successful planetary observers have proved defective when it came to glimpsing small satellites and faint stars. Seemingly, two different visual processes are involved, and it is clearly sensible to make observations in the field in which one's talents lie. One excellent idea is to train each eye separately, one for planetary work, the other for the observation of very dim objects.

PREPARATION. Spasmodic and ill-considered observation is not only useless, but also wastes energy and enthusiasm. The observer should know exactly what he intends to observe before he goes to the telescope; assuming that he has reached the stage of knowing the main constellations and the positions of the planets, it is possible to cover a wide field and yet work systematically. The best education of all, because it exercises the eye and also teaches one the layout of the night sky, is to survey each constellation in turn for its more prominent features: colored stars, double stars, star clusters, and nebulae. If a list is prepared in advance, they can be found, or "swept up," in conjunction with *Norton's*, and notes made on their appearance. After several hundred of these objects have been identified, the observer will have a vastly different attitude toward the apparent chaos of the sky, which will now have resolved itself into familiar and unforgettable patterns.

It is the same with lunar and planetary work. If it is decided to observe a certain region on the moon, or a particular planet, it must be viewed on every possible occasion when conditions are good enough. The aim should be to produce a series of observations that may be compared with one another, and this can only be done by being prepared whenever an opportunity presents itself.

RELAXATION. It is useless to try to make a critical observation under conditions of strain. The problem of muscular comfort can usually be solved by using some form of observing steps, but the effects of cold are less easy to combat; the only practical agent of warmth on the coldest nights is the observer's own enthusiasm! Once one becomes really uncomfortable, it is far better to close down, or at least go indoors for a while, since the reliability of observations made under such conditions is open to question, and no observation at all is better than a spurious one.

When observing, the idle eye should be kept open; the strain involved in shutting the lid has a prejudicial effect on one's vision. This takes a certain amount of practice; at first, the half-seen images seem to obliterate rather than improve the view, but after a time they can be ignored. Another physiological habit that must be mastered is the tendency of the eye, when fatigued or when trying to catch a detail at the limit of its powers, to focus on something near at hand rather than to relax at infinity. When this happens, the critical sharpness of the image may be destroyed without the observer realizing it. Therefore, it pays to pause every five minutes or so, throw the eyepiece out of focus, and then refocus carefully.

PART

II

THE SOLAR SYSTEM

6

The Moon

The moon is without doubt the favorite object for observation by the beginner with a small telescope. The reasons for this are really too obvious to need mentioning: It is bright, it is visible regularly every month, and it is so close that a fantastic amount of detail can be made out. The fascination of seeing so literally unearthly a world in close-up never palls, and it takes no astronomical knowledge to be impressed by the way every rock casts its own sharp shadow, with the great craters looming up in indescribable magnificence.

The moon is the earth's natural satellite, and, whether or not it once formed part of the earth (few scientists support this theory nowadays), we can at least be sure that they were both formed at around the same time—some 4,500 million years ago. The moon is a small world, however, only 2,163 miles across, and it lost its internal heat and its atmosphere relatively quickly. These factors, together with the fierce scarring of its surface, have produced a world bearing absolutely no resemblance to the earth. Where there is no air there can be neither rain nor wind, and consequently no erosion; hence, the lunar topography, most of which is probably older than the most ancient mountain chains on our own planet, has been preserved in its original form, as if in some celestial museum. Only with the brief impact of *Luna 2* in September, 1959, did the moon begin to awaken to the present.

The fact that the moon is so unchanging—on the large scale, at least—means that modern observations can be compared directly with those of our predecessors. This is not so with the other planets of our solar system, whose surfaces show slow or rapid changes as the case may be. The dark lava plains, known generally as *maria* (seas), have made up the face of the "man in the moon" since the beginning, and a telescope not only brings these out in great detail, but also shows the vast walled plains, craters, craterlets, and mountain ranges that were first seen in 1609, following the invention of the telescope. Galileo compared them to the "eyes" in a peacock's tail, while an early British observer, Sir William Lower, described how he saw the "mountaintops shining like stars." Indeed, the first proper map of the moon was produced by Hevelius in 1647, and the formations that he charted can be clearly recognized.

Although there is now very little scope for useful lunar observation with amateur instruments, the moon is a marvelous medley of tangled detail even with a 3-inch refractor, and no observer will dismiss it before he has wandered many times with his eye over the stark surface. He will come to know its craters and mountains as well as he knows the constellations; but on no two nights will these unchanging features ever appear the same, because the angle at which the sunlight strikes the lunar surface is varying all the time. To begin with, therefore, we should examine the motion of the moon and the way its appearance changes.

Movements and phases

The moon revolves around the earth in an elliptical orbit; its mean distance is 238,866 miles, and this varies from around 228,000 at perigee to around 252,000 at apogee. Since most of the orbits in the solar system lie in roughly the same plane, we shall not be far wrong in drawing them all in plan view on a sheet of paper. Figure 23 shows a schematic view of the moon's orbit around the earth. It must be remembered, however, that the earth itself is revolving around the sun in the same direction (counterclockwise if viewed from the north).

The moon, like all the planets and satellites in the solar system, has no light of its own; in position *A*, therefore, where it lies roughly between the earth and the sun, it is invisible, since its night hemisphere is turned toward us. If the lunar orbit lay in exactly the same plane as that of the earth, it would always pass centrally across the sun and produce a solar eclipse; but

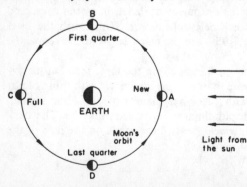

Figure 23. *Phases of the moon.*

there is a tilt of about 5°, which means that the *new moon*, as it is called in position *A*, generally lies either north or south of the sun in the sky, and only rarely does an eclipse occur. The usual new moon, masked by the solar glare, is utterly invisible.

The moon's counterclockwise motion carries it eastward around the earth at the rate of about 12° per day, so that in a couple of days' time a thin crescent can be seen in the evening sky after sunset. As the days pass, this crescent widens as more of the sunlit hemisphere is turned toward us and the moon moves farther away from the sun, until at position *B*, or *first quarter*, it forms a right angle with the sun and appears to us as a perfect half. The *terminator*, the line separating lunar day from night, is now straight, and the features across which it runs cast long shadows and are seen with great clarity. It is obvious that as the terminator sweeps across the lunar disk, different regions will be seen under low-lighting conditions.

The moon takes about 29½ days—the lunar month, or synodic period— to complete its cycle of phases, so that the period from new to first quarter is about 7¼ days. After this, the moon passes through the second quarter of its orbit. The phase becomes *gibbous*, or intermediate between half and full, with the terminator convex instead of concave; after a total of 14½ days it has reached position *C*. It is now on the far side of the earth from the sun, and appears fully illuminated. This is *full moon*, and, referring to the diagram, it follows that at this stage it appears as more or less opposite the sun in the sky, rises at sunset, and remains above the horizon all night. If the line-up is nearly perfect, it passes through the earth's shadow and suffers eclipse.

The full moon appears glaringly bright not only because the whole disk is illuminated, but because the sunlight is now shining down squarely on its surface. Just as a photographer places his camera so that the illumination on the model's face is slightly slanted, so for the most spectacular views we should see the moon when the sun is to one side. But at full moon, since the sun is behind our backs and no shadows are visible, the lunar disk appears as a blinding, patchy blur of light.

The moon now moves on toward *D*, or *last quarter*, and the evening terminator passes across the surface. The formations are now seen in their evening, as against morning, aspect; by the time the moon's last quarter is reached, half the disk has been lost in night. The moon has now moved to the west of the sun, so that it rises late in the night. Finally it shrinks to a crescent, visible just before dawn, and after the completion of the synodic period it is back at new again.

It is clear from this that the best time at which to observe any particular feature of the moon depends on its position. For example, if it is near the right-hand (western) edge, or *limb*, we shall have low illumination soon after new (sunrise), and again soon after full (sunset). On the other hand, a crater on the disk's meridian will be well seen at first and last quarters. It should not be thought that observations can be carried out only when the object is near the terminator, but there is no doubt that the most spectacular views are to be had at this time.

The two hemispheres

The moon presents its features in the same relative positions from night to night; we enjoy this conveniently standard state of affairs because it keeps roughly the same face turned toward us. When it was young, it was much closer to the earth than it is today, and gradually the earth's gravitational pull slowed down its spin until one of its hemispheres remained in the "captured" state. At the same time the moon slowly receded; it is still spiralling outwards today, although unimaginably slowly, and it will be thousands of millions of years before it appears appreciably smaller to us. At all events, observers have only just over half the lunar surface at their disposal. (The map on page 94 shows the main features permanently visible on the disk.) We can, however, see part of the averted hemisphere because the moon's eastern and western limbs alternately advance and recede during the lunar month in a slightly swinging motion known as *libration*. The eccentricity of the moon's orbit is the reason for this. The moon itself, because of its captured rotation, spins on its axis at a uniform speed, but its orbital speed is not constant. When near perigee it is moving faster than when near apogee, so that the two motions move slightly out of synchronization, with a fortnightly presentation of each margin. Furthermore, the lunar axis is slightly tilted, which means that we can also peer a little way beyond the north and south poles. Altogether, in terms of area, 59 per cent of the moon's surface is at one time or another accessible to our gaze, and the splendid success of the Orbiter, Surveyor, and Apollo spacecraft has given us a key to most of the hidden side.

Lunar observation

Though the moon is a bright and regular visitor, exhibiting the same formations month after month (except in the limb regions) and going through the same cycle of phases, lunar observation is far from being the simple matter that it may seem at first sight. There is little or no color on the moon; its surface tints are merely grays of various depths and textures, and we are highly dependent on shadows to mark out features and reveal tiny details. But all the time the terminator is creeping along, not only from night to night but also from hour to hour; in addition to this, the general angle of the sunlight is somewhat different with each revolution, or *lunation*. Should a particular formation be under examination, and a vital night be missed because of clouds or poor seeing, it may be months before conditions are right again. Roughly comparable conditions occur at intervals of 2 and 15 lunations (59 days 1½ hours and 442 days 23 hours), but this does not take the effects of libration into account; these greatly affect one of the hunting grounds of the amateur—the marginal regions. The lunar observer must be systematic, must plan ahead, and must seize his chances as they are presented.

For obvious reasons, professional interest in the moon has stepped up in the last few years. Until the recent launching of the Apollo project, which suddenly made it necessary for a large-scale "official" map to be constructed, our satellite was considered as more of a nuisance than a benefit; astronomers, anxious to photograph faint stars and galaxies that are of far more cosmic importance than our diminutive solar system, have more or less to stop work during the fortnight around full moon, when the bright light fogs their sensitive plates. Formerly, only amateurs bothered to take the moon seriously; one of the pioneer lunar maps was produced in 1837 by two German astronomers, Johann von Mädler and Wilhelm Beer, using a 3¾-inch refractor. The next great advance occurred in 1878, when another German, Julius Schmidt, who observed from Athens, produced a chart more than six feet across. Shortly after this, in 1890, the British Astronomical Association (B.A.A.) was formed, and amateur astronomy became established on a more coordinated footing. Three different directors of the B.A.A. Lunar Section published charts, in 1895, 1910, and 1946. The spate of amateur observations became even more marked when, just after World War II, American observers banded together to found the Association of Lunar and Planetary Observers (A.L.P.O.), at University Park, New Mexico.

However, the position has changed. Photographic techniques have improved tremendously, and some professional observatories, such as the Pic du Midi in the Pyrenees, are taking photographs showing details considerably beyond the range of a small telescope. The publication in 1960 of the *Lunar Photographic Atlas*, compiled by Gerard Kuiper, was an event of great importance. Add to this the wonderful detailed coverage of almost the whole lunar surface by the Orbiter and Apollo missions, and the amateur cartographer cannot avoid the hard fact that his scientifically useful days are over. On the other hand, there is nothing to stop him from drawing lunar formations just for the joy of it, and as a compensation he can refer his finished product to a closeup photograph for verification!

Paradoxically, the current situation can only enhance the value of any lunar observation that may be undertaken. The position closely parallels the attitude of observers in the period following the appearance of Mädler and Beer's great map, which was generally accepted as the last word on the subject, effectively discouraging any further effort. This was a desolate period for lunar work, and the same is true today. Therefore, should anything remarkable occur, such as a great meteorite fall, or a major landslip, or even one of the curious glows that have been reported from time to time, there may be no eyes to witness it. So we must hope that amateurs will continue to let their telescopes rove over the scarred surface of our neighbor world.

The lunar features

The first step in becoming a lunar observer is to learn the moon's principal features. An outline map shows the main maria and the largest craters, and a few nights with a telescope (or even binoculars) will bring a good deal of order out of the initial confusion. Indeed, the surface presents a certain amount of harmony in the sense that, although it is covered with a tangle of objects, these can be organized into a number of different classes.

MARIA, OR SEAS. These form the lunar lowlands. The Latin name for "seas" (singular, mare) was bestowed on them by the early telescopic observers, who thought they really were liquid bodies; we now know there is no water anywhere on the moon. The maria, which differ in appearance from the highlands in being both darker and smoother, are evidently colossal lava plains, perhaps covered with a thin layer of dust. Winding ridges are common, and so are small hillocks and craterlets ("small" on the telescopic scale, of course), but there are few prominent craters in these regions. Here and there we find very low rings, often of considerable size; these "ghosts," so called because of their faintness, are evidently the walls of ancient craters that were overlaid by molten lava when the seas were formed.

Most of the maria run into each other, extending from the Mare Foecunditatis in the S.E., or Fourth, Quadrant, via the Maria Tranquillitatis and Serenitatis (N.E., or First, Quadrant), into the beautiful Mare Imbrium (N.W., or Second, Quadrant), and ending in the region of the Mare Nubium and Oceanus Procellarum, the latter being the largest sea, although somewhat ill-defined. All these maria have areas of a hundred thousand square miles or more. In addition, there are smaller dark areas which may be known as a lake (lacus), marsh (palus), or bay (sinus), all of which are offshoots of the maria.

Some maria are quite detached. The most prominent of these is Mare Crisium (First Quadrant), while four other examples lie along the eastern limb. One of these, Mare Australe, extends on to the averted hemisphere.

WALLED PLAINS. Although it is convenient to refer to all circular lunar formations as "craters," they actually fall into definite categories. The walled plains are the largest features, ranging from about 60 to 180 miles across, with a mountainous ring several thousand feet high overlooking an inner floor that may itself have some fair-sized craters embedded in its surface. Some are in good condition, with their walls entirely preserved, whereas others have been so broken down by later upheavals and remelting of the crust that the reduced ridges can be well seen only when near the terminator. Judging from the way in which so few are perfect, it seems probable that these immense formations are among the oldest of the moon's visible features, presumably originating when the crater-forming activity was at its most violent.

RING PLAINS. These are the most impressive of the lunar features. They are smaller than the walled plains, with diameters of between about 30 and 60 miles. The walls are relatively high, sometimes rising 10,000 feet or more, and the interior floor may be depressed below the level of the external terrain. These walls are often magnificent in themselves; they may be ridged or terraced with great complexity, especially on the inner slope. The floor itself often contains a central mountain mass that may rise to a height comparable to that of the wall, and this mass is often divided into separate peaks.

CRATERS. Similar in cross section to the ring plains, except that the walls are much less complex, are narrower, and the central mountain is either small or totally absent. They are smaller than the ring plains, with diameters as small as 5 miles, and many form the center of *ray systems*, deposits of bright matter that seem to radiate away from the crater across the outside surface. Craters, together with the smaller craterlets, occur all over the surface and are often plastered over much larger formations.

CRATER CONES AND CRATER PITS. These are very small features, requiring a moderate telescope for their detection. Crater cones are roughly conical, though not steep, formations, bearing tiny orifices at their summits. Crater pits are simply minute depressions in the lunar surface without recognizable walls; they are well shown in the Orbiter photographs.

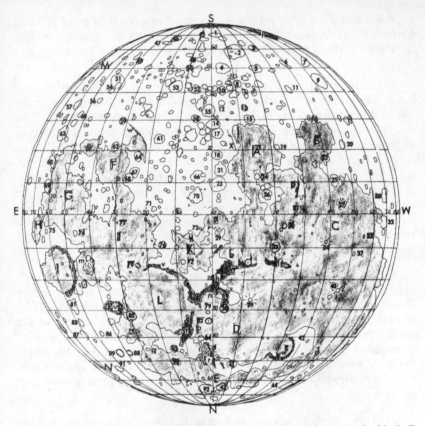

Map of Lunar Features. (From *The Telescope Handbook and Star Atlas* by Neale E. Howard, Thomas Y. Crowell Company)

A. Mare Nubium (Sea of Clouds)
B. Mare Humorum (Sea of Moisture)
C. Oceanus Procellarum (Ocean of Storms)
D. Mare Imbrium (Sea of Rains)
E. Mare Frigoris (Sea of Cold)
F. Mare Nectaris (Sea of Nectar)
G. Mare Foecunditatis (Sea of Fertility)

H. Mare Spumans (Foaming Sea)
I. Mare Crisium (Sea of Crises)
J. Mare Tranquillitatis (Sea of Tranquillity)
K. Mare Vaporum (Sea of Vapors)
L. Mare Serenitatis (Sea of Serenity)
M. Mare Australe (Southern Sea)
N. Mare Humboltianum (Humbolt Sea)

1. Moretus
2. Clavius
3. Scheiner
4. Maginus
5. Longomontanus
6. Schiller
7. Phocylides
8. Tycho
9. Schickard
10. Orontius
11. Hainzel
12. Lexell
13. Wurzelbauer
14. Regiomontanus
15. Pitatus
16. Doppelmayer
17. Purbach
18. Arzachel
19. Bullialdus
20. Mersenius
21. Alphonsus
22. Gassendi
23. Ptolemaeus
24. Guericke
25. Letronne
26. Fra Mauro
27. Flamsteed
28. Grimaldi
29. Pallas
30. Reinhold
31. Landsberg

32. Hevelius
33. Copernicus
34. Kepler
35. Reiner
36. Eratosthenes
37. Marius
38. Archimedes
39. Timocharis
40. Aristarchus
41. Plato
42. Mairon
43. Birmingham
44. Pythagoras
45. Curtius
46. Maginus
47. Mutus
48. Lilius
49. Vlacq
50. Cuvier
51. Janssen
52. Stöfler
53. Maurolycus
54. Fabricius
55. Walter
56. Rheita
57. Furnerius
58. Aliancensis
59. Piccolomini
60. Stevinus
61. Sacrobosco
62. Frascastorius

63. Petavius
64. Catharina
65. Vendelinus
66. Albategnius
67. Cyrillus
68. Theophilus
69. Langrenus
70. Hipparchus
71. Delambre
72. Triesnecker
73. Maskelyne
74. Taruntius
75. Appolonius
76. Julius Caesar
77. Manilius
78. Plinius
79. Autolycus
80. Posidonius
81. Cleomedes
82. Aristillus
83. Geminus
84. Cassini
85. Eudoxus
86. Franklin
87. Messala
88. Hercules
89. Atlas
90. Aristoteles
91. Endymion
92. W. C. Bond

a. Sinus Medii (Central Bay)
b. Sinus Aestuum (Seething Bay)
c. Sinus Iridum (Rainbow Bay)
d. Sinus Roris (Bay of Dew)
e. Leibnitz Mountains
f. Doerfel Mountains
g. Riphaean Mountains
h. Apennine Mountains
i. Carpathian Mountains
k. Jura Mountains
m. Palus Somnii (Marsh of Sleep)
n. Palus Putredinis (Marsh of Decay)

o. Lacus Somniorum (Lake of Dreams)
p. Palus Nebularum (Marsh of Mists)
q. Lacus Mortis (Lake of the Dead)
r. Alps (mountains)
s. Alpine Valley
t. Caucasian Mountains
u. Taurus Mountains
v. Haemus Mountains
w. Altai Scarp
x. Straight Wall
y. Hyginus Cleft

CRATER CHAINS.   Many of these features are distributed in chains. The walled plains exhibit some impressive line-ups, such as the great series running roughly along the meridian in the southern hemisphere. The same is true of the craters and craterlets, which often occur in pairs and occasionally in long strings. However, with the exception of the very small objects only a mile or two across, there is no evidence that these alignments are other than would be expected from a random distribution of craters produced by interplanetary particles striking the surface. There are great differences between the moon and a volcanically active body such as Jupiter's satellite Io.

DOMES.   These curious features, like low swellings or hillocks, are of heights up to 1,000 feet and diameters up to two miles; they can be seen well only when near the terminator. More than a hundred are known, but it is certain that many remain to be found, since they were first identified only forty years ago. Dome-charting used to be a popular amateur activity, but it has, of course, been vitiated by the mapping work carried out from spacecraft.

VALLEYS AND CLEFTS.   Sometimes called *rills,* these range from colossal ravines more than a hundred miles long and several miles wide, to hairline cracks at the limit of telescopic visibility. The more delicate clefts are evidently true cracks in the surface, but some of the more prominent ones are seen to contain the remains of craterlets along their length. Great clefts often pass across the floors of walled plains, frequently breaking through the mountain wall and extending across the ground outside.

MOUNTAIN RANGES.   It is not surprising that a world showing such drastic primeval activity has thrown up superb mountain ranges. One of the finest, the Apennine Mountains, runs across the meridian in the northern hemisphere, while the highest peaks on the moon, some of which reach 30,000 feet, occur near the south pole. These, the Leibnitz Mountains, actually run along the limb, so that under suitable conditions of libration they can be seen in profile, their glittering summits poking out against the sky. Other mountain ranges also run along the western limb, as well as marking off the Mare Imbrium from the Mare Serenitatis.

NAMES.   The system of crater naming follows the plan suggested in 1651 by the Italian observer Giovanni Riccioli, who called them after famous scientists (reserving a fine formation for himself) and other prominent figures. The moon therefore houses such illustrious personages as Plato, Newton, Archimedes, and Copernicus. Riccioli himself supplied about two hundred names for the craters he had plotted, but subsequent work has naturally added many more; there are now more than a thousand listed formations on the earth-turned hemisphere, and about five hundred on the far side. These names have been internationally agreed upon, based on recommendations by the International Astronomical Union working group on lunar nomenclature, whose original map of the earth-turned hemisphere, drawn up in 1932, has required extensive revision as a result of the intensive mapping programs of the past two decades. In addition to the fifteen hundred or so "primary" named craters, smaller features are usually referred to by the adjacent primary crater's name followed by a capital letter, such as Archimedes A, Hadley B, and so on.

It should be noted that using the classical east and west orientation (i.e., with the eastern limb following and the western limb preceding the moon's daily motion across the sky), the sun rises in the lunar west and sets in the east. With the coming of manned exploration it was decided to reverse this convention, a change that must be borne in mind when referring to earlier descriptions of the surface. The Mare Crisium, therefore, is near the "modern" east limb. Lunar latitude is positive for the northern hemisphere and negative for the southern; lunar longitude is positive for objects east of the central meridian, negative for those west of it.

The best way of getting a general picture of the lunar surface is to follow the moon through its phases from crescent to full. In this way, all its major features will be seen as they are revealed by the slow progress of sunrise over its surface; and it will be noted how soon they lose their distinctiveness as the sun rises high over them and the shadows disappear. With this end in view, let us now assume two weeks of favored weather, and view the moon at two-day intervals throughout the first half of a lunation.

The thin crescent

It is interesting to see just how soon the young moon becomes visible with the naked eye in the evening sky. Conditions for northern observers are most favorable in the first three months of the year, the moon being highest in the evening sky at this time. Late June to August is the best time for southern observers to look. If the old moon is being sought in the dawn sky, these times should be reversed. Observers in lower latitudes have the best opportunity of seeing the moon near the sun, because it rises and sets at a greater angle to the horizon, and therefore is at a greater altitude for a given elongation from the sun. W. F. Denning, in his 1891 classic work *Telescopic Work for Starlight Evenings,* mentions seeing the crescent from southern England when it was only 20 hours 38 minutes old, and quotes an observer in Belgium seeing it only 18 hours 22 minutes before new. In the southern United States, an observer saw a crescent only 16 hours 17 minutes before new on

January 27, 1979, and it was seen again on the evening of the following day, only 17 hours 10 minutes old, using binoculars. Challenges such as this, although of no particular scientific benefit, form an interesting test of vision and application.

Three days old

The most obvious feature of the young crescent is the Mare Crisium (Sea of Crises). This is one of the smaller lava plains, measuring 355 by 280 miles; but, although it is elongated in an east-west direction, foreshortening makes it appear to be extended along the north-south axis, and it compensates for its small size by the majesty of its mountainous surroundings. At this phase and, similarly, two or three days after full, the rugged border can be seen very well with a small telescope. There are also three small craters on its surface (Picard, Pierce, and Graham); these are just coming into view. Another interesting feature is a "quadrangle" of white streaks and tiny craterlets near the southwest corner. This can be seen easily with a 3-inch refractor when the sun is high, but it somehow escaped notice until thirty years ago. Several well-known amateurs have observed that some small features on the mare's surface appear unaccountably obscure at times; but these observations must be taken with a certain amount of reserve, for it is desperately easy to be misled by different conditions of seeing and illumination.

There are four other maria located in the crescent (Australe, Smythii, Marginis, and Humboldtianum), running from south to north, but these lie on the extreme edge of the disk and can be well seen only at conditions of eastern libration, when this part of the moon has swung its maximum extent toward the earth. Even so, they always appear extremely foreshortened, and our first good view of them was had from the *Luna 3* photographs.

The crescent phase reveals some fine walled plains. The southernmost, Furnerius, has been reduced by subsequent activity, but its walls still rise to 11,000 feet above the 80-mile floor. To the north is the magnificent crater Petavius. This is 100 miles across, and has a splendid central mountain, from which a great cleft runs to the southwest wall. This cleft is easily seen with a 3-inch, as is much other interior detail. North again is Vendelinus, as large as Petavius but badly destroyed. The fourth member of this chain, Langrenus, is 85 miles across and has a central mountain; its walls are only a little lower than those of Petavius. It is worth remembering that although these craters, and indeed almost all the craters on the moon, are roughly circular, they appear elliptical when they are near the limb because of the foreshortening effect.

Two 78-mile walled plains in the northern half of the crescent are worthy

of attention. The first, Cleomedes, lies immediately north of the Mare Crisium, and has a fine central mountain; parts of its wall rise to 16,000 feet.
Much nearer the north pole is Endymion, which has an unusually dark floor
that seems to change in tone as the lunation progresses. Several craters show
this effect, and, although it is one of contrast, it is important as a warning to
those who think that any real changes of this nature should be readily
apparent.

When viewing the crescent moon, it will be noticed that the entire disk is
often visible, glowing faintly against the sky. This effect, known as *earthshine,* can be seen with the naked eye on some occasions, and a low magnification will reveal some of the darkest and brightest features on the night
side. Earthshine is caused by the cloudy terrestrial atmosphere reflecting
sunlight onto the moon, so that its distinctness is a measure of atmospheric
conditions. It is strongest when the moon is near new, since at this point the
earth would appear "full" to an observer on its surface.

Five days old

One of the major seas, Mare Foecunditatis (Sea of Fertility), is now
revealed. It contains a few small craters dotted here and there, but none is
more interesting than the "non-identical" twins Messier and Pickering.
These lie close together in the center of the mare, and are easily found
because a curious, double white streak projects from Pickering in a roughly
westerly direction. Although they are small, about eight miles across, they
are well worth observing. Beer and Mädler, in their 1847 map, described the
two craters as exactly alike; Walter Goodacre, a well-known British observer, measured their diameters in 1932 and found Messier to be the larger of
the two; and many amateurs often see Messier as smaller than Pickering!
These "changes" are, of course, entirely optical effects, probably due to the
different depths of the craters. R. M. Baum, a contemporary observer well
known for his lunar work, has commented: "The apparent shallowness of
Messier leads to internal reflection, and thus possibly, at the right phase, to
some dilution of shadow intensity. By contrast, Pickering seems deeper and
able to hold shadow longer." This example shows how easy it is to be misled
by the subtle and varying play of light. Another instance was the fiasco of
the supposed "bridge" discovered near the eastern shore of Mare Crisium in
1954, caused by unusual lighting conditions whose effects deceived an observer. The moon may be an almost dead world, but the progress of sunrise
and sunset certainly breathes a ghostly life across its surface.

Mare Nectaris (Sea of Nectar), a smaller sea to the west of Mare Foecunditatis, has its interesting features. The nearby chain, Theophilus, Cyrillus, and Catharina, is indeed the most striking sight at this phase. These

three objects are superb, between 65 and 70 miles across. Theophilus is the best preserved, displaying a bright central mountain mass, a depressed floor, and terraced walls rising to 18,000 feet. It is interesting, too, to see how Theophilus has broken into its neighbor Cyrillus, indicating that it is of more recent origin, although it appears to be of about the same age. There are a number of cases of such overlapping craters, and the general rule is that the larger is overlapped by the smaller. On the whole this is to be expected, since, if the craters are impact features, the larger bodies would tend to be rarer than the smaller ones and would become proportionally rarer still as they were destroyed in impacts. A glance at photographs of the surface shows that the largest impact features are usually much older than most of the smaller craters, with ruined walls, reduced central mountains (or none at all), and other signs of age. It seems clear that, after the main epoch of crater formation, some areas of the moon suffered extensive remelting that not only caused the maria but also obliterated the terraces and inner peaks of the older craters.

Further proof that the moon was once a lava-strewn inferno comes from the nearby crater Fracastorius, on the southern shore of Mare Nectaris. Here, we see that what was once its northern wall has disappeared almost entirely under the molten lava of the mare, which long ago flooded its interior and converted it into a great bay. There are many examples of flooded and partly drowned craters, suggesting that the seas were formed in the later stages of the moon's active history.

At this stage, Mare Tranquillitatis (Sea of Tranquillity) is also coming into view, but the most distinctive feature of the northern part of the crescent is the 62-mile crater Posidonius, on the western shore of Mare Serenitatis. Posidonius has low, rather narrow walls, and a great amount of interior detail. North of Posidonius are the twin craters Atlas and Hercules. Atlas, 55 miles across, is the larger of the two; its floor contains some interesting dark patches which, like the floor of nearby Endymion, seem to vary in tone.

First quarter

Half moon is the most spectacular phase. When the terminator runs like a knife down the central meridian, and the great walled plains are thrown up into plan view, the observer feels convinced that he is hanging in space over this black-and-silver world.

All over the southern hemisphere cluster the sunlit walls and shadow-filled interiors of hundreds of craters. Some immediately catch the eye: the great walled plain Stöfler, which has a smaller crater, Faraday, on its southwestern margin; Aliacensis and Werner to the north; and, north again and near the center of the disk, the 80-mile Albategnius and the

84-mile Hipparchus. Hipparchus must once have been a magnificent object, but it is now sadly reduced; its walls are low, and have been completely breached in the northeast quarter. In fact, near the time of full moon, when Hipparchus is experiencing noon, it is not too easy to make out the crater at all. Similarly, if we now look near the eastern limb we shall find Vendelinus and Furnerius hard to distinguish from their glaring surroundings, while the mountainous border of the Mare Crisium shows no sign of relief. Only very bright and very dark craters can be distinguished easily when the sun shines vertically upon them.

Mare Vaporum (Sea of Vapors), an ill-defined sea in the center of the disk, contains a small but well-known object. This is the 4-mile crater Hyginus, which lies in the middle of a famous cleft 150 miles long. The cleft is only a mile wide, but it can easily be seen with a 3-inch refractor because it is distinctly lighter than the surrounding surface. The Hyginus cleft is particularly interesting because it is not a true fault at all, but a string of tiny craterlets. A moderate telescope is required to show this well, but there can be no doubt at all of its nature; it is hard to see how these craterlets can be anything but volcanic. Hyginus forms a right-angled triangle with the bright crater Manilius at one corner and the dark-floored Boscovitch at the apex, and these form a convenient guide since they are distinct under all conditions of illumination.

Mare Serenitatis (Sea of Serenity), one of the finest of the large seas, is now well seen. Its southern and western borders are marked by the Haemus and Caucasus ranges, while the magnificent Apennine range now leads off into the dark hemisphere. The mare's surface is relatively featureless, except under very low illumination, when a number of ridges become obvious. There is only one distinctive crater, Bessel, which lies on a bright streak that passes centrally across the huge expanse. Farther west toward the gap between the Caucasus and Apennine ranges, a conspicuous white spot marks the position of the controversial crater Linné.

The Linné furor started in 1866, when Julius Schmidt, who was then working on his famous lunar map, announced that the crater had either disappeared or else changed beyond all recognition. Observing it on October 16 of that year, he found that instead of appearing as a regular crater, it seemed to be just a small bright patch. This flew in the face of observations he himself had made between 1841 and 1843, and of those made by other earlier observers, Johann von Mädler among them. On previous occasions, when Linné lay near the terminator, they had all seen a distinct shadow cast on the crater's floor by the surrounding walls. Yet in 1866, and indeed a century later, this impression was never regained. With powerful instruments, Linné appears as a craterlet about a mile across, surrounded by a whitish patch.

The matter was actually cleared up later by Mädler himself, who stated that he did not consider the crater to have changed its aspect from what it was in the 1830s; but this statement was largely ignored by other observers, and today there are still people who believe that the Linné affair has a basis of reality. Nevertheless, the evidence overwhelmingly supports some sort of optical deception. When conditions are suitable, tiny Linné does give the impression of being larger than it really is; and the fact that observers of the caliber of Mädler and Schmidt could be misled only emphasizes how very careful one must be about jumping to conclusions. The Linné affair is a valuable lesson, and one should not forget it.

To the north of Mare Serenitatis lie two conspicuous craters, Aristotle and Eudoxus. They both have bright floors and are easily seen at full.

Nine days old

It is well worth looking at the moon just a day after the quarter, for at this time a magnificent string of walled plains comes into view. These are Ptolemaeus, Alphonsus, and Arzachel, which lie slightly to the west of the Hipparchus-Albategnius pair. The 90-mile Ptolemaeus must once have been a magnificent object, but it has been somewhat reduced by subsequent activity, although not so drastically as Hipparchus. It is still an imposing sight when caught on the terminator, and there is a conspicuous small crater, Lyot, on the floor.

Its southern companion Alphonsus hit the publicity spotlight in November, 1958, when the Soviet astronomer Nikolai Kozyrev reported detecting an emission of gas, accompanied by a reddish glow, at the base of its central mountain. Kozyrev was using a large telescope—the 50-inch reflector of the Crimea Observatory—but previous observers had also reported seeing faint glows there from time to time. Moreover, the floor contains several dark patches that have been reported as showing variations during the lunation. Although a moderate aperture is necessary for a proper study, a 3-inch will reveal these patches plainly enough. Clearly, Alphonsus is one of the most interesting craters on the moon.

A great number of objects come into view two days after first quarter, but there is no room here to touch on more than a few of them. One of the most remarkable craters on the moon, 54-mile Tycho, lies on the southern region of the terminator. It has a splendid central mountain and is in a good state of preservation; it will already have been noticed that a number of bright streaks crossing the southeastern region of the moon appear to converge on it. These are just some of the far-flung members of Tycho's ray system. It is by far the most extensive on the moon, and by full the whole disk is dominated by these radiating white streaks, some of which extend for more than a thousand miles. They are obscure when near the terminator, but shine out under high illumination.

Tycho's ray system is by no means the only one. Copernicus, a magnificent crater in Oceanus Procellarum (Ocean of Storms), is a prominent ray center, and there are many other examples. The rays are a surface deposit of very small glassy globules, evidently caused by almost instantaneous vaporization and condensation of silica squirted out at the moment of impact when the crater was formed.

Some distance to the south of Tycho lies Clavius, second largest crater on the moon. This is a colossal walled plain 145 miles across, and the distinctive string of craters across its floor makes it a most conspicuous object. To the north, in the rather ill-defined Mare Nubium (Sea of Clouds), lies the half-drowned Guerické, with the group of other old craters Parry, Bonpland, and Fra Mauro nearby. All these features have been partly submerged by the wave of melting that produced the huge western maria. The fine 40-mile crater Bullialdus, also on Mare Nubium, now stands out near the terminator. Nearby lies the curious fault known as the Straight Wall, which is an almost straight ridge some 60 miles long, with the eastern ground about 800 feet higher than that to the western side. It is situated near the prominent crater Thebit, and its southern end runs into a branching group of hills appropriately known as the Stag's Horn Mountains.

To the north of the lunar equator, and near the southern shore of Mare Imbrium (Sea of Showers), the crater Copernicus rears up in magnificent relief. Because of its relatively level surroundings, we see Copernicus undisturbed by later activity. In its superb isolation, it is perhaps the finest crater on the moon. It is 56 miles across, with walls rising to 17,000 feet and bearing on their inner slopes an intricate system of terracing. Impressive, too, is the central mountain mass, which contains three separate peaks. When caught right on the terminator, so that the interior is filled with shadow, Copernicus really seems to jut up toward the observer. The region nearby contains a large number of domes.

Many craters appear impressively deep when seen near the terminator, but it must always be remembered that low sunlight exaggerates. Car headlights can amplify harmless dips into seemingly deep ravines; the same is true of sunrise and sunset conditions on the moon. Relatively speaking, Copernicus and its companions are far shallower than saucers; in a sense, they are more impressive as seen from the earth than they would be from close at hand. For example, someone standing near the center of the walled plain Ptolemaeus would see no mountain border at all, for the peaks would be below the horizon! The moon, being a smaller world than the earth, has a more sharply curving surface, and the horizon is only about two miles away from the observer.

Mare Imbrium is a fine sea, with many interesting features. Its eastern shore is bounded by the magnificent Apennine range, in which one peak, Mount Huygens, rises to almost 20,000 feet. To the north lie the Alps, whose less impressive appearance is compensated for by the curious valley, as straight as if cut with a knife, that slices its way through the mountains. This can be seen very well immediately after the half phase.

Before the eye crosses Mare Imbrium it is a good idea to look at Eratosthenes, a smaller version of nearby Copernicus, which lies at the southern tip of the Apennines. The floor of Eratosthenes contains some well-known dark patches that give the impression of varying in size during the lunation, although, again, it is certainly no more than a curious optical effect.

The dark-floored walled plain, Plato, on the northern shore, may well be the most studied object on the moon. It is 60 miles across, with relatively low walls and a level floor, and appears noticeably elliptical because of its closeness to the north pole. Its darkness makes it easily distinguishable at full. What makes Plato so interesting is some of the tiny craterlets scattered about its interior. Figure 24 shows a chart of the major features, and although these craterlets are relatively small, there is no doubt at all of their existence. Yet, on occasions, some or all are strangely difficult to see, when they should, by all accounts, be obvious. On the other hand, as in the case of Linné, it is easy to jump to conclusions; the angle of illumination has so drastic an effect on the appearance of the lunar surface that without a really thorough survey no definite conclusions can be drawn. Proper study of the crater requires a telescope of at least 10 inches aperture, although some of the spots have been made out with a 3½-inch refractor.

A little way south of Plato, on the surface of the mare, is Pico, a bright, isolated mountain. To the west, the impressive Sinus Iridum (Bay of Rainbows) is creeping out of the darkness. It is an unforgettable sight when the terminator passes through it; the semicircular mountain range is lit up like a great gleaming scimitar, while the "bay" itself is still lost in night.

Eleven days old

This is the time for catching two interesting features: the brightest crater on the moon, Aristarchus, and the nearby Schröter's Valley.

In 1783, Herschel made a startling announcement. He had been observing the dark, earthlit portion of the moon, on which, as often happens, the very dark and very light features could be dimly made out, when he noticed the central peak of Aristarchus glowing quite brightly. Moreover, as he reported to the Royal Society, "all the adjacent parts of the volcanic mountain seemed to be faintly illuminated by the eruption, and were gradually more obscure as they lay at a greater distance from the crater."

We know more now than was known in Herschel's day, and the idea of volcanic activity on a scale sufficient to produce such a glow seems distinctly unlikely. On the other hand, an observer of Herschel's experience

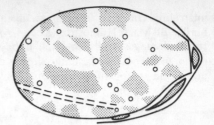

Figure 24. *Plato. This shows some of the more prominent craterlets and light streaks on the floor, as charted by A. Stanley Williams, a famous amateur lunar observer of the last century.*

would not report so remarkable an observation without being sure of his facts, so we are left seeking alternative explanations. One is that the dark side of the moon was illuminated by exceptionally strong conditions of earthshine, causing Aristarchus to glow more brightly than usual.

Aristarchus itself is a normal enough crater, 29 miles across and with a central peak, lying in the rather barren northern reaches of Oceanus Procellarum. It is the center of a small but bright ray system, and the walls and floor are coated with a white deposit that makes it noticeably bright even when it lies on the terminator. The central peak is even brighter than the rest, and by full moon it is so glaring that it is difficult to examine with a large telescope unless a neutral filter is used to cut down the light. Violet hues have been reported in the area, but the most famous observation was made as recently as October 30, 1963, when two United States Air Force lunar mappers were observing from Lowell Observatory in Arizona. They recorded three orange glows in the area, lasting for twenty-five minutes, and were convinced that some form of activity had taken place. However, these observations are made extremely difficult by the great brilliance of the region; the eye is easily deceived, and the fact that a refracting telescope was used makes any color estimates somewhat uncertain.

Aristarchus is also the prototype of a class of lunar features known as *banded craters*. Soon after the sun has risen over Aristarchus, even a small telescope will show two or three bands running from the central mountain toward the western wall; these appear to extend and darken during the lunar day, even passing over the wall and onto the outside plain. Once again, we cannot be sure whether or not this is an optical effect, although the development appears definite enough. A 3-inch refractor will show the main bands easily, and they provide an interesting field of investigation.

The much darker crater Herodotus lies closely southeast. It is slightly smaller than Aristarchus, and acts as the terminus for a curious U-shaped valley named after its discoverer, Johann Schröter. Schröter's Valley, which runs north for some 30 miles, then doubles back in a southwesterly direction, is a true fault, with nothing of the crater-chain nature of the Hyginus cleft. It is easily seen with a small telescope, although an instrument of at least 8 to 12 inches aperture is required for a proper investigation. Another interesting crater to be seen at this phase is Gassendi, on the northern border of Mare Humorum (Sea of Moisture). It is 55 miles across, with rather damaged walls, and the floor contains a most intricate system of clefts. Gassendi has recently shown evidence of bursts of activity on its floor, and is one of the regions being watched with colored filters (see page 111).

Thirteen days old

With the moon now moving toward full, the great formations running down the west limb are coming into view; at this phase, depending on the libration conditions, some superb objects become visible. It must, of course, be borne in mind that many of these walled plains are so near the limb that they appear extremely foreshortened.

By far the most conspicuous of these formations is Grimaldi, a colossal walled plain 120 miles across, with a rather low mountain ring and a floor that is even darker than Plato's; parts of it are often cited as exhibiting the deepest tint on the lunar surface. Under low magnifications, the floor appears relatively featureless, but the curve of the lunar surface can be clearly seen. Its companion, Riccioli, is only slightly smaller, and it too has a dark floor. Somewhat to the south, and running right along the limb, are the Rook Mountains, some of whose peaks rise to 20,000 feet. Nearby is the 20-mile crater Sirsalis, through which a prominent cleft runs from north to south.

Farther south still is the immense formation Schickard. This is only slightly smaller than Clavius, but the walls are rather lower. It is chiefly interesting because of nearby Wargentin, which the Reverend T. W. Webb, a well-known observer of the last century, likened to a slice of cheese! Wargentin is a 54-mile crater that for some reason has become filled to the brim with lava, so that it is a true lunar plateau. We occasionally find other instances of the same phenomenon, but on a much smaller scale, so Wargentin is something of a freak.

Between Schickard and Clavius lies an interesting example of two large craters that have virtually coalesced into one. This is Schiller, which is about 110 miles long and only 60 wide, with a slight "neck" where the walls of the two original formations have joined. Another well-known Siamese-twin feature is Palitzsch, near Petavius, best seen a day or two after full.

The northern quadrant of the west limb is rather less spectacular, much of it being taken up with the ill-defined border of Oceanus Procellarum. There is, however, an immense plain named after Otto Struve, the famous Russian astronomer. (It is hardly a true crater, being formed by the confluence of two mountain ranges.) The 85-mile Pythagoras, near Sinus Iridum, would be a noble object were it better placed. At conditions of western libration, however, many features creep into view.

Amateur work

This tour of the moon has necessarily been sketchy, but it has at least brought to light some of the characteristic features of the lunar surface, and the observer will now have some idea of when the more prominent features are best placed for observation. What is the next step in becoming a lunar observer? Many people turn their telescope to the moon, perhaps to see if any spectacular views are presented that night, or to give a friend a peep, and then pass on to the planets or double stars; but such cavalier treatment of our satellite hardly deserves the term "observation."

Given that lunar cartography by earth-based observers is now superfluous, we can view our satellite as the perfect training-ground for visual observers, no matter where their ultimate interest may lie. A telescope of the smallest kind, or even binoculars, will show enough detail to be worth committing to paper, and in this act of representing with a pencil what is seen in the eyepiece, the tyro is taking a very practical step forward in becoming a useful observer. It will be a good idea, at the outset, to prepare a list of interesting formations—not necessarily large ones, for they may be too complex to draw properly—scattered about the surface so that one is bound to be well placed when the weather is favorable. Ideally, a number of sketches of each feature should be made on different nights, showing its appearance at various phases. Thus, a picture will be built up of its real form rather than the two-dimensional aspect it may wear at an isolated viewing. Much interest attaches to noting how the shadows move with the sun, and how they vary at corresponding phases of different lunations. Knowing the altitude of the sun in the lunar sky, the slope of a wall or mountain can be estimated by noting when the shadow vanishes. Shadow edges may be rippled by evanescent folds in the surface, effects so fleeting that they may never be recaptured by the same observer. All this is available to the student of the moon, be his aperture large or small—only the scale varies with the power of his instrument.

Making a drawing is by far the best way of exercising the eye. We may scrutinize a feature and imagine that we see all that there is to be seen; but, the moment details begin to be committed to paper, a host of fresh and finer detail springs into evidence, so that the initial sketch soon turns out to be a mere outline of what is really visible. This happens so frequently that the observer tends to give up in despair; the eye learns faster than the hand can follow, and it is here that the natural draftsman is at an advantage, in being able to translate the eye's messages to paper almost without thought.

A sketchbook of good-quality drawing paper makes the best lunar observation book. Sketches are commonly made too small; a scale of about twenty miles to the inch is a good standard, for the finer features can then be added without complicating the drawing. Some gifted observers have produced really beautiful lunar representations, using ink and pencil, but a simpler outline sketch can be just as useful if written notes are made of intensities and other features that are too subtle to be rendered visually. Each observer has his own method, but it is a good idea to draw in the shadows first, adding the crater walls, peaks, and finer touches as the observation progresses. If the outline of the formation can be traced from a suitable photograph, this will insure basic accuracy. It is necessary to work fairly fast, for when the sun is low the shadows show perceptible movement in the course of an hour.

If atmospheric conditions are good, the highest possible magnification should be used when drawing in the finest features, whereas a somewhat lower power may be better for sketching in the initial outlines. It will be found that high powers can be used on the moon more often than on most of the planets; this is one of those charming idiosyncrasies that so often turn up in amateur astronomy and that are quite unintelligible to the non-observer. Assuming, of course, that conditions are good, both the moon and Saturn show up well under a powerful eyepiece, whereas the other planets tend to give a clearer image with rather less magnification (say, × 200 instead of × 300 with a 6-inch). Mars, however, which shows a small disk, requires a powerful eyepiece if its surface markings are to be seen clearly. There is certainly a distinct difference between the telescopic images of Jupiter and Saturn, Jupiter appearing somewhat vague under powers that show Saturn's disk and rings perfectly clear-cut. It is, of course, necessary to experiment with different planets on the same night, so that seeing conditions are comparable, and it must not be forgotten that steadiness improves with altitude. For this reason a formation near the moon's western limb is better observed just after full (i.e., under sunset illumination) than at the crescent stage, since the full moon is visible much higher in the sky. On the other hand, should it be necessary for some reason to make an observation at local sunrise, the inherent difficulties of the observation must be overcome somehow; one way is to observe during the day when the crescent is high, using a deep-red filter to cut down the sky light. When making a thorough survey of any particular feature, it must not be forgotten that high-light observations are just as important as those made under the more spectacular angles of illumination.

Leaving aside the tremendous amount of statistical research still to be done on crater distribution, ray systems, twin craters, ridges, etc., for which ample material exists in the Orbiter photographs, the amateur with a telescope of between 6 and 12 inches aperture who wishes to carry out serious lunar research will find himself almost entirely confined to the study of temporary surface obscurations or changes, or colored glows.

Surface phenomena

Although occasional colored glows have been reported for many years, and are now considered established phenomena, though fleeting in their occurrence, there have been very few well-documented cases during a century and a half of critical examination. The most famous is Kozyrev's observation of Alphonsus, and it is worth remembering that it was made with a very large telescope. Any amateur expecting to see a crater formed or spirited away before his eyes is not only highly optimistic, but is also unlikely to make a reliable observer! It is no use "expecting" to see something unusual, for the eye will almost certainly oblige. An example of the fallibility of even experienced observers was afforded by the impact of *Luna 2* in 1959, when a number of "sightings" were reported—all in different places! Nevertheless, it may be of interest to list some of the regions in which unusual phenomena have been observed by reliable witnesses on more than one occasion.

MARE CRISIUM.   The case of the "quadrangle" in the southwestern corner was undoubtedly due to the oversight of early observers, but the same region is dotted with a great number of tiny craterlets that were also missed, and that are still, on occasions, strangely obscure or even invisible. There are even cases of Graham, the smallest crater on the mare, being missed when it should have been obvious.

ALPHONSUS.   There seems to be no doubt at all that some sort of activity occurred here in 1958. The region of eruption lay to the south of the central mountain, and was seen by Kozyrev as a reddish spot. He was, however, using a 50-inch reflector, and it is unlikely that a small instrument could have picked it up. Some observers claim to have seen reddish areas on other parts of the moon; Mädler, for instance, reported reddish tinges near Lichtenberg, a small crater near Aristarchus, and others have made similar observations. Unfortunately, there are many traps for the unwary; the moon is so bright that even the best achromatic eyepieces can sometimes produce false color effects, and it may be significant that the Alphonsus affair heralded an outbreak of color reports.

PLATO.   Some of the tiny craterlets in the interior have often been missed under good conditions, but they are so small that a 3-inch refractor will only show them as tiny spots. Plato's floor is covered with streaks of different tint, and these, together with its apparent darkening, make it one of the most interesting areas on the moon.

ARISTARCHUS. A great many transient phenomena have been reported in and around this crater: both red and bluish tints have been seen, while the area has sometimes appeared unusually brilliant. Kozyrev claimed that an outburst similar to that seen by him in Alphonsus occurred here in 1969.

Obscurations have also been reported in Schickard, Tycho, and around Schröter's Valley, but these are less well documented. In any case, it cannot be emphasized strongly enough that no one should expect to see anything unusual until he has become thoroughly used to lunar observation and has observed the formation in question under all conditions of illumination. There seems little point in *searching* for many of these phenomena, which have been noticed only incidentally during other work.

Transient lunar phenomena (TLP)

The so-called TLPs, however, have been placed by some observers in a special category. Their current popularity stems from the Alphonsus phenomenon of 1958 and the Aristarchus glows of 1963, already mentioned. In 1968, a NASA Technical Report listed 579 reported events that occurred between November 26, 1540 (a naked-eye observation!), and October 19, 1967. A further 134 events, covering the period from November 15, 1967, to May 4, 1971, were later published by P. A. Moore. Of these, no less than 54 refer to Aristarchus and its immediate area, although only 11 events (10 in Aristarchus, and one in Alphonsus) were considered "very reliable"—i.e., independently confirmed by experienced observers using adequate equipment. Most of these TLP reports refer to unusual brightenings of the surface, or reddish or bluish patches, lasting for anything from a few minutes to several hours. In some cases, a crater wall has appeared to pulsate in brightness.

It is only fair to say that some lunar observers are skeptical of these reports. For example, during the particular interest and excitement of the first manned lunar landing, amateurs paid close attention to those craters likely to be noticed by the astronauts as they orbited close to the surface. No less than 12 events, in six different craters, were reported on July 19, 1969, alone! Probably not the most enthusiastic TLP observer would accept all these reports as definite. But if 90 percent or even 99 percent are dismissed, we are still left with the two photographs by Kozyrev showing evidence of carbon and nitrogen emission, and a few independent observations, agreeing remarkably well in timing and description.

Aristarchus, the brightest crater on the lunar surface, has been the site of many TLP reports, but its particular brilliance makes it extremely hard to observe. For one thing, there is no "standard" region against which to match it, since it already outshines the other bright spots. Its great brilliance also very easily gives rise to false color effects. Most so-called achromatic refractors will give a bluish fringe to the brightest parts of Aristarchus; but a reflecting telescope is no perfect solution either, since most eyepieces will also produce some false color, although very much less than that from an object glass unless the mirror has an unsuitably short focal length. Aristarchus does, on occasions, appear remarkably blue, but only the consistent observer will be in a position to judge whether the effect is atmospheric, ocular, or lunar!

On February 24, 1975, I was testing a recently completed 4-inch Newtonian on the planet Saturn, and decided to look also at the almost-full moon. At once, Aristarchus was seen to be surrounded by blue flare, and the effect was mentioned to a friend at the time, though with no more than a joking reference to surface activity. On the following day came reports that a number of regular lunar observers had noticed blue or violet tints in Aristarchus that night! On the next clear night, the twenty-seventh, the crater looked virtually white. Had I been in the habit of observing the moon with that particular telescope, the appearance on the twenty-fourth would have attracted serious rather than casual attention. This little instance illustrates the great importance of consistent work, not only in lunar observation but in almost all fields of study. The amateur can only rarely refer to a yardstick and be confident that the same yard is always being measured. Photometry and colorimetry of the moon, in particular, must be largely relative and subjective.

The two-color "moon-blink" device was introduced, initially in the United States, in an attempt to reduce these influences and make the work more objective. Most of the lunar surface is virtually colorless; in other words, it reflects sunlight more or less evenly across the spectrum. This means that the relative intensities of different regions will appear the same no matter what colored filter is interposed in the telescope's light path. But if a region is preferentially colored, it will appear relatively brighter if viewed through a filter of its own tint, and relatively darker if the filter is of a different tint. At its simplest, the moon-blink consists of two filters, one red and the other blue, as well as a clear patch, which can be alternated rapidly in front of the field lens of the eyepiece. A reddish glow, for example, will appear alternately lighter and darker in the red and blue views respectively, even though it may not be obviously colored to the eye. There are, however, difficulties in

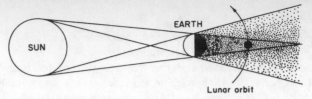

Figure 25. *Eclipse of the moon. (Not to scale.)*

comparing relative intensities in different colors, and not all observers find this device easy to use. In the original device, both colored views were converted to monochrome, to permit of easier comparison. Without this facility, a color-blind person might have more success than one of normal visual response! The Kodak Wratten filters No. 25 (red) and 80B (blue) have been widely used for the purpose.

It will be remembered that the *Apollo 17* astronauts came across patches of orange soil, and it is important to recognize those areas of the moon's surface that are sufficiently tinted to reveal a blink on all occasions. Fracastorius is the most obvious, but the floor of Plato also contains colored patches. The regular observer will learn to recognize these. But some TLPs have been recognized and confirmed by others by color alone, without any blink device, such as the reddish glows seen inside Gassendi by several British observers on April 30 and May 1, 1966.

Any observer interested in this work should first of all compile a short list of "likely" craters. Aristarchus, which has also been reported as appearing unusually bright under earthshine conditions, is the best studied, though probably not the easiest to observe. Alphonsus and Gassendi have also produced a number of claims as have Theophilus and Plato. The next step is to organize a small network, or join an existing group, of observers working on a mutual alert system. An unconfirmed TLP report, interesting and convincing though it may be to the observer, will carry little chance of acceptance. To make the confirmation worthwhile, however, it is essential that the other observers be given no clue other than the crater or region in which activity is suspected. If they then proceed to agree on its appearance and position, confidence in its objective existence must be strengthened. For this reason alone (quite apart from the fact that any local circumstance or telescopic effect is eliminated), confirmation from separate observers carries far more weight than does agreement among a group using the same instrument together.

If TLPs exist, what are they? The evidence of the spectrograms obtained by Kozyrev indicates that some venting of gas from beneath the surface must occur. Possibly this venting raises a temporary cloud (some observers have reported obscuration of surface detail associated with colored patches), which is either highly reflective or glows through electrostatic discharges. The continued work of amateurs, while it may not clear up the mystery of their nature, must do more to establish whether there is any periodicity or other significance in their occurrence.

Eclipses of the moon

Just as the moon casts a shadow which at new may fall upon the earth and produce a total eclipse of the sun, so the moon at full sometimes passes through the shadow cast by the earth (figure 25). The most striking observational difference between the two phenomena is that, whereas a solar eclipse is total over a very restricted region of the earth's surface, a lunar eclipse appears the same wherever the moon is above the horizon. The net result is that, for any particular site, lunar eclipses are rather more common than solar eclipses, and about two hundred times as common as *total* solar eclipses!

Since the sun is larger than the earth, the shadow the earth casts is in the form of a cone. The total length of the cone is about 860,000 miles, and at the average distance of the moon it is about 5,700 miles across. Around this central *umbra* (not to be confused with sunspot terminology) is the *penumbra*, about 6,000 miles wide, the region in which the sun is only partly cut off by the earth. Therefore, as the moon passes into the penumbra it experiences only a very slight darkening until it reaches the edge of the central shadow.

Usually, two full moons a year pass through the earth's shadow.* Not all of these are total, for the moon can simply pass through the north or south edge of the umbra; it may, indeed, just encounter the penumbra, in which case the dimming will be so slight that nothing unusual may be seen. At a total eclipse, the penumbral dimming becomes noticeable about half an hour before the umbra is reached; it takes the moon about an hour to pass fully into the shadow, and it may remain totally eclipsed for up to one and three quarter hours, so that the observable duration of the eclipse may be almost five hours.

Lunar eclipses are of great interest. They have none of the drama of a solar eclipse; but useful observations can be made. One never knows what course the eclipse will take, since the darkness is affected by the earth's meteorological conditions.

It may seem curious that the earth can influence the moon in this way, but the explanation is simple enough. The umbra of the earth's shadow is not perfectly black, because the dense atmosphere, acting as a colossal lens, refracts red light into the shadow. If the earth were airless, a lunar eclipse would be quite dark; but, as things are, the eclipsed moon usually glows with a dull coppery tint; and an observer on the moon would see the earth as a great black ball, with its atmosphere forming a reddish halo.

*A list of forthcoming lunar eclipses is given in Appendix I.

Because the light of the umbra has to pass through the earth's atmosphere, it is clear that meteorological conditions can affect the resultant brightness.

These conditions change from year to year. The total eclipse of May 13, 1957, was quite normal; but during totality on June 25, 1964, the moon was so dark that it was hard to see without a telescope, although conditions had cleared a little for the second eclipse of that year, on December 19. Searching through the records, we find that the eclipses of 1620 and 1642 were invisible with the naked eye, and the Swedish astronomer Pehr Wilhelm Wargentin reported that he could not see the totally eclipsed moon of 1761 even through a telescope! On other occasions, when the earth's atmosphere has been exceptionally transparent, the moon has been quite bright at mid-eclipse.

It is interesting to try to relate these effects to geophysical events. The explosion of Krakatau in 1883 undoubtedly caused the eclipse of the following year to be unusually dark, for it scattered a huge volume of dust and ash throughout the upper atmosphere, and this would certainly have cut down the transmitted light. Similarly, the eruption of Bali's volcano, Mt. Agung, in 1963 may have contributed to the darkness of the 1964 eclipses.

A number of interesting observations can be made during the course of a total lunar eclipse. They include: (a) the timing of the instant at which craters pass into and out of the shadow; (b) the observation of eclipsed features, to compare their appearance with that of normal visibility; (c) estimates of the total brightness of the moon during the eclipse.

Shadow timings can give information about the true diameter of the umbra at the distance of the moon. Its theoretical diameter can easily be calculated, but the average observed diameter is some 2 percent larger than this. Despite the fact that the umbra and penumbra merge into each other with no sharp edge, the eye does experience the illusion of a fairly well-defined boundary provided only low magnifications are used, and its passage across craters can be timed to within a minute or less. Such observations should be passed on to the national amateur society for reduction. Timings of the four "contacts" (i.e., first entry into the umbra, commencement of totality, end of totality, and final exit from umbra) can also be made, although their accuracy will be lower than that of crater timings.

It is also interesting to see how long before first contact the first trace of penumbral shading can be seen at the western (following) limb. The moon takes about an hour to pass through the penumbral zone, but the first hazy shading at the moon's limb may not be seen until some 20–30 minutes have passed. It is worth noting that the exit from the penumbra may be observed more satisfactorily than the entry into it, because the bright highlands of the eastern limb offer a better contrast with the shadow than does the much darker Oceanus Procellarum.

Observation of eclipsed features was a popular pursuit when some regions of the moon's surface (such as dark patches in Eratosthenes and Endymion) were believed to show a real variation during the lunar day. Since the surface temperature drops during a total eclipse by some 200°C, followed by an equally sharp rise when the shadow moves away, we might expect to see some greatly accelerated change taking place. The results of many decades of eclipse observation have been negative or inconclusive. Some features, however, are nearly always prominent in the shadow, particularly the bright craters Aristarchus, Copernicus, and Tycho.

Estimates of overall brightness are always useful and interesting, and can be made in two ways. The classical scale is that of Danjon, introduced in 1920, based on the visual appearance of the totally eclipsed disk, as follows:

0   A very dark eclipse. The moon is nearly invisible, particularly at the middle of totality.
1   A dark eclipse with gray or brownish coloring. Surface details are difficult to distinguish.
2   A dark red or rust-colored eclipse, with a darker region at the center of the shadow, and the edge of the shadow rather bright.
3   A brick-red eclipse, the shadow often being surrounded by a bright or yellowish border.
4   A very bright eclipse, copper or orange, with a very bright bluish border to the shadow.

On Danjon's scale, the total eclipses of December 30, 1963, and June 25, 1964, were both rated at 0, much the darkest of recent times. Danjon introduced his scale in an attempt at correlating the brightness of different eclipses with solar activity, but such a relationship is difficult to establish, since the moon does not pass equally far into the umbra at different eclipses; and, as we have seen, atmospheric transparency can vary considerably from one year to the next. Some mean estimates for recent total eclipses are as follows:

Date	Danjon Scale
1979 Sept. 6	3
1978 Sept. 16	2½
1978 Mar. 24	2
1975 Nov. 18/19	3
1975 May 25	2

An alternative way of measuring the intensity is to compare the total brightness with that of a star or planet (see page 255 for a discussion of stellar brightness or *magnitude*). Since the moon shows a disk, some means must be found of shrinking the lunar disk to a point. One way is to observe the moon's image in a bright convex surface, such as a Christmas tree ball;

or hold a low-power eyepiece at arm's length and view it through that. The moon and a comparison star or planet are viewed alternately with the device and a magnitude estimate using standard variable-star methods is made. For example, at the eclipse of November 18/19, 1975, I found the moon to be 0.5 of a magnitude brighter than the planet Jupiter, or −2.9 (about 1/10,000 of the brightness of a normal full moon). Observations of the very dark eclipse of December 30, 1963, however, obtained a magnitude of +4, some 600 times fainter still.

Lunar occultations

Since the moon is much closer to us than any other celestial object, it will naturally pass in front of and block out any star or planet that happens to lie in its path. It revolves around the sky in a month, and naked-eye observation reveals its drift; if it is close to a star one night, it will be about 12° east of that star at the same time on the following night. This means that in the course of an hour it appears to move, relative to the stars, across a space equal to its own diameter, so that several of these *occultations*, as they are called, may occur in one night if it is moving across a rich region of the sky.

However, the moon cannot stray over the whole sky. Its orbit is fixed to one plane, which is more or less the same as the plane of the planetary orbits (including the earth's), so that they all keep to a certain band of the sky.* This band is known as the *zodiac*, and it is clear that only stars lying within the zone can be occulted. It is about 18° wide, and runs through the twelve zodiacal constellations: Aries (the Ram); Taurus (the Bull); Gemini (the Twins); Cancer (the Crab); Leo (the Lion); Virgo (the Virgin); Libra (the Scales); Scorpio (the Scorpion); Sagittarius (the Archer); Capricornus (the Goat); Aquarius (the Water Bearer); and Pisces (the Fishes). There is a banal but useful rhyme for remembering the order of the zodiacal constellations:

> *The Ram, the Bull, the Heavenly Twins,*
> *And, next the Crab, the Lion shines,*
> *The Virgin, and the Scales.*
> *Scorpion, Archer, and the Goat;*
> *The Man who pours the Water out,*
> *And Fish with glittering tails.*

Of these, Taurus, Leo, Virgo, and Scorpio each contain a bright star

---

*Because of the large inclination of their orbits, both Mercury and Pluto can sometimes lie outside the recognized limits of the zodiac. The same is true of some of the minor planets.

that is occasionally occulted. Even in these cases, however, the glare of the moon (except when a thin crescent) is such that binoculars or a telescope are required for the observation.

Depending on the phase, disappearance, or *immersion*, occurs either at the moon's dark limb (before full) or at the bright limb (after full); conditions for reappearance, or *emersion*, are the opposite. An occultation of a bright star at the moon's dark limb, provided it is not illuminated by earthshine, is startling: One moment it is shining steadily; an instant later, it is blotted out. The very suddenness of disappearance offers one proof that the moon cannot possess a considerable atmosphere, for a depth of air would produce a gradual fading and reddening of the star's light. But nothing like this is ever seen.

All occultations are interesting to watch, but they also offer the amateur the chance of making some really useful observations. If the instant of immersion or emersion is timed and the observer's location on the Earth's surface is known sufficiently accurately, the relationship between his telescope, the occulting point of the moon's limb, and the star is uniquely determined. Either the moon's position along its orbit, or the star's position on the celestial sphere, can be calculated. Until quite recently, it was taken for granted that the stars' positions are known more accurately than the moon's, so that the moon's position was the one being checked. But nowadays the moon's position in its orbit is known to within a few tenths of meters at any instant as it speeds round the Earth at a rate of about 1 kilometer per second. Therefore, even if the time of occultation is correct to within 0.1 second, the moon's position has only been defined to within about 100 meters. But this 100 meters represents a movement of the moon in front of the stars of only $0''.05$. Although many stars have had their positions on the celestial sphere determined to within a tenth of this amount, the majority of the fainter ones have not. Therefore an occultation can lead to an improved determination of the star's position.

This argument assumes that the moon is a perfectly smooth sphere. But it is not, since surface irregularities can amount to several miles. Furthermore, the east-west and north-south librations are continually changing the profile as seen from the earth. It follows from this that the instant of occultation may be uncertain by at least a second and possibly more. In this case, the observation may well be supplying information about the altitude or depression of the feature on the lunar limb that occults the star. Some regions of the limb (especially those near the equator and in lower latitudes generally) are well known, and their irregularities can be allowed for, while others are poorly known or have not been observed at all.

Occultation predictions are given in such publications as the *Astronomical Almanac*, the *Observer's Handbook* of the Royal Astronomical Society of Canada, and the *Handbook* of the British Astronomical Association (see bibliography).

The visibility of a star near occultation depends critically upon atmospheric transparency. The slightest haze, when back-lit by the moon, can make even a bright naked-eye star difficult to see with a small telescope. Near full, the extra brilliance adds to the difficulty, while events at the bright limb are harder to observe than those at the dark limb. The most favorable circumstances for a first-class observation occur when the occulting limb is earth-lit, in the crescent phase. In general, occultations of stars down to about magnitude 7 or 8 can be handled by a 6-inch telescope if conditions are favorable, but near full the limit is brighter than this.

Disappearances (immersions) are much easier than reappearances (emersions) to observe properly. If a reappearance is being timed, the point on the lunar limb (which may be invisible) must be determined beforehand, using the data in the prediction, and the telescope kept accurately tracking on the spot so that it remains at the center of the field. Even with a driven equatorial instrument, it is not easy to determine the position of reappearance with precision, and if the eye is not looking at just the right place, it will be missed. Disappearances, on the other hand, can be observed perfectly well with an altazimuth. The star is located some minutes before the event is due to occur, and its drift through the field is timed. Using a low power (the lower the better, although a higher magnification may be necessary if the star is faint), this duration may be a couple of minutes. Divide the duration by two, and arrange the telescope so that the star is just entering the field this interval before the event is due to occur. If all is well, the occultation will occur with the star near the center of the field of view. This method has the additional benefit of giving the observer some warning of when the critical period is approaching. Of course, if an assistant can call out a "countdown," this is better still.

A good stopwatch is the best method of timing because it is the simplest. A tape-recorder used in conjunction with a radio tuned to one of the various international time-signal transmissions can also be used, but with the result that there is more to go wrong. This is, however, the best method if observations have to be made from a remote site and there may be multiple events, as with a graze occultation (see below). All that the backyard observer needs to do is to start the watch when the event occurs, then dial the telephone time signal and stop the watch against a convenient instant. Of course, the watch needs to be rated (i.e., run for several minutes to check its error, which must be allowed for). The common stopwatch, with a second hand making one revolution per minute, is not sufficiently accurate to do justice to a good observation, since the stopping and starting mechanism has an error of at least 0.2 second. A 30-second dial is preferable, and some watches make one revolution every 10 seconds.

When reducing the observation, the observer must allow for two errors that always occur: error of observation and personal equation. Between any cause and its effect, there must be a delay. First there is delayed perception, which may be long or short depending upon the circumstances. If a star is clearly seen, its disappearance will be apprehended almost simultaneously; if it is glimpsed or intermittently seen, more time will have to elapse before the observer can be sure that it has vanished. This perception error can vary from virtually zero to several tenths of a second (it could be longer than this under very poor conditions, but in most cases such an observation would be practically worthless). The second error, personal equation, is the physiological delay between the brain signaling the reaction and the muscles obeying the message. A number of tests have shown that, under normal conditions, a delay of about 0.2–0.3 of a second elapses between an observer seeing a star disappear and his finger starting the watch. Personal equation is reasonably constant, and can be assessed by covering the left half of the stopwatch dial with a piece of paper whose edge coincides with (say) the zero second line on the face. As soon as the hand is seen to emerge from beneath the paper, the watch is stopped, and the error can be read directly on the dial. The personal equation, and the estimated perception error (which will, if the observation is a good one, be negligible) must be deducted from the actual instant as recorded on the watch. There should be a negligible error in stopping the watch against the time signal, since the preliminary seconds' beat that leads up to the marker signal allows the button to be pressed as close to the time instant as makes no difference.

The geographical position of the telescope needs to be known to within about 1″ of latitude and longitude (equivalent to about 30 meters). Height above sea level must also be known to within this accuracy. Such precision is obtainable from large-scale survey maps available in many libraries. Observations should be submitted to the national society.

On rare occasions, a star does not snap out suddenly but shows a perceptible fading before it disappears. This is almost certainly due to its being a close double. The eye is sensitive to a fade as short as a tenth of a second, equivalent to an angular separation of as little as 0″.05, a better resolution than can be achieved visually with any telescope. Some observers have reported a fading effect for stars that appear to be single, but there is no obvious explanation for such a phenomenon.

In recent years, there has been a great deal of interest in so-called *grazing* occultations. These are occultations where the extreme north or south limb of the moon passes in front of the star. Graze occultations are important on two counts. First, they permit the moon's declination or celestial latitude to be determined with greater precision than do normal occultations, which give more information about its position *along* its orbit. Second, they impart information about the profile of a section of the limb rather than just the height of a single point. Perhaps the main reason, however, is that graze occultations usually require an expedition to the appointed site, are fun to organize and take part in, and give the flavor of serious, cooperative observing.

The predicted track of a graze occultation usually indicates the line on the earth's surface along which observers would see the star exactly coincide with the mean lunar limb (in other words, the edge of a perfectly smooth moon). The procedure is to arrange observing sites at intervals of about 100 yards both inside and outside this line. When the occultation occurs, therefore, each site will see the star pass through a different "slice" tangential to the limb, and the timings of the disappearances and reappearances at the various stations will enable a profile to be drawn up. Some sites, particularly the ones that are located "outside" the mean limb, may see the star disappear and reappear more than once as it passes behind individual peaks.

The uncertainty of what may be seen, and the possibility of multiple events having to be timed, means that the flexibility of a tape-recorded commentary is to be preferred to a row of stopwatches! If a radio time-signal transmission is also recorded on the tape, the timings can be reduced later at leisure. Continuous seconds beats are transmitted from Fort Collins, U.S.A., (WWV) on 20, 30, 60, and 120 meters; from Ottawa, Canada, (CHU) on 20.5, 41, and 90 meters; and from Nauen, East Germany, (Y3S, formerly DIZ) on 66.3 meters. Observers in the United Kingdom can use either Y3S or the long-wave transmissions from Rugby on 60 kilohertz.

7

The Sun

The sun is 864,600 miles across, or, as near as makes no difference, a hundred times the diameter of the earth. It is a great sphere of hydrogen gas that is shining at a fiercely hot temperature—about 6,000°C at the surface, and several millions of degrees in the interior. Through atomic reactions, this hydrogen is gradually being converted into helium, so that the sun is really a vast nuclear furnace.

The sun is essentially no different from any of the stars that crowd the night sky, but it appears very different to us because it is so close. It is the center of our solar system, which comprises the planets, comets, and other bodies that revolve around it; and some 5,000 million years ago, so astrophysicists tell us, the sun gave birth to these bodies, including the earth, collecting the material for their formation from a vast cloud of gas and dust far out in space.

In all this time, however, the sun has probably changed very little. The earth itself, which revolves around it once a year at a distance of about 93 million miles, has cooled from its molten state, filled its surface hollows with oceans, seen the development of life from the primitive sea creatures to men who have already traveled to the moon and back. During this whole development it has received the sun's steady radiation, and there is no reason to suppose that our star will not continue shining in its present manner for many thousands of millions of years. Its resources are so unimaginably vast that its prodigious emission of light and heat seems to affect it not at all.

Since all the other stars are so far away that we cannot observe their surfaces at all, the sun is of special interest to professional astronomers. Every day, observatories all over the world take photographs of its surface and examine the various features with special equipment that is far beyond the amateur's range, and it must be admitted that there is little chance of the observer making a startling discovery. However, this is no reason for dismissing solar work out of hand, for useful observations can still be made. It is both interesting and instructive to make drawings or take photographs of whatever sunspots may be visible; since they often change rapidly and unpredictably from day to day, one never knows quite what to expect.

Direct observation

It must be re-emphasized that *the sun is a lethal object*. Many a beginner, underestimating its power, has peeped at it through a small telescope and as a result suffered permanent partial blindness. Rumor has it that Galileo's eventual blindness was caused by his unwise solar observations, and the danger is greater now, since telescopes are correspondingly more powerful.

Perhaps the greatest disservice done to astronomers by unthinking opticians has been the invention of the sun cap, a thick, heavily dyed circle of red, blue, or green glass designed to screw over the ocular's eye lens, and intended to make direct observation safe. Unfortunately, the makers have not allowed for the fact that telescopes vary in aperture; thus, while such a filter may (if it is dense enough) provide safety when used for a period of half a minute or so with a 2-inch refractor, it is emphatically *not safe* when used with anything larger. In any case, it is virtually impossible to see anything through a sun cap because of the thickness of the glass and the correspondingly restricted field of view, since the eye cannot get close to the eye lens.

If the sun is to be observed directly, the correct place for the filter is at the upper end of the tube, so that the radiation is reflected or absorbed before it enters the telescope. There are two reasons for this: It minimizes heating of the air within the tube, which can create tube currents, and as we have seen, any filter placed near the focus is subjected to concentrated heat, and sooner or later will be damaged. The drawback to a full-aperture filter used to be expense, since an optically worked glass disk was needed as a support for the coating. During the past decade or so, this problem has been solved by the production of thinly aluminized mylar, a plastic material so diaphanous that it has no detectable effect upon telescopic performance. The best-known filter of this type, sold under the trade name of Solar-Skreen, has been tested by various authorities and found to be quite safe for visual use. It turns the sun a metallic blue color.

Fogged and developed photographic film has been used for small-aperture filters (to fit over binocular objectives, for example), but this material is treacherous. The sun may appear dim, but this does not mean that all the radiation is being suppressed, since both ultraviolet (shortwave) and infrared (longwave or heat) radiation can pass through and damage the retina without their presence being suspected. Some types of film are relatively safe, others are not. Fogged color film, in particular, is almost transparent to infrared waves, and terrible damage to the retina can result. With aluminized mylar so readily available, there is no need to seek a dangerous substitute.

A full-aperture filter, used with an aperture of 6 inches or so, will give a beautiful view of the sun in good conditions. The fine structure of the spots can be observed, and the photosphere is seen to be mottled. In steady seeing, with a magnification of about 300, a telescope of this aperture will reveal the granulation of the photosphere, and a large spot group is a magnificent sight. It will be noticed that the best conditions for solar work often occur early in the day, before the ground has warmed up sufficiently to produce bad turbulence; good views are rarely to be had around noon, despite the higher altitude.

Projection

While direct observation shows the greatest detail, one of the principal pastimes enjoyed by the amateur is the day-to-day pursuit of the spot groups as they are carried across the disk by the sun's steady rotation. The positions of the spots (noting whether they are north or south of the equator, and in what longitude) are just as interesting as the spots themselves, since at different times different regions of the sun are active. The projection method is the best way of recording them accurately. Figure 26 shows a very simple apparatus that can be made in an hour. Essentially, it consists of a holder that fits over the drawtube, carrying an arm that holds a square board rigidly a foot or so behind the eyepiece. A shield is also needed to

Figure 26. *Solar projection apparatus.*

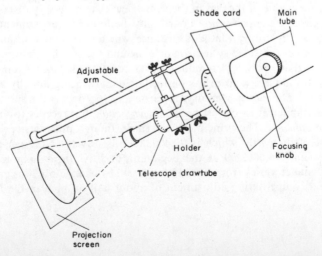

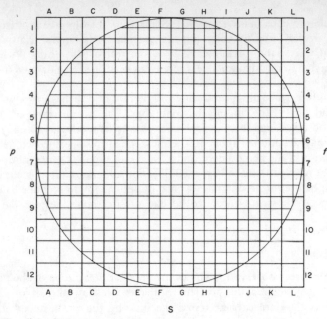

Figure 27. *Solar projection grid.*

protect the projected image from direct solar rays, and a sheet of cardboard or light metal, pierced with a central hole to fit over the object-glass end of the telescope, will serve this function and also help to balance the tube. In the case of a reflector, which is altogether less convenient for solar work because of the heating effects, no shield is required, since the image is not facing the sun. A useful disk diameter is 6 inches, which will require about a ¾-inch eyepiece if a 3-inch refractor is being used; it must, of course, include the whole of the sun in the same view. The necessary distance from the eyepiece to the projection board must be found by experiment.

It is worth noting that this distance will not remain unchanged. All the planets in the solar system revolve around the sun in slightly elliptical orbits, so that instead of remaining at a constant distance from the sun, as they would were their orbits perfectly circular, they gradually approach and recede. The planet Mercury, for instance, approaches the sun to within 36 million miles at its closest point, or *perihelion*, but recedes to 48 million miles at *aphelion*. The principle is the same, though much less extreme, in the case of the earth, which is 91,400,000 miles away at the beginning of January, and 94,600,000 at the beginning of July. In angular terms, the sun's diameter varies from 31′ 31″ to 32′ 38″, a change obvious enough to require a compensating adjustment of about half an inch in the length of the arm.

A piece of clean white paper, bearing a 6-inch circle crossed with a fine grid, is attached to the board, and the solar image is accurately focused on it. This grid (figure 27) must be drawn with a hard, sharp pencil, since coarse lines may obscure some of the finest detail. It is a matter of individual taste just how many lines are drawn in, but ⅓-inch squares are convenient and allow of considerable accuracy in copying the sunspot positions. Since these are copied onto a sheet of thin paper placed over a similar grid, it is helpful if they can be identified in some way. One good idea is to draw every other line slightly bolder (thus forming larger quartered squares) and marking them with reference letters and numbers, as shown. The second grid should be drawn with ink, so that it shows through the paper.

Before making a drawing, the disk must be orientated. This is done by leaving the telescope stationary and letting the image drift across the screen, twisting the attachment until a sunspot trails along one of the E-W lines. Due to the earth's rotation, the drift is from right to left, with the west limb of the sun leading, or *preceding*, and the east limb *following*—two terms frequently used to describe the apparent motion of a celestial object across the sky. The north and south points are at the top and bottom, respectively. (In the southern hemisphere these directions are reversed.)

Getting accurate positions with an altazimuth mounting is nowhere near as easy as with an equatorial, but it can be done. Very tiny spots, or *pores*, are seen more easily if a sheet of blank paper is placed over the grid so that the disk is seen projected without any confusing lines. Once their general position is established, they can be marked in by reference to the grid. If any bright clouds, or *faculae*, are noticed, these should also be recorded. The main thing to remember is that positions are far more important than details; any attempt to make the result look "realistic" will almost certainly result in exaggerated spot sizes. The secret of accuracy

Figure 28. *Different views of the sun.*

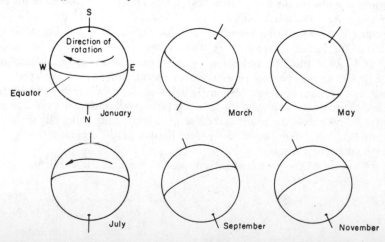

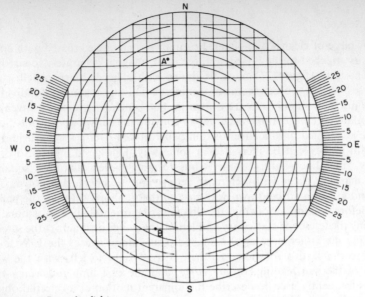

Figure 29. *Porter's disk.*

is to wait until the solar disk is exactly central on the screen, and then to memorize the position of a spot in relation to the grid lines. It is then drawn in on the blank.

Heliographic coordinates

We now have a representation of the sun's earth-facing hemisphere, orientated with respect to the earth's cardinal points. This does not mean, unfortunately, that the north pole of the sun is at the top and the south pole at the bottom. From January to June, the sun's north pole is inclined somewhat westwards (i.e., to the left), whereas during the rest of the year it swings out toward the east. Furthermore, from the end of May until the end of December, the north pole is inclined toward the earth, the south pole coming into view during the other months. Figure 28 shows the projected view of the sun at different times of the year.

It is clear that the reduction of the positions to solar latitude and longitude, or *heliographic coordinates*, is not as simple as might appear at first sight. It can, however, be done in a few minutes by a technique devised more than twenty years ago by J. G. Porter, a well-known British astronomer. *Porter's Method* requires the use of a hand calculator, but since the formulas are very simple this is no great hardship; it is a great pity that few textbooks give it mention.

A sketch of the projected solar image is made in the usual way, great care being taken to mark the east and west points accurately. This is then laid over a Porter's Disk (see figure 29), which must, of course, have the same diameter. Using the table in the British Astronomical Association *Handbook*, the axial inclination (*P*) is found and the sketch rotated until the E-W line is inclined at the correct angle. It is worth remembering that the rotation from the zero position is clockwise if the value of *P* is positive, and counterclockwise if it is negative. For instance, if a drawing were being made in late September, when $P = + 26°$, the east point on the drawing would be turned to coincide with the lower 26° mark at the right-hand side of the disk.

When the sketch is orientated, there only remains the task of estimating the spot's position in terms of horizontal position (*x*), and vertical position (*y*). The *x* factor is reckoned as positive if to the west of the meridian, negative if to the east; whereas *y* is positive if to the north, and negative if to the south. Each division of the grid corresponds to a value of 0·1. Thus, in the case of spot *A*, the values would be $x = + 0·14$ and $y = + 0·64$; for spot *B* we should have $x = + 0·25$ and $y = - 0·60$.

It is also necessary to make an estimate of the distance (*d*) from the center of the disk. This is done with the aid of the concentric lines; in both cases it works out at about 0·65.

When these values are obtained, the *latitude* of the spot (*B*) can be found from the formula

$$\sin B = y + \text{correction}$$

the correction being found by reference to Table I, following. The value of *d* is already known, while B_0, which is the latitude of the center of the disk, can be found in the B.A.A. *Handbook*, or it can be found in Appendix VII.

TABLE I. Correction to the value of y

d	$B_0 =$	0°	1°	2°	3°	4°	5°	6°	7°
0·0		0·00	0·02	0·03	0·05	0·07	0·09	0·10	0·12
0·1		00	02	03	05	07	09	10	12
0·2		00	02	03	05	07	09	10	12
0·3		00	02	03	05	07	08	10	12
0·4		00	02	03	05	06	08	10	11
0·5		00	02	03	05	06	08	09	11
0·6		00	01	03	04	06	07	08	10
0·7		00	01	02	04	05	06	07	09
0·8		00	01	02	03	04	05	06	07
0·9		00	01	02	02	03	04	05	05
1·0		0·00	0·00	0·00	0·00	0·00	0·00	0·00	0·00

The sign of the correction is always the same as that of B_0 (see Appendix VII).

Once the value of B is known, the *longitude* of the spot (L) can be found from the formula

$$\sin (L_0 - L) \times \sec B$$

Tables in the *Handbook* of the British Astronomical Association, the *Astronomical Almanac,* and other annual publications, enable the longitude of the sun's central meridian, L_0, to be determined for any hour of the day. It should be noted that L_0 decreases with time, so that features in the western hemisphere (i.e., past the meridian) have a larger value of L than do features to the east of the meridian. The rate of change of L_0 at a latitude of 15° N or S is 0.55° per hour, or 13.2° per day. At other latitudes, it is approximately as follows:

0°	13.4° per day	20°	13.1° per day	35°	12.5° per day
5	13.4	25	12.9		
10	13.3	30	12.7		

Individual sunspots and their characteristics

Some spots, especially the small ones, are quiescent; they either die out as quickly and quietly as they formed, or remain more or less unchanged from day to day. Large groups, however, can be much more active, and may show striking changes in just a few hours. At these times, if the atmosphere is steady enough, it is interesting to make a large-scale drawing of the details of the group, once again by projection.

A large disk is required if the finest details are to be seen, and since the image will appear faint unless protected from the light in some way, the answer is to make a *projection box*. This is made from thin plywood, about 5 inches square and 10 inches long, with the interior painted mat black. The drawtube fits into one end; at the other end is a piece of white paper or Bristol board to act as the screen. The lower half of one of the sides is omitted, so that the image can be examined. Under these conditions, well shielded from daylight, the solar surface is seen with great brilliance. Once the drifting E–W line has been established against the grid drawn on the screen, the details of the group can be copied in the same way as with whole-disk projection. To produce the necessary enlargement, of course, a high-power eyepiece must be used.

The sunspots themselves are titanic disturbances on the visible surface or *photosphere*. Characteristically, a spot consists of a central dark *umbra* surrounded by a lighter *penumbra;* patches of this penumbral matter may also be found scattered over the adjacent photosphere. A complex spot may contain several separate umbrae, and a common sight is that of two spots having roughly the same latitude, one following the other across the disk at a distance of anything up to 50,000 miles.

Although a spot umbra appears dark, it is certainly not black. This is something that can be proved when the moon passes across the disk during a solar eclipse; against the absolutely dark lunar silhouette, a sunspot appears as distinctly brownish. Evidently, then, it is an effect of contrast. The interior of a sunspot, being some 1,100°C cooler than the photosphere, radiates light less efficiently; yet if we could see it divorced from its over-powering surroundings, it would appear as dazzlingly bright!

Sunspot positions are of interest on three counts, and it is worth saying a few words about each of these.

ROTATION PERIOD.   The sun, like the giant planets Jupiter and Saturn, does not rotate as a solid body. Instead, its period of rotation relative to a fixed point, or *sidereal period*, varies from about 25 days at the equator to about 27½ days at a latitude of 40°, with more polar regions (if one can imagine "polar" regions on the sun!) rotating even more slowly. Since the published values for the longitude of the central meridian are based on a mean period, we find higher-latitude spots gradually dropping behind. It should be remembered, however, the the earth's own revolution around the sun makes it appear to rotate more slowly; the *synodic period*, or period relative to the earth, for the equator is about 27¼ days, and about 30 days in the "temperate" regions.

PROPER MOTIONS.   As well as showing a drift according to their latitude, individual spots sometimes have movements of their own. The leading spot of a freshly formed pair frequently gains on its follower at the rate of about 1° a day, but it quickly slows down again to a more orderly rate. These *proper motions*, as they are called, are not always predictable, but they need accurate longitude observations if they are to be made out.

DISTRIBUTION.   Sunspots do not occur erratically all over the solar surface. The main areas of occurrence are the 5° to 40° latitude limits in both hemispheres; only rarely do they appear at the equator, and hardly any have been observed in latitudes higher than 45°.

The lifetimes of spots are exceedingly variable. The majority of groups never get beyond the "pore" stage; these require careful observation to be made out at all, and they die out in a day or two. If two distinct umbrae develop, however, and a major group occurs, it usually reaches its greatest development in about ten days, at which time it is well elongated in longitude. After this, the trailer dies away and the leader becomes small and round, although it may persist for weeks; it will, of course, be temporarily out of view while on the averted hemisphere. In 1943, a spot was followed for almost two hundred days, but this was most exceptional.

The solar cycle

The frequency with which large and small groups form varies greatly. In some years, the sun throws up many groups, while at other times the disk may be spotless for days or weeks on end. The discovery of the *solar cycle*, as it is called, was made by an amateur—a German apothecary of Dessau, Hofrath Schwabe. In 1826, he bought a small telescope and started making daily observations of the sun, drawing whatever spots happened to be visible. By 1843, some sort of periodicity seemed to be emerging, but it was not until 1851 that he considered the fluctuations to be confirmed—an example of devoted persistence that can have few equals. We now know that the cycle has an average period of just over eleven years, and that the rise to maximum is faster than the subsequent decline; but the period is elastic to the extent of several years, while some maxima are much more active than others. A recent maximum, which reached its peak in December, 1957, was the most active ever recorded; but the largest sunspot ever seen appeared during the previous maximum, in April, 1947. This group covered an area of about six million square miles, and was easily visible with the naked eye when protected by a dense photographic negative or smoked glass. A group larger than about 25,000 miles can usually be seen without a telescope.

It should not be thought from this that the sun is interesting only around the time of maximum activity. Near minimum the disk is more often blank, but one never knows when a large group may appear; in September, 1963, for example, when the sun was only a year away from minimum activity, I watched a splendid group develop until it could be seen with the naked eye. At the time of writing (1981), the sun has passed through an extended maximum, documented by the graph in figure 30. This shows the average number of active areas, or centers of activity, seen on each day of observation since 1975. This data was obtained using my 3½-inch refractor, using a full-aperture filter, and a magnification of × 70. It is fascinating to watch new spots break out and old ones decay—and one never knows just what is

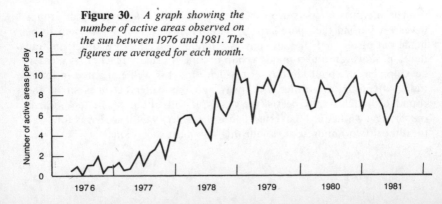

Figure 30. *A graph showing the number of active areas observed on the sun between 1976 and 1981. The figures are averaged for each month.*

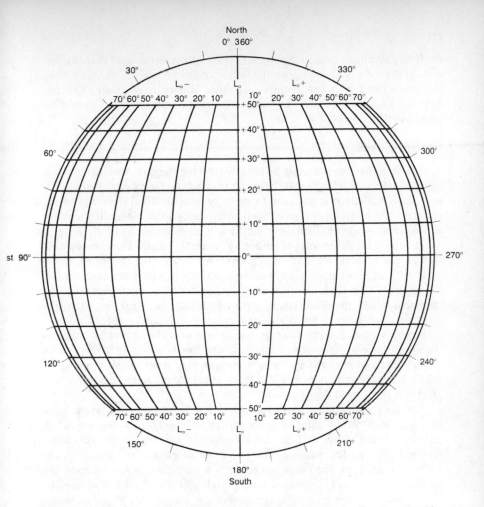

Figure 31. *The presentation of heliocentric latitude and longitude when* $B_o = 0°$. *The sun spins from east to west, so that the longitude of the central meridian* (L_o) *decreases with time.*

going to appear around the eastern limb! If an exceptionally large group is carried into the averted hemisphere, it is worth keeping a special lookout two weeks later to see if it has reappeared at the eastern limb.

An interesting by-product of the solar cycle is the way it influences sunspot latitudes. The first spots of a new cycle always occur in the higher latitudes, the active regions moving toward the equator as the cycle develops. The sun's recent new phase of activity was heralded by the appearance of a small, high-latitude spot in November, 1974, at the same time that the last spots of the old cycle were appearing in the equatorial regions, so that during this phase there were two distinct spot-zones in each hemisphere. This behavior is known as Spörer's Law, after the nineteenth-century German astronomer who observed it.

Sunspot counts

The fact that the sun must be observed during the day is something of a mixed blessing. While the solar observer is saved the frosty dark vigils of his nocturnal brother, it is not easy to make regular detailed drawings of the surface when he is out at work! This is particularly so in winter. If the sun is active, an accurate observation can take half an hour or so.

But there are other ways of gathering interesting data. The simplest is to make a daily count of active areas (AAs). An active area is defined, somewhat arbitrarily, as any region of sunspot activity in which the main centers lie within 10° on the solar surface of each other. Two spots farther apart than this would therefore count as two active areas. Near the limb, the foreshortening is so severe that it may become difficult to judge the separateness (or closeness) of two spots, but to help with this, figure 31 indicates the appearance of lines of solar latitude and longitude at 10° intervals. A typical observation of the sun near maximum may reveal 12 or more active areas, but near minimum the disk may appear completely devoid of any markings for several days at a time.

The great attraction of this approach is that an adequate observation can be made in a few minutes. No knowledge of the coordinates is needed, although, if the position of the solar equator is known, the record can be refined by giving the number of active areas in each hemisphere. As the record accumulates, the usual procedure is to calculate monthly means: the number of active areas observed during each calendar month is divided by the number of days on which an observation was made. Monthly means formed the basis of the graph in figure 30.

This simple method of counting is open to the obvious objection that no allowance is made for the relative importance of different groups. A well-developed group may contain two large spots and a dozen or more small ones scattered over the immediate vicinity, but it still counts as a single active area. The international Wolf Number system attempts to remedy this by including both active areas and individual spots in its rating. If f is the number of individual spots that can be made out, and g is the number of active areas, then the Wolf Number or Relative Sunspot Number R is given by

$$R = 10g + f.$$

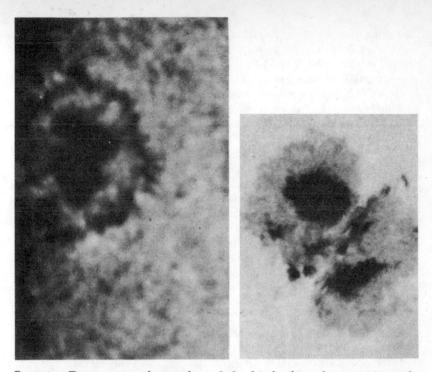

Sunspots. *Two amateur photographs. Left, 3-inch object-glass, negative scale 5½ inches to the sun's diameter, subsequently enlarged. The granulation of the solar surface can be seen clearly. (H. N. D. Wright, London.) Right, 4-inch O.G., negative scale 60 inches to the sun's diameter. This shows the relative contrast between photosphere, penumbra, and umbra. (W. M. Baxter, London.)*

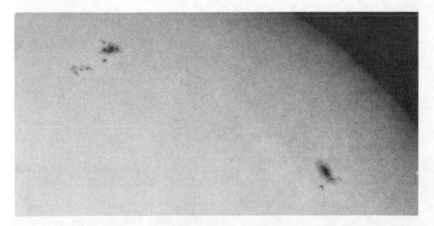

Two small sunspot groups photographed near the limb by Lee S. Najman, using a 12½-inch (32cm) Newtonian protected by a Solar-Skreen filter. The exposure was 1/1000th of a second on H & W control film.

The value of R will tend to increase with increasing aperture, since more small spots and pores will be seen. The observer can apply a constant to his derived value to bring it into better line with the standard Sunspot Numbers that are published monthly in *Sky & Telescope*.

It is also helpful to make an estimate of the position of each active area. I took up regular solar observation some years ago, using a very simple technique which at least allows the different spots to be identified from one day to the next. The telescope is a 3½-inch altazimuth refractor with a full-aperture solar filter. The eyepiece unit contains an ex-government roof prism, with a crosshair engraved on one surface. The eyepiece is focused on the cross. The roof prism gives an erect image of the sun, so that it corresponds to the naked-eye view. The first stage in the observation is to twist the drawtube until a sunspot trails accurately along one of the cross-hairs, which must then indicate the east-west direction. Bringing the solar image central in the view, and starting from the north point in an anticlockwise direction, the position angle (P.A.) and the distance from the sun's center in terms of its radius are estimated for each active area. Position angle is reckoned in degrees from north (0° or 360°), east (90°), south (180°), and west (270°). Simple eye estimates can derive positions to within 10° in P.A. and 0.1 in radius.

To derive the heliographic coordinates of each active area, the values of P, B_0 and L_0 must be obtained for the time of the observation. Additionally, for each active area we must determine the following two values:

A: The position angle measured from the sun's north pole. In other words, the observed position angle $- P$.

ρ: Sin ρ is given by the distance of the feature from the center of the disk, in terms of the sun's radius.

Then,

$$\sin B = \cos \rho \sin B_0 + \sin \rho \cos B_0 \cos A;$$

$$\sin (L_0 - L) = \frac{\sin A \sin \rho}{\cos B}.$$

These values for each active area can be obtained in a few seconds, using a pocket calculator. This method of deriving positions can, of course, be applied equally well (and with greater accuracy) to a drawing, but even eye estimates can give positions to within 2–3° in latitude and 5–10° in longitude, depending on how close the feature is to the limb.

The Wilson effect

Some British amateur solar observers have recently been active in investigating a phenomenon known as the *Wilson effect*. This was first noticed in

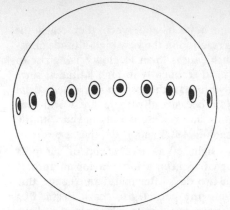

Figure 32. *The Wilson effect. The effect is greatly exaggerated in this diagram.*

1769, when Alexander Wilson, Professor of Astronomy at Glasgow, followed a large sunspot as it approached the western limb. He noticed that as it neared the limb, the umbra became steadily more and more displaced toward the center of the sun, as shown schematically in figure 32. Wilson realized the significance of this: The spot must have been shallow, with the umbra lying below the level of the photosphere—rather like a saucer that is viewed from a more and more sharply inclined angle. His supposition was confirmed when the spot, which proved to be a long-lived one, reappeared at the eastern limb with its umbra once more displaced, becoming central as it moved toward the center of the disk.

This was all very well until other observers noticed that only a relatively few spots showed the Wilson effect, whereas a small number produced an umbral deflection in the *opposite* direction, inferring that the spot in question was convex! To investigate this curious state of affairs, members of the Solar Section of the British Astronomical Association, which is an entirely amateur body, took and measured photographs of 79 spots that showed either a positive or negative displacement. The results were most interesting: It was discovered that three times as many spots were concave as convex, and that the degree of depth or elevation of the center was much less than had been widely believed—a spot 20,000 miles across would, on average, show a difference of level of about 500 miles, though larger variations were sometimes found. Moreover, fluctuations occur; one spot was observed to progress from convex to concave. It is clear that a great deal remains to be done before our knowledge of the Wilson effect and its consequences is at all satisfactory. This is a field in which other amateurs may be able to do useful work.

Sunspots and auroras

Sunspots are intensely magnetic, and active groups emit great quantities of what is known as *corpuscular radiation*. These atomic particles are

sprayed out almost like water from a hose, and when they reach the
vicinity of the earth they trigger off a reaction in the very high-altitude molecules in the earth's atmosphere, which causes them to glow, giving rise to
auroral displays. Auroras are seen most frequently in high latitudes, since
the earth's own magnetic field usually prevents the disturbing particles from
reaching equatorial regions; but just occasionally there are world-wide
displays. It is not surprising to find that auroras are closely linked with the
solar cycle; although they were common in 1968 and 1969, they are rare at
the present time. Nevertheless, the sight of a large group of sunspots
passing the sun's meridian is the signal to keep a lookout for an auroral
display; it will occur, if at all, about two days after meridian passage, this
being the time taken for the slow-moving particles to reach Earth. For
regular auroral observation, the observer must have a latitude of at least
50° to 55°, and the higher it is, the better.

Faculae and flares

Closely associated with sunspots are the *faculae*, clouds of glowing
vapor that float at an altitude of several hundred miles above the photosphere. In some mysterious way, faculae seem to herald the outbreak of a
new spot group; they also linger after the group has decayed—rather like
solar vultures! If a group of faculae is seen, it is wise to keep an eye on
that particular region and be ready for the appearance of spots.

Faculae can ordinarily be seen only when they are near the limb; this is
explained by the extra-effective thickness of the solar atmosphere, or
chromosphere, at this point. The chromosphere is several thousand miles
deep, and acts as a division between the dense photosphere and the tenuous
outer atmosphere. The chromosphere itself shines brightly, but its light is
overpowered by the brilliance of the photosphere and generally cannot be
seen except when the sun is totally eclipsed by the moon. It absorbs so
much light from the limb region that the edge of the photosphere appears
much dimmer and redder than the center, an effect well seen on the projected image. The faculae are really of about the same luminosity as the
photosphere, but since they are floating at a considerable altitude they are
less affected by this dimming, and so stand out clearly.

Faculae are quiescent phenomena and rarely spring surprises; but once
in a lifetime, if he is attentive, the systematic solar observer may witness
something like the following sight:

> *A very brilliant star of light, much brighter than the sun's surface,
> most dazzling to the protected eye, illuminating the upper edges of the
> adjacent spots and streaks, not unlike in effect the edging of the clouds
> at sunset.*

Such was the appearance of the first solar *flare* ever observed, witnessed independently by two British observers, Richard Carrington and Richard Hodgson, on September 1, 1859. Around the time of sunspot maximum, faint flares are often observed photographically, but very rarely is one sufficiently bright to be seen visually, even if an observer happens to be watching at the time. Flares are tremendously powerful outbursts of radiation that last for just a few minutes, but that often produce disastrous effects on terrestial communications, temporarily upsetting the ionospheric layer high in the atmosphere and interfering with shortwave reception. Not surprisingly, flares and accompanying "fade-outs" are associated with spot groups, a hazard the *Apollo* astronauts were forced to face when they made their first lunar missions, which took place during a particularly active period of maximum solar activity. The danger is present because flares also emit certain short-wave radiations that are filtered out by the earth's atmosphere, but that for man out in space might have a destructive effect on unprotected human tissues. Flares are not only seen when the sun is near the peak of its cycle; on July 4, 1974, when it was near minimum activity, several observers in the United States observed a brief, brilliant flare in the very active spot that was crossing the disk at the time.

Solar photography

Although four chapters of this book are devoted to astrophotography, solar photography is a field of its own. These notes should, however, be read in conjunction with Chapter 21. Although it might at first sight appear a simple matter to record so bright an object, the fact remains that first-class photographs showing anywhere near all the details that the telescope is capable of recording are few and far between. Seeing conditions in the daytime are often bad. On the other hand, exposures can be so short that a motor-driven telescope is not needed, and very fine-grain film, permitting considerable enlargement, can be used.

For most work, a full-aperture solar filter will pass sufficient light for very short exposures to be used. If it should prove too dense, it is preferable to use a faster film than to resort to one or more neutral-density filters, which are likely to be less optically perfect and must, because of their small size, be located near the focus of the objective. If a reflector is to be used exclusively for direct visual and photographic work on the sun, removing the coating from the main mirror will reduce the image intensity by a factor of about 20 times. If the diagonal also is uncoated, the overall reflection of the system will be reduced to about 0.25 percent. This compares with a visual transmission of about 0.01 percent for Solar-Skreen.

The size of the image formed of an object in the focal plane is a function of the objective's focal length. In the case of the sun and moon, which appear to be of about the same diameter, it is about 1/110 the focal length. Therefore with a typical f/8, 6-inch reflector (focal length 48 inches). the image will be about 0.44 inch across. The working dimensions of a 35mm negative are 24 × 36mm, and therefore the image size needs to be adjusted to a diameter of about 20mm, if it is to fit comfortably into the frame. A Barlow lens is the most convenient way of achieving this. If a telescope larger than 6 inches in aperture is being used, there is little point in employing the full aperture, since the full resolving power of the objective will not be achieved in so small an image, even with the fine-grain film that can be used in solar work. Instead, a large mirror should be stopped down by placing a cardboard disk with an offset half-aperture stop cut out of it, as shown in figure 33. An eccentric stop of this kind eliminates the diffraction effects around the diagonal and gives excellent images, comparable with those produced by a first-class refractor.

Many photographs, otherwise excellent, have been spoiled by poor focusing. It is difficult or impossible to focus directly on the screen of most single-lens reflex cameras, although it can be tried. Astronomical images tend to lack the crispness and bite of a normal view, and there is a margin of uncertainty in the focusing where the image continues to look "all right," whereas on the developed negative it may look all wrong! The enthusiast who intends undertaking regular work would be well advised to invest in a special optical viewfinder that can be attached to some types of camera for technical purposes. However, if only a short length of film is being exposed, or if the adjustment, once made, is reasonably permanent, the simple way to be described is as good as any. Open the back of the camera, take a short piece of razor blade, and carefully tape it across the middle of the film guide— making sure that it is flat—so that the sharp edge crosses where the center of the film frame will be. This edge is then virtually in the plane that will be

Figure 33.   *How to stop down a reflecting telescope for solar work. The same method can be used with a catadioptric telescope.*

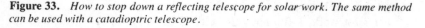
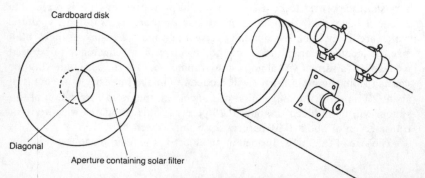

Cardboard disk

Diagonal

Aperture containing solar filter

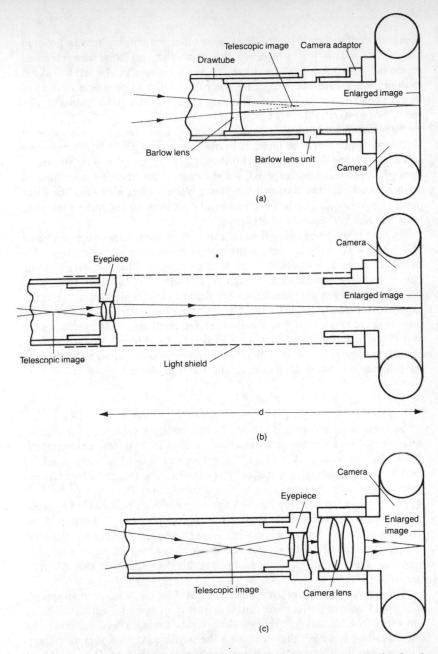

Figure 34. *Photography using a lens to magnify the primary image.* (a) *A Barlow lens is normally supplied in a unit that fits into the drawtube, since it is placed inside the telescopic focus.* (b) *If an ordinary eyepiece is used to enlarge the image, a considerable extension must be allowed for.* (c) *If the camera lens remains in position, it must be placed as close to the eyepiece as possible, both telescope and camera being focused on infinity.*

occupied by the film. With the Barlow or other magnifying lens in position (figure 34) mount the camera on the drawtube and point the telescope (complete with solar filter) so that the sun's limb is in the field. Take a medium-power eyepiece and bring the edge of the razor blade to a crisp focus. Then adjust the telescope focusing until the limb is simultaneously in focus. The system is then ready for use.

Exposure must be determined by practical tests. Using a slow color or black-and-white film (the latter is preferable on the grounds of cheapness and ease of home development), run through the range of shutter speeds. If the negatives are all overexposed, a light neutral-density filter may have to be interposed near the Barlow lens. Never place a filter very near the focal point of a telescope, since dust specks and defects on its surface may be projected sharply onto the photograph.

To take photographs of individual sunspots, a whole-disk image diameter of 4–6 inches is suitable. There is little point in having a higher enlargement than this, since a slow film will be able to show all the fine detail in the image; very big enlargements make it difficult both to focus the image sharply and to keep it accurately in the field of view, besides prolonging the exposure or necessitating a lighter filter. Since a Barlow lens cannot magnify more than about three times, a well-corrected medium-power eyepiece such as a Plossl or Orthoscopic should be used. The focal length of the eyepiece being f, and the required amplification of the enlarging system being E, then the distance d between the eyepiece and the film plane is given by

$$d = f(E + 1).$$

Suppose that a 4-inch diameter image, using a 6-inch, f/8 reflector, is required. Since the prime focal image is 0.44 inch across, an amplification of × 9 is needed. If a 15mm focus eyepiece is used, d corresponds to $15(9 + 1)$mm, 150mm, or 6 inches. The camera must therefore be mounted on a bracket with the film plane at this distance from the eyepiece, and all light other than that coming through the eyepiece must be excluded by some means. The use of an air-bulb shutter release minimizes any vibration of the telescope when the exposure is made. Since exposures of the order of 1/500 of a second are used, vibration is much less serious a problem than with other branches of astronomical photography, but this does not mean that it can be ignored.

If the atmosphere were always perfectly steady, the solar photographer's task would be straightforward: There is plenty of light by which to focus, and exposures are so short that guiding errors cannot affect the result. In compensation, however, the air during the daytime is often very turbulent. This can result either from the intermingling of warm and cool air masses high in the atmosphere (characteristic of the passage of fronts) or from warm air rising from the heated ground in the immediate vicinity of the telescope. Additionally, a hot telescope tube will create its own turbulence.

The solar photographer should, therefore, bear in mind the following points:

(a) Days of settled weather offer better high seeing than do periods of changeable, showery conditions. An anticyclone, even though the sky may be slightly hazy, is often excellent.

(b) It is useless to expect steady images when the surrounding ground is hot. At the very least, situate the telescope on a grassy lawn rather than on concrete or stones. Raising the telescope even a few feet higher above the ground can improve the seeing. Ground seeing is at a minimum in the early morning and late evening, or when the sky clears after a cloudy period.

(c) The telescope must be kept cool. A cardboard shade fixed across the upper end of the tube will keep the rest in shadow. A white-painted instrument will absorb less heat than will one with dark paint or a tarnished brass tube.

Some of the best seeing, in the British Isles at least, occurs during a winter anticyclone, when the sunshine is not strong enough to warm the ground. A snow covering helps to blanket in the heat. At such times, the visual observer will see the spots and limb looking as hard and sharp as an engraving: a rare and beautiful sight.

Solar eclipses

A total eclipse of the sun, produced when the moon passes directly across the solar disk, blotting out the brilliant photosphere, is the most impressive of all natural phenomena. At such a time, for a few seconds or minutes, the chromosphere and *corona* (the sun's outer atmosphere) glow in the darkened heavens like some cosmic eye, and it requires no special astronomical knowledge to thrill to a spectacle that awed and terrified our ancestors. It is worth journeying many thousands of miles to see such an event, and a journey will almost certainly be necessary, for total eclipses are visible over a very restricted region of the earth's surface.

The reason for this is not hard to understand, for the moon, at its average distance from the earth, is only just large enough to cover the sun's disk. If an eclipse should occur when the moon is at its greatest distance, or *apogee*, it actually appears too small and produces an *annular* eclipse, in which a thin ring of sunlight remains visible at the central phase. Conversely, if the moon is near *perigee*, when it appears largest, the eclipse will last for a longer period. The maximum possible duration, however, is only 7½ minutes, and most eclipses are far briefer.

When the moon passes in front of the sun it casts a small black shadow

Solar camera. *An ordinary plate camera attached to H. N. D. Wright's 3-inch refractor is used to take his solar photographs. The white disk attached to the camera receives the projected solar image from the finder and indicates when the telescope is pointing to the desired region of the sun.*

on the earth's surface. At a favorable eclipse (i.e., one occurring with the moon near perigee), this shadow can be almost 170 miles wide, and would be seen by an interplanetary observer as a black spot. The moon's orbital velocity carries the shadow over the surface at a velocity of more than 1,000 miles per hour, so that when the eclipse is plotted in advance we find out the path of totality. Observers on either side of this path will see only a partial eclipse; care must therefore be taken to establish the equipment on the central line. Also, it is worth remembering that most eclipse paths are considerably less than one hundred miles across.

The standard work on eclipse data is Oppolzer's *Canon der Finsternisse*, giving relevant details of 8,000 solar and 5,000 lunar eclipses from 1205 B.C. to A.D. 2152, and including maps showing the path of each total or annular eclipse. This was first published in 1887, but was reprinted by Dover Books in 1962 under the title *Canon of Eclipses*, translated by O. Gingerich, and so is easily available. Another work, *Canon of Solar Eclipses* by Meens, Grosjean, and Vanderleen, covers eclipses of the sun from 1898 to 2510 with even greater precision than Oppolzer. However, more accurate predictions for the year in question are published in the *Astronomical Ephemeris*, as well as in the B.A.A. *Handbook*, *Sky and Telescope*, and similar publications. A list of forthcoming solar eclipses is given in Appendix II.

A total eclipse of the sun is a superb sight. It begins innocuously enough, an hour before totality is due, when the moon makes a dark nick in the west (right-hand) limb of the sun. It grows rapidly to begin with, then spreads more slowly; but before long the sun is reduced to a thick crescent. The professional astronomers, who have spent months preparing for the vital moments of totality and the last week in anxious anticipation of the weather, make a final examination of their apparatus. The crescent is narrowing, and events happen quickly; there must be no mistakes, for another total eclipse may not come their way for years. Now the tiny fragments of sunlight cast under trees are also crescent-shaped, the gaps in the leaves acting as tiny pinholes. At the same time, some clouds low in the south begin to glow with sunset hues. In the west, from whence the shadow will come, there is the darkness of a thunderstorm.

Only a hairline of sunlight remains. Suddenly, the landscape is covered with trembling, rippling lines, the *shadow bands* that are caused by atmospheric refraction of the vanishing crescent. Now the crescent breaks up into irradiating points of light as the sunlight shines through irregular gaps in the moon's limb. A great wave of darkness sweeps across the ground; the last pinpoint dies; and now, in absolute silence, we see the black lunar disk set in a circle of rosy light—the chromosphere—while, beyond it and far more extensive, the pearly *corona*, or outer atmosphere, glows against the dark sky. The bright planet Venus shines brilliantly; perhaps we also glimpse Mercury or Jupiter or a few bright stars. At three or four points around the moon's limb we see bright spots of red light; these are the *prominences*, masses of glowing gas that surge up from the photosphere for tens of thousands of miles, and that may last for hours before collapsing back again.

But now the western part of the chromosphere is brightening. The moon has almost completed its transit, and totality will soon be over. Suddenly, a blinding spark of light appears, as the first fragment of sunlight shines through a lunar valley; a few seconds more, and the edge of the photosphere flashes into view. Gone are the corona and chromosphere; gone are the planets; the sky rapidly lightens, and the eerie silence is replaced by excited chatter and the return of birdsong. The eclipse has ended, and everyone who watched it carries away his own memories of those unforgettable seconds.

Observing a total eclipse

There is so much to observe during a total eclipse of the sun that no single observer can cover everything. Some of the features are terrestrial as much as astronomical—for instance, the temperature falls through several degrees during the ten minutes around totality—but the more easily observable phenomena occur in the following order:

FLORA AND FAUNA.   Some flowers close during totality, and birdsong is diminished.

THE SHADOW BANDS.   These may be seen just before totality, being more or less distinct depending on the meteorological conditions. They seem to be caused by erratic refraction of the thin crescent in the atmosphere, and the rippling effect makes them extremely difficult to photograph. The distance between the bands—usually a foot or two—varies at different eclipses.

APPEARANCE OF STARS AND PLANETS.   Bright planets, such as Venus and Jupiter, may appear before the onset of totality. It is best, before the eclipse occurs, to work out just where in the sky they will be seen. During totality, if the sky is transparent, fainter objects, such as Mercury and the brighter stars, may be detected if their positions are known beforehand.

THE MOON'S SHADOW.   According to where on the earth's surface the eclipse takes place, the moon's shadow sweeps across the ground at a velocity of from 1,000 to 5,000 miles per hour. When the moon is low in the sky (as during an eclipse occurring near sunrise or sunset), the effective velocity of the shadow will be much higher than if it is overhead. Just before totality occurs, the lunar shadow may be seen in the west as a black veil blotting out the landscape.

BAILY'S BEADS.   Named after the English astronomer who first observed them in 1836, these are the bright fragments of the photosphere left shining through irregularities of the moon's edge after the moon has passed fully onto the sun. These cannot be predicted with much accuracy, since the lunar silhouette is different at every eclipse; their effect is to make true totality occur slightly later than the theoretical time. For the same reason, Baily's Beads may hasten the end of the eclipse. Sometimes one particularly bright bead is left shining after the others have vanished, combining with the already revealed inner chromosphere to give the so-called diamond-ring effect. Although the fragments of light are really very small, they appear large because of irradiation.

CHROMOSPHERE AND PROMINENCES.   The chromosphere, when seen shining in its own right, has a characteristic rosy light. Most prominences are relatively small, but just occasionally very large ones are seen, extending for several hundred thousand miles away from the sun. They are really in rapid or even explosive motion, although the brief minutes of totality are insufficient to reveal any movement.

THE CORONA.   This is undoubtedly the most beautiful sight of all—the sun's outer atmosphere, extending for millions of miles and glowing with a distinctive pearly hue. Its observed size depends on the clarity of the atmosphere; under exceptionally favorable circumstances, it has been seen to a distance of five or six solar diameters. The form also changes according to the state of the solar cycle. Near maximum, it is more or less evenly distributed all round the limb, while near minimum the polar extensions are short and the equatorial streamers somewhat winged.

Solar corona. *This photograph was taken at the total eclipse of February 15, 1961, by H. C. Hunt, who carried his 3-inch refractor to Pisa, Italy. An H.P.3 plate was used at the direct focus of the objective; the exposure time was 3 seconds, and the telescope was not driven in any way.*

When planning to observe an eclipse, it is best to decide on a definite work program. Photography may well prove rewarding; the solar corona shown here is a fine example of amateur work. This was taken by a British observer, H. C. Hunt, who took his 3-inch refractor to Pisa, Italy, to observe the eclipse of February 15, 1961. The photograph was taken at the direct focus of the object glass, using an H.P.3 plate, with an exposure of three seconds; the telescope was not guided in any way. Circumstances were favorable, for the eclipse occurred not long after sunrise, and the air was very transparent. The structure of the inner corona can be clearly seen.

If the eclipse lasts long enough, it should be possible to experiment with different exposures. A fraction of a second is enough to record the chromosphere and prominences, but the inner corona requires at least a second, an exposure that will lose the other features in the glare. In recording the tenuous outer corona, a plate-camera lens of about f/6 gives much greater concentration of light and may produce good results. The fields for experiment are many; the time, pitifully short.

On the other hand, as often happens, too many potential eclipse observers accord photography priority in their plans. Since no photograph can show as much detail as the eye can detect, because of the great disparity of brightness between the inner and outer corona, it seems manifestly absurd to restrict ocular observation for the sake of the camera! In a sense it is understandable to want to return home with a tangible souvenir, but a photograph of one eclipse is very much like that of another, and they all, whether color or black and white, fail to do justice to the phenomenon. This is not to say that photography should be set aside altogether, but it should be restricted to a short, pre-planned section of the totality, so that the rest of the time can be spent in visual observation, the best plan being to photograph the first part and observe the last part, when the eyes have become somewhat dark-adapted. Probably the ideal instrument would be a pair of large binoculars, such as 15×80, with which the fine streamers of the corona and the details of the prominences can be made out; never forgetting, however, that no unprotected visual instrument of any sort, however small, should be pointed at the sun until it is *completely eclipsed*. In this way the observer will appreciate the extraordinary range of colors in the eclipse, particularly the plum-colored prominences standing out against an almost turquoise inner corona. He will also escape

Figure 35. *The principal lines of the solar spectrum. The* h, f, F *and* C *lines are due to hydrogen, the* C *(H$_X$) line being the brightest visually.*

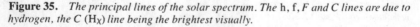

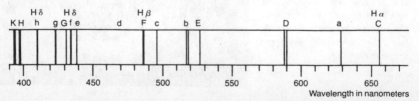

joining the ranks of so many camera-laden observers who, when asked what the eclipse was like, can only reply: "Oh, I didn't have time to *watch* it!"

Monochromatic observation

The solar photosphere emits what is known as continuous radiation. In other words, it is not releasing energy at discrete wavelengths or colors. Discrete emission can happen only when a gas is so thin that its atoms are relatively free from mutual interference. If a thin gas is heated until it shines, the light it emits is confined to a few narrow wavebands. A common example is the low-pressure sodium street lamp, which emits an intense yellow light. It is loosely termed *monochromatic* or single-colored light, although in fact the sodium atom emits several separate wavelengths. Similarly, thin hot hydrogen shines by the light of several narrow lines of color spread along the visible spectrum. The brightest of these, as far as the eye is concerned, is the red C line or Hα line, with a wavelength of 656.3 nanometers (one nanometer is equal to one millionth of a millimeter). The denser hydrogen of the photosphere, however, emits a continuous spread of colors, which together combine to form slightly yellowish sunlight.

Above the photosphere, where the chromosphere or inner atmosphere and the prominences are to be found, hydrogen is sufficiently rarefied to emit its characteristic Hα line. This explains why both the prominences and the chromosphere, when seen during a total eclipse, have a red color. The prominences that project from the sun's limb against the sky are much fainter than the photospheric light scattered by the atmosphere, and therefore can normally only be seen when the sunlight is blocked out by the moon. But if a filter that passes only Hα light is used, then the bright atmospheric halo is practically cut out, and the prominence can be seen shining against a dark background.

Monochromatic observation does not only reveal prominences; it can show the swirling clouds of hydrogen that float above the photosphere, as well as flares. It is the method *par excellence* for observing the dynamic features of the sun. Until quite recently, filters capable of passing so narrow a band of light did not exist, and the job could be done only by using a bulky prismatic arrangement known as a spectrohelioscope. But now it is possible to buy glass filters that transmit a waveband less than one nanometer wide, and using one of these, built into a suitable small refracting telescope of about 3 inches aperture (anything larger is unnecessary), it is possible to observe prominences around the sun's limb on any fairly haze-free day. A prominence telescope of this type was described by H. E. Dall in the *Journal* of the British Astronomical Association (Vol. 77, No. 2, 1967).

8

The Planets and
Their Movements

Of the nine major planets that form our solar system, only four, Venus, Mars, Jupiter, and Saturn, are really suitable for amateur observation. Venus and Mars, although relatively small—of the same order of size as the earth—are near enough to show disks in a small telescope. Though Jupiter and Saturn are much farther away, they are very large. Of the others, Mercury is extremely difficult to observe because it is always close to the sun in the sky, while the outer planets (Uranus, Neptune, and Pluto) are very remote. Uranus can just barely be seen with the naked eye, and shows only a tiny telescopic disk; Neptune is dimmer still, and Pluto is so faint that it takes a moderate telescope to make it out at all.

There is no room here to discuss any but the most important facts we know about the planets. The writer's *Stars and Planets* covers the field in greater detail. However, before examining each planet from the observational angle, we should look at the solar system as a whole. The major planets (called "major" to distinguish them from the thousands of tiny "minor" planets that lie between the orbits of Mars and Jupiter) are listed in Table II, in order of increasing distance from the sun.

The planets fall neatly into two families. The four inner bodies are relatively small and close to the sun; these are known as the *terrestrial* planets. The next four, the *giant* planets, are all much larger than the earth and so far away from the sun that they are intensely cold. At the frontier of the solar system we find Pluto, a tiny, isolated world that is of no interest to the amateur.

The two planetary groups also differ markedly in their physical makeup. The terrestrial planets are solid and rocky, with a hard crust; we can see the true surfaces of Mercury and Mars, while Venus is enveloped in thick cloud.

TABLE II. The Major Planets

PLANET	DIAMETER (MILES)	MEAN DISTANCE FROM SUN (MILLIONS OF MILES)	LENGTH OF DAY	LENGTH OF YEAR
Mercury	3,030	36	59 days	88 days
Venus	7,520	67	243 days	224 days
Earth	7,920	93	24 hours	365¼ days
Mars	4,220	141½	$24^h 37\frac{1}{2}^m$	687 days
Jupiter	88,700	483	$9^h 50^m$	12 years
Saturn	74,500	886	$10^h 14^m$	29½ years
Uranus	32,300	1,783	16^h-28^h	84 years
Neptune	30,100	2,793	18^h-20^h	165 years
Pluto	1,900	3,666	$6^d 9^h$	248 years

The giant planets, on the other hand, consist principally of frozen gases, and seem to possess no truly solid surface at all, so that observation is confined to the upper layers of their dense and turbulent atmospheres. We can draw a chart of Mars, but the features visible on Venus and the giant planets are constantly changing.

The planets all revolve around the sun in the same direction (counterclockwise, if the solar system is viewed from the north), and in more or less the same plane, so that it is possible to draw a plan of their orbits on a sheet of paper. Most of the orbits, too, are almost circular. Mercury and Pluto are the chief exceptions in this respect, and to a lesser extent Mars; but the observer is more concerned with their distance from the earth, since the closer they come, the better the view. This information is embodied in Table III.

TABLE III. The Major Planets

PLANET	DISTANCE FROM THE EARTH (MILLIONS OF MILES) Max.	Min.	APPARENT DIAMETER Max.	Min.	MAGNIFICATION TO APPEAR SIZE OF MOON (MEAN DIST.)
Mercury	136	50	13″	4½″	× 200
Venus	160	26	66	9½	× 70
Mars	248	35	25½	3½	× 120
Jupiter	413	366	50	30½	× 45
Saturn	1,032	741	21	15	× 100
Uranus	1,961	1,605	3½	3	× 600
Neptune	2,911	2,675	2¼	2	× 900
Pluto	4,660	2,672	0.1?		—

The earth, lying third in the sequence from the sun, automatically divides the planets into the *inferior* and *superior* classes. The inferior planets,

Mercury and Venus, always remain near the sun in the sky, whereas the others circle the zodiac quite independently. Figures 28 and 29 explain why this is so.

Figure 36 shows the orbit of an inferior planet. In position A, when it lies more or less between the earth and the sun, like the new moon, its night hemisphere is turned toward the earth. This position is known as *inferior conjunction*. Its orbital motion then carries it westward, toward position B, so that a narrow crescent becomes visible, swelling until it has become a perfect half. This position is known as *elongation*, because the planet then appears at its maximum angular distance from the sun. In the case of Venus, it can appear up to 48° away from the sun, whereas the greatest possible elongation of Mercury is only 28°.

After elongation (in this case, western elongation), the planet seems to move toward the sun again. The phase becomes gibbous, and at position C, when it is lost in the solar rays, the sunlit hemisphere is turned fully toward the earth. At this *superior conjunction*, the planet is shrunken and totally unobservable until it reappears on the eastern side of the sun, passes through eastern elongation and position D, and finally returns to inferior conjunction. The inferior planets are therefore alternately visible on either side of the sun, either rising before dawn when west of it, or visible in the evening after sunset when to the east. In the *Iliad*, Homer mentions Hesperus and Phosphorus, the morning and evening stars, which, 2,500 years later, are still visible. They are, in fact, the brilliant planet Venus, seen near one of its elongations, when it appears brighter than any other object in the sky aside from the sun and the moon. At these times, many people notice Venus without realizing what it is.

The two inferior planets, then, present their own observational problems. They can be seen with the naked eye only near sunrise or sunset; and when closest to the earth and therefore largest, they appear as a thin crescent. At this time, Venus appears so large that some people claim to have seen the phase with the naked eye, and it is certainly perceptible with

Figure 36.   *Movements of an inferior planet. (Not to scale.)*

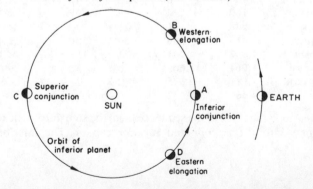

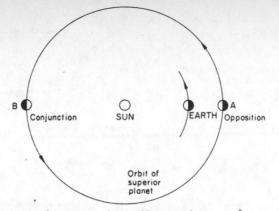

Figure 37. *Movements of a superior planet.* (*Not to scale.*)

binoculars. But the tiny, full disk it displays near superior conjunction gives little delight to observers with small telescopes.

The superior planets are, in many ways, more convenient for viewing. Since the earth lies between the planet and the sun (figure 37), the planet when closest (at position *A*) appears opposite the sun in the sky, a position known as *opposition*. What is more, the disk is fully illuminated, for the case is similar to that of the full moon. The planet then circles in the usual counterclockwise direction until it has found refuge on the opposite side of the sun, at *B*, in the position known as *conjunction*. The disk still appears full, but at this position in its orbit it is clearly more distant from the earth, and its proximity to the sun makes observation impossible for a time. Nevertheless, a superior planet is decidedly more cooperative than the two inferior ones, and presents fewer problems to the observer.

Sidereal and synodic periods

The interval between successive inferior conjunctions or successive oppositions is called the *synodic period*, an important quantity when we wish to find out how regularly a planet will be well placed for observation. To find the synodic period, account must also be taken of the earth's own orbital movement. For example, the period of revolution around the sun, or *sidereal period*, of Venus is a little more than 224 earth-days long, so that if we take inferior conjunction as the starting point (position *A* in figure 36), it will have returned to that point after 224 days. During this interval, however, the earth has completed more than half its own orbit, so that Venus has to travel on for some considerable distance before the three bodies come in line again and another inferior conjunction occurs. This interval, or synodic period, is actually about 584 days, so that Venus is not so regularly placed for observation as might at first appear. Since the earth is moving in

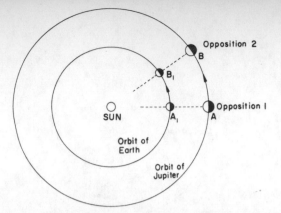

Figure 38. *Successive oppositions of Jupiter. It must be remembered that the earth completes one circuit of its orbit before passing on to B₁. (Not to scale.)*

the same direction, its effect is to make a planet's apparent movement around the sun seem slower, in the same way that a car traveling at 60 miles per hour seems to be moving much more slowly when we are following it in another car.

Since both Venus and Mars are relatively close to the earth and have sidereal periods, or years, comparable with our own, their synodic periods are very long; the average interval between oppositions of Mars is as much as 780 days. But in the case of the giant planets, which move more slowly, the earth does not have to move very far past its starting point to catch up with them again and produce another opposition (figure 38). Jupiter's synodic period is only 13 months, while Pluto moves so slowly that the earth has to travel on for only one and a half days after completing a lap before another opposition is reached. Mercury is convenient for the opposite reason: It moves around the sun so quickly—in 88 days—that it takes little extra time for it to catch up with the earth again. Its synodic period is only 116 days, shorter than that of any other planet.

Northern and southern declinations

We have said that the paths of the planets keep to the zodiac, and this is mainly true, although Mercury and Pluto, because of the exceptional tilts of their orbits, can actually stray beyond its limits. But the zodiac itself is inclined toward the equator, which means that a planet can appear in a more northerly or southerly latitude, or *declination*, depending on its position along the zodiac. The zodiac's tilt is not difficult to understand when we remember that the planets all revolve around the sun in more or less the same plane. Since the earth's axis of rotation is tilted with respect to this plane (it is 23½° off the vertical), it follows that all the planets' orbits appear as tilted, relative to the equator, by the same amount. It is precisely this tilt of the earth's axis that gives rise to our seasons. Northern midsummer occurs in June, when the sun appears farthest north; southern midsummer occurs in December, when it appears farthest south. The planets follow more or less the same path, so that they too appear to move up and down in the sky. The exact track traced by the sun during its annual revolution around the sky is called the *ecliptic*, and this is centered on the zodiac.

The movements of Mars afford a convenient illustration. Oppositions occur at intervals of somewhat more than two years, and in 1954 and 1956 the planet was well south of the equator and therefore low in the sky for northern observers. Conditions improved in 1958, and the 1961 opposition occurred with it in the most northern part of the ecliptic (in the constellation Taurus), so that to observers in Europe and the United States it appeared at a high altitude. In 1963 and 1965, it had started to sink southwards again, while the opposition of 1969 will show it in its greatest southern declination; after that, conditions will once more improve for northern observers. It is important to observe a star or planet when it is high in the sky, because seeing conditions show marked deterioration at low altitudes.

Orbital eccentricity

It is clear that the planets' movements are not quite so simple as might at first have appeared, and there is yet a third effect to be taken into account before the amateur can know just when is the best time to observe. This is the planet's orbital *eccentricity*, or divergence from a circular path, of which Mars is a striking example.

Figure 39 shows the orbits of the earth and Mars drawn to scale. The earth's orbit is only very slightly eccentric, and for most purposes can be considered as circular; but it will be seen that Mars swings quite appreciably first toward, then away from, the sun. At perihelion, or its closest point, its distance from the sun is only $128\frac{1}{2}$ million miles; at aphelion, or farthest distance from the sun, it is as much as $154\frac{1}{2}$ million. It is clear that oppositions occurring near the time of perihelion are much more favorable for viewing, the minimum distance being about 35 million miles from the earth, as against 63 million miles when they occur near aphelion.

The last perihelic opposition was in 1971, when Mars appeared as a glorious orange object in the night sky; but previously, as the diagram shows, its distance was much greater. The opposition of 1965 was aphelic, the planet glowing dimly in Leo and showing little detail in small instruments. At present, opposition distances are decreasing in advance of the next favorable oppositions in 1986 and 1988.

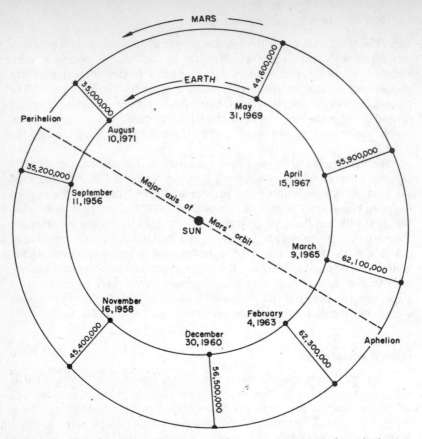

Figure 39. *The orbit of Mars. The earth completes over two circuits of its orbit between successive oppositions. (The orbits of the two planets are to scale.)*

Jupiter and Saturn are so far away and their obits are so nearly circular that their eccentricity counts for little. Venus, too, has an almost circular orbit. Mercury's is very eccentric, however, its distance from the sun varying from 29 to 43 million miles. This effect is most noticeable at the time of elongation. Mercury's elongations vary from 18° to 28°, depending on whether they occur near perihelion or aphelion, and the farther Mercury appears from the sun, the better the view.

Retrograde motion

At the beginning of an apparition, a superior planet reappears from conjunction to the west of the sun. This is because the sun, due to the earth's orbital motion, appears to be moving eastward more quickly than the

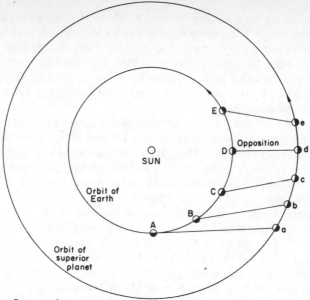

Figure 40. *Retrograde motion.*

planet, and so leaves it behind. Consequently, the planet appears in the morning sky, rising shortly before dawn. As the weeks pass, it rises earlier and earlier, until by the time opposition is reached, when it appears opposite the sun in the sky, it rises at sunset. After that, it becomes more and more of an evening object until the sun has once more caught up with it, and it disappears in the western glow shortly before conjunction.

All this time, the planet has been moving eastward in front of the stars. But some weeks before opposition is due, a curious phenomenon occurs: The eastward drift slows down, pauses, and then reverses into a westward back-tracking. Opposition passes, and some weeks later still, the backward motion comes to a halt and the planet reverts once more to its orthodox eastward progress. This reversal, which occurs with all the planets, is known as *retrograde motion*.

Figure 40 explains what happens in the case of a superior planet. At position *A*, when opposition is approaching, the earth is moving more or less directly toward the other planet. Since we see the planet's movement not in the three-dimensional aspect, but as a two-dimensional projection against the sky, it appears to be moving eastward at its own orbital velocity.

The earth, because of its smaller orbit and higher orbital velocity (18·5 miles per second), moves through a much greater arc in a given time than the other planet. At position *B*, the planet has reached only *b*, so that the earth is catching up with it quite quickly. Moreover, since the two planets are moving in the same direction, the earth's velocity seems to make the other planet move more slowly than it really does. By the time the earth reaches C, with the other planet at c, their two velocities are effectively matched and the other planet appears not to be moving at all, just as two cars racing side by side at the same speed show no relative motion. This is called the *stationary point*.

After the earth sweeps past, the other planet appears to be moving backwards, retrograding fastest at opposition (D and d), when the two planets are moving in parallel paths. Then, as the earth swings away again, its superior velocity is in effect cut down. The retrograding decelerates, another stationary point is reached at E and e; after this, we see the other planet continue its interrupted easterly progress.

Retrograde motion is most obvious in the case of Mars, which moves around the sky more quickly than the outer planets, and so covers a larger arc in a given time. At the opposition of 1965, for instance, Mars reappeared in the morning sky in the constellation Leo, and proceeded to advance into Virgo, reaching its stationary point on January 29. It then retrograded back into Leo, reaching opposition on March 9 and pausing once more on April 21. After that, it drifted back into Virgo, and continued on the zodiacal path through Libra and Scorpio.

Planetary motions are extraordinarily complicated, and it is in no small way a tribute to our mathematicians that the recent planetary probes have been so successful. Fortunately, the demands on the amateur observer are much less stringent. Mercury, Venus, Mars, Jupiter, and Saturn all shine in the sky like well-known friends, and he needs no almanac to identify them. Soon the newcomer to astronomy becomes familiar with their surfaces and their seasons, and the planets become companions of whose faces he never tires.

9

Mercury

Mercury is an intriguing planet, and the fact that we know so little about it makes it more fascinating still. There is obvious reason for its neglect, since it is very small, remains close to the sun in the sky, and is normally only visible for periods of about a fortnight on three or four occasions a year. Few useful observations can be made with a telescope of less than 12 inches aperture. Even so, it is satisfying to glimpse the innermost planet, and a dusky shading or two can be seen, on favorable occasions, with a 3-inch.

That Mercury is an elusive little world is evidenced by the fact that Herschel, a keen follower of planets as well as stars, paid it scant attention. The first systematic observations were made by Johann Schröter, his contemporary, who used one of Herschel's 6-inch reflectors as well as two considerably larger but probably inferior telescopes. He observed Mercury from about 1780 until 1801, managing to make out a few dusky streaks, and to notice that when the planet is in the crescent phase, its south horn appears somewhat blunter than the north. This feature, which has been noticed many times since Schröter's day, is probably caused by a rather dark patch on the disk. Unfortunately, Schröter had a rather wild imagination; he conjectured that it might be the shadow cast by a mountain eleven miles high! From this sensational deduction, he went on to suggest a Mercurian rotation period of about 24 hours 4 minutes. It seems that Schröter had a complex about mountain peaks—perhaps inspired by his lunar work?—since he also bestowed one on cloud-covered Venus, an observation that drew unusually acid comment from his friend Herschel.

With the tragic death of Johann Schröter, three years after the French army occupied Bremen in 1813 and destroyed his observatory, Mercurian observation languished. The next systematic attack was made by an Italian astronomer, Giovanni Schiaparelli, better known for his work on Mars. Using an 8½-inch refractor stationed in Milan, he started observing Mercury in 1882 and came to the conclusion that it keeps the same face to the sun. This announcement caused a good deal of surprise in the astronomical world, but it seemed confirmed by the work of Percival Lowell, also better known for his Martian studies, who observed the planet with the 24-inch refractor of his private observatory at Flagstaff, Arizona. Both observers decided that the dusky patches remained in the same position relative to the terminator. Had Mercury been spinning rapidly, they would be carried out of view into the dark hemisphere and reinstated some time later.

Schiaparelli's observing technique was most interesting: He observed in broad daylight, when the sun (and therefore Mercury) was high in the sky. It is not difficult to find a bright star or planet in daylight, provided the sky is really transparent and the object's position is accurately known. The advantage of observing a bright planet against the blue sky is that the glare is greatly reduced. In the case of Mercury, which is always near the sun, this is the only time when it is high in the sky and the seeing conditions are good.

Schiaparelli had no great difficulty in finding Mercury, for his telescope was mounted equatorially and could be pointed directly to its position as marked in the almanac. Positions in the sky are marked out in terms similar to latitude and longitude, in just the same way as geographical locations. We have the celestial equator and the north and south celestial poles; and the grid is laid down with lines of *declination* (latitude) and right ascension (longitude). (A fuller description of the *celestial sphere*, as it is called, is given on page 265.) It is clear that if the telescope's axis is lined up with that of the earth, it is a simple matter to point it in any given direction. With an instrument on an altazimuth stand, the approximate position must be estimated with the naked eye, and the region swept with a low power.

Schiaparelli found the markings so definite that he was able to draw up a chart. The features he found seemed to be confirmed by the Greek astronomer E. M. Antoniadi, who observed between 1920 and 1940 and issued a revised map. It therefore came as something of a shock to observers of the planet when radar observations, carried out at Cornell University Observatory in April, 1965, suggested that the planet did not keep the same face to the sun at all, but rotated with a period of about 59 days. With the arrival of the *Mariner 10* probe in 1974, which photographed much of the crater-covered surface in great detail, Mercury was virtually removed from the realm of the serious amateur.

Yet the fact that no useful observations may be possible has not lessened the attraction of this exotic planet. No world is more eagerly sought during its rapid rise and fall through the twilight sky. To see and identify this tiny planet for the first time is one of the beginner's greatest triumphs.

Telescopic appearance

Mercury looks like a tiny moon, its phases changing from almost full to a narrow crescent as it moves around the sun; at the actual times of inferior and superior conjunction, it is totally invisible. It becomes visible in the twilight after it has moved between 10° and 15° away from the sun. Since its synodic period is only 116 days, there are six elongations a year: three in the evening (eastern) and three in the morning (western). Mercury moves very quickly through the sky and its position changes perceptibly from night to night, although its path is not easy to follow because of the lack of comparison stars in the bright sky. Should Venus happen to be near, it affords a helpful guide.

This sounds as though Mercury can be seen regularly every other month, but unfortunately there are two snags. First of all, half the elongations occur with the planet south of the sun in the sky, so that north temperate observers have a very poor view. Secondly, the most favorable elongations, which occur when the planet is near aphelion and so appears about 28° away from the sun, take place when it is in southern declination. Observers in the southern hemisphere clearly have the best of it, and it is perhaps to their discredit that most of the important observations have been made by northern observers. In temperate latitudes (between about 45°N and 55°N) the planet can never be seen with the naked eye on more than fifteen or twenty occasions in the year. Should the enthusiast carry his telescope down to Florida, or the Canary Islands, he would notice a distinct improvement. There the sun rises and sets at a much steeper angle, so that Mercury is carried higher above the horizon and is a far more

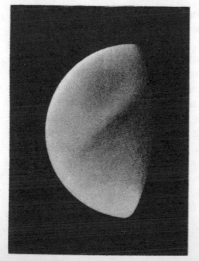

Mercury. *A drawing made by R. M. Baum in March 1952. A 6½-inch reflector was used, with a magnification of ×216. The northern cusp is clearly brighter than the southern one.*

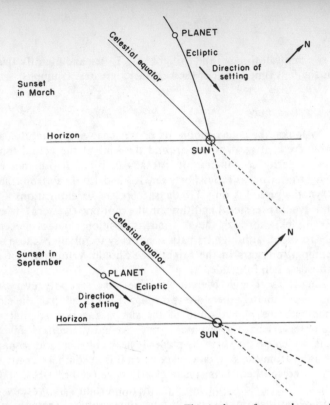

Figure 41. *Spring and autumn elongations. This is drawn for an observer in latitude 45°N. The lower the observer's latitude, the greater the angle the celestial equator makes with the horizon.*

regular visitor to the twilight sky. For all planetary observation, in fact, a low latitude is a considerable advantage, since the zodiac and ecliptic are higher in the sky.

There are definite seasons for morning and evening observation of both Mercury and Venus. Figure 41 explains why. It shows the western sky as seen at sunset in March and September, with an inferior planet at elongation. Both the sun and the planet lie on or near the ecliptic, but at the March elongation the planet is much higher above the horizon than is the case in September, because in spring the ecliptic, which is caused by the inclination of the earth's axis, makes a greater angle with the horizon. On March 21 the sun is exactly on the celestial equator, and is traveling north. Consequently, the ecliptic lying ahead of its present position is inclined to the equator, at an angle of $23\frac{1}{2}°$—the axial tilt of the earth. This means that the planet is situated well north of the celestial equator. But in September, when the sun is once more on the equator but traveling southwards, the ecliptic is passing down toward its midwinter position, so that the planet now appears very low in the sky. An evening elongation is therefore best observed when occurring in the spring, while a morning elongation is most favorable in the autumn. Conditions are reversed, of course, for observers in the southern hemisphere.

Observing Mercury

A high magnification is essential for observing Mercury. A power of at least × 100 is required to make out the phase, while nothing less than × 250 will show a reasonable disk, and × 350 is better still. Denning, in a series of observations made eighty years ago, used powers of × 250 and × 312 on his 10-inch reflector. It is optimistic to expect to see much with smaller apertures, although the blunting of the southern cusp can be glimpsed with a 3-inch. Daylight observation requires moderate apertures, but the tube must be protected from the direct sunlight, or the air inside it will become so turbulent that nothing can be seen. A great deal, too, depends on the transparency of the sky; the slightest haze will produce so much glare around the sun that the planet is lost from view even in a large telescope, and the success of Schiaparelli in seeing markings with an 8½-inch refractor is testimony not only to his own eyesight, but also to the clarity of the Italian sky.

Many textbooks say that the disk appears pinkish. Certainly, the planet appears distinctly warm-colored when viewed with the naked eye close to the horizon (though on these occasions the seeing will be far too unsteady for serious work); but this tint may be due, in part at least, to the twilight glow. Denning saw it as leaden in hue; and I once observed it with a 3½-inch refractor when it appeared so close to Venus in the sky that both planets were included in the same high-power field. On this occasion, Venus shone with a brilliant, pure white luster, and Mercury appeared to be a dull reddish-gray. Mercury's bare surface is a poor reflector of sunlight and the cloud-covered Venus dazzles, so it is not surprising that the one overpowers the other. Mercury appears brightest as an evening star between 10 and 14 days before elongation, and during the same period after elongation, when it is a morning star. At such a time, its phase is about 80 per cent.

The elusiveness of Mercury's markings, combined with its rare favorable appearances, have helped to explain the earlier belief that one face is turned toward the sun. The solar day on Mercury lasts for exactly two of its years, or 176 days, so that one hemisphere always points toward the sun at perihelion, and another faces the sun at aphelion. It so happens that the most favorable elongations for northern observers occur with the planet near peri-

helion (elongation about 18°), while observers in the southern hemisphere
see the aphelic elongations best (distance from the sun about 28°). A single
observer will, therefore, tend always to see the same illuminated hemisphere
at the times of best presentation.

Transits

Just occasionally, when the line-up is perfect, the planet at inferior
conjunction passes across the solar disk and can be seen as a tiny black
spot. This is known as a *transit*; the last one occurred on November 9,
1973, and only two more will be seen in this century; on November 12,
1986, and November 14, 1999. The path of the 1960 transit is shown in
figure 42.

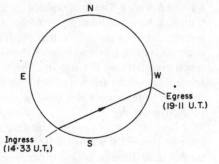

Figure 42. *Transit of Mercury, 1960.*
This shows the path of Mercury across
the solar disk on November 6, 1960.

Transits of Mercury are interesting phenomena, but it must be emphasized that unless a proper solar filter (not a sun cap) is used, the only safe
way of observing one is by projection. The planet appears as a tiny black
spot, like a sunspot without a penumbra, and takes about five hours to cross
the disk. Some observers, using direct vision, have noticed curious effects of
light and color around the disk as it moves on to the sun, but these are
certainly caused by contrast.

10

Venus

Venus is a most suitable object for study with a small telescope. Like Mercury, it is an inferior planet and shows phases, but its disk is usually much larger, and it is so bright that it cannot possibly be missed. Not only has it been seen in daylight with the naked eye, but there are records of its casting a shadow on those occasions when it can be seen against a dark sky.

Yet Venus, despite its shining visibility, is far from easy to observe; in fact, it is precisely its brilliance that makes it so easy to be deceived by false effects. The only answer is to observe it either during broad daylight or around the time of sunrise or sunset. It is easy to find even with an altazimuth mounting, and, apart from the time spent near superior conjunction, when its disk is not much larger than Mercury's, it can be followed almost continuously. Systematic work is essential, for the features are so vague and changing that they must be observed as consistently as possible.

The true surface of Venus can never be seen. It is the earth's twin in size, with a diameter of 7,550 miles, but it seems unlikely that there are any other great resemblances. The surface is hotter than that of Mercury, with a temperature of about 500°C, and the carbon dioxide atmosphere has a surface pressure about a hundred times that of the earth: equivalent to the pressure 1,000 meters below the surface of the sea.

Of course, this is of academic interest to the amateur observer, who can see only the surface of the cloudy atmosphere. The curious thing is that, although the solid globe has a rotation period of 243 days (longer than the planet's 224-day year), the cloud markings suggest a much faster rotation period of about 4 days. It is these markings that have occasionally been recorded by amateur observers, though so fleetingly that doubts about the rotation of the atmosphere persisted until the planet was photographed by space probes some ten years ago.

The markings of Venus

The first reliable observations of markings on Venus were made just three centuries ago by the Italian observer J. D. Cassini, who also discovered the main division in Saturn's rings and made other observations of note. Later, other observers, not all of them as reputable, reported seeing dusky features that seemed to be moving across the disk, and the popular view was that Venus' day is of about the same length as the earth's. However, the fact that Giovanni Schiaparelli's work in 1877–78 suggested to him that Venus keeps the same face toward the sun shows just how unreliable such estimates are. The occasional shadings are hopelessly faint and ill-defined, and it is worth noting that Johann Schröter, a model of honesty and patience if not of observational skill, watched the planet for *nine years* before he felt convinced that he had observed a definite marking! This should be remembered when considering sketches made by observers using small telescopes who record features each time they examine the planet.

However, the situation offers some hope. Although the disk often appears blank, apart from a perceptible falling-off of light toward the terminator, most elongations produce one or two shadings that have an element of reality. Of these, the most frequently seen are the *cusp caps,* which can appear during either the gibbous or crescent stage. These are bright areas near the apparent poles of the planet. Sometimes they even appear to have a dark border, or collar, although this is probably the result of a contrast effect. There are two cusp caps that have been widely recorded in recent years. Some observers put them down to optical deception, but the fact that they are not always visible seems to disprove this, and most regular students of the planet consider them to be real phenomena. The photographs taken by *Mariner 10* and other probes have shown beyond doubt that the clouds are swept or brushed in longitude, giving the impression of a polar cap and collar. However, it is easy to fall into the trap of justifying an observation simply because it happens to agree with what we think *ought* to be seen!

Many years ago, it was discovered that the cloud markings of Venus show up more clearly in ultraviolet light than at the wavelengths to which the eye is most sensitive. Since most eyes have little or no response to ultraviolet radiation, these markings were investigated photographically, first by Ross in 1927 and later by Boyer, at Paris, who in 1953 first proposed the 4-day rotation period. Space photographs of the planet taken through a yellow filter (the color to which the eye is most sensitive) showed a featureless disk, although many amateurs have recorded markings using a yellow filter to improve the planet's contrast with the twilight sky! Must we assume, then, that observers of Venus are recording what they think they ought to see?

Bias and preconcept.ons must influence *any* observation. Two observers, viewing the same object under identical conditions, will "see" it in a different way. The suspicion has been raised more than once that drawings of Venus began to look slightly different after *Mariner* took its close look! But this cannot be the whole story. To infer that a yellow-light photograph must automatically duplicate the visual appearance through a yellow filter implies that the filters must be identical and that the retina and the film must respond to color in the same way. Probably neither is true. Furthermore, although the average eye has very poor sensitivity at the extreme blue end of the spectrum, there are exceptional people who can detect radiation into the ultraviolet. It is obvious that the slightest difference in visual sensitivity may have far-reaching consequences when features as elusive as the Venusian shadings are being observed. It would be interesting to measure the ultraviolet response of different observers of the planet, and to see if there is any correlation between this and the drawings they produce.

The "wishful thinking" theory does not explain the occasional independent observation of an unusually definite marking, such as the dark feature seen independently near the center of the gibbous disk by two British observers on May 30 and June 2, 1978.

Apart from the cusp caps and the occasional dark shadings, Venus sometimes shows slight irregularities in the smooth curve of the terminator, particularly when it is in the crescent phase. These, like the shadings, have been recorded for three centuries, but it would be rash to call all the observations reliable. On many occasions of poor seeing, the limb of a planet appears irregular or serrated; under such conditions an unreliable observer might well record nonexistent deformities. A bad observation of any planet will, to a very large extent, prejudice the value of the good ones that have preceded it, and it is far better to leave the telescope covered and to wait for another more favorable occasion. Bad seeing is particularly treacherous in the case of Venus. With such planets as Mars and Jupiter, its effect is to erase detail; with Venus, it seems to inspire markings. "The better, the blanker" might well be the slogan of regular observers of the planet.

Terminator deformities, which certainly seem to occur from time to time, are probably contrast effects. A dark shading near the terminator can give the impression of an indentation at that point, while a bright cusp cap can seem to project slightly due to irradiation. If such an effect is noticed, it should be followed most carefully to see if there is any change of position from day to day, or even from hour to hour.

If any additional proof were needed that Venus has a dense atmosphere, it occurs when the planet is visible as a very thin crescent either just before or just after inferior conjunction. At this time, the horns can be seen not terminating sharply at the cusps, but extending in a vague, glimmering arc that may almost completely encircle the planet. This twilight arc can be seen with a 3-inch refractor, since at this time the planet, being closest to the earth, appears very large. But since it is also within a very few degrees of the sun, it must naturally be observed in broad daylight. If its position relative to the sun is known, it can be found easily enough if a low-power, wide-field eyepiece is used, but the greatest care must be taken not to accidentally sweep across the sun itself. Venus is furthest from the sun at inferior conjunction when it occurs in March or September.

Phase effects

It has been known for many years that the theoretical and observed instants of half-phase, or *dichotomy,* do not agree. Schröter was the first observer to point this out, and it has been referred to as *Schröter's effect.* It is caused by the considerable falling off of brightness at the terminator, with the result that, like Mercury's, the observed phase of Venus appears always less than computed.

Dichotomy offers the most convenient moment for checking the phase, since it is not too difficult to time the instant at which the terminator appears perfectly straight—although there may well be an uncertainty of two or three days. Thus, dichotomy is always early during an evening apparition, since the phase is lessening, and late during a morning one, when the planet is waxing. Strangely enough, the discrepancy seems to be greater at evening apparitions, when dichotomy may occur as much as 8 to 10 days early, while at morning elongation it is about 4 to 6 days late. Making estimates of these discrepancies is interesting, since the difference seems to vary somewhat, as is shown by the following dichotomy dates observed by members of the Mercury and Venus section of the British Astronomical Association:

PREDICTED DATE		ELONGATION	DISCREPANCY (DAYS)
1971	Jan. 25	Western	+4.3
1972	Apr. 4	Eastern	−5.75
1972	Aug. 31	Western	+4.55
1973	Nov. 7	Eastern	−5.55
1974	Apr. 6	Western	+0.9
1975	June 18	Eastern	−7.7
1975	Nov. 6	Western	+7.2
1977	Jan. 27	Eastern	−9.8
1977	June 16	Western	+4.0
1978	Aug. 27	Eastern	−7.3
1979	Jan. 18	Western	+4.9

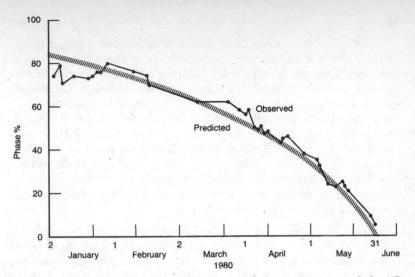

Figure 43. *Phase estimates of Venus. This series of observations was made by A.R. Hutchings, using a 4-inch refractor, during the evening elongation of 1980.*

It is, of course, possible to measure the phase effect at other times than dichotomy. Probably the most accurate method is to make a drawing, and measure that. Figure 43 shows how the phase measurements of one observer, made during the eastern elongation of 1980, compare with the theoretical curve.

The Ashen Light

A phenomenon which has given rise to much dispute, even though its occurrence seems established by the weight of observation, is the occasional very faint luminosity of the dark side, aptly termed the *Ashen Light.* Some good observers, using large telescopes, have never seen a trace of it and so remain skeptical; but many regular observers of Venus have occasionally recorded it with instruments ranging from 3 to 12 inches in aperture. This tenuous gray veil, spread between the horns, is far too elusive to photograph, so the evidence rests almost entirely on the work of amateurs who have patiently followed the planet during its crescent phases. On the whole, independent confirmation is good; for instance, many British observers recorded the Ashen Light during the first half of January, 1958, whereas sightings since that time have been sparse. The first observation was made as long ago as 1643, by Riccioli; since that time it has been reported by many observers of repute. One of them, the Reverend T. W. Webb, the famous nineteenth-century amateur observer, made his first observation on January 31, 1878:

> *Though a frequent observer of the planet Venus through a long series of*
> *years, I have never till yesterday evening seen the unilluminated side, which*
> *presented itself rather unexpectedly, as I had not been thinking particularly*
> *about it, and was not making it an object of special examination. The air was*
> *frosty and hazy, and definition tremulous, and the planet low; my*
> *beautiful 9⅓-inch With reflector brought nevertheless the horns to sharp*
> *points. I had noticed nothing remarkable with a low Kellner eyepiece, but*
> *on changing it for 212 I perceived the phenomenon pretty distinctly at*
> *intervals; it was much overpowered by the splendid light of the planet,*
> *but came out occasionally rather paler and browner than the back-*
> *ground of the twilight sky ... and was equally perceptible when the*
> *planet was hidden by a bar in the field.*

What is the Ashen Light? We do not know and can therefore only theorize, but it seems possible that it could be caused by intense auroras in Venus' atmosphere. We must remember that Venus is relatively close to the sun, and so receives much more radiation than does the earth. If Ashen Light sightings could be tied in with solar activity, the evidence would be conclusive.

It is treacherously easy to imagine seeing the dark side, for some ocular effect makes the area between the horns appear lighter in tone than the outside sky, and a careless observer might well record this as a "sighting." The only way of vindicating the observation is to get rid of the bright crescent by using an *occulting bar*. A very simple one, which has been found to give good service, can be produced in five minutes by gluing a scrap of paper across the field stop of the eyepiece. With most types (apart from the Huygenian and Ramsden, which are useless for Venusian observation because of the glare), the stop lies just in front of the field lens, and it is an easy matter to fix the strip of paper so that it appears sharply in focus. This allows the crescent to be hidden behind the bar, so that if the dark side is seen projecting, it is clearly real. An extra refinement is to cut a scoop out of the bar to match the curve of the terminator, but this is a delicate operation. Good results are also obtained by using 15-amp fuse-wire, with an induced kink to block out the crescent.

Because of its faintness, the Light is usually best seen when the sky is fairly dark. Some observers have reported seeing the entire disk by day, appearing *darker* than the sky, but this is most probably an optical deception, although some observers believe it may be due to silhouetting of the planet against the sun's outer corona.

Observing Venus

Venus rarely behaves according to form, if, indeed, it may be said to have any form at all, and its unpredictability demands an open mind. The planet's cloudy surface can efface itself in a strange way; one evening some feature is recorded with fair certainty, and the next evening the disk appears totally blank. A reliable observer has to perform a difficult task: He must eliminate from his thinking all expectation of seeing a dark marking, a cusp cap, or the Ashen Light, and come to the telescope with an open mind. If reason says that a feature *ought* to be there because it was there the previous night, it has an uncanny habit of appearing!

The planet is also demanding in an instrumental sense. Because of the great brilliancy of the disk, it shows up telescopic flaws that seem of negligible account for other work. Both object glass and eyepiece suffer a severe test of achromatism when Venus shines in the sky. The best object glass is not perfect; no matter how carefully computed and figured, there is always a very slight bluish halo remaining, known as the *secondary spectrum;* and a bright object naturally exaggerates this halo. However, since it is usually most unwise to observe Venus in a dark sky, when it produces flares and false color effects, the secondary spectrum does not present any serious problem. Generally speaking, the best time to observe the planet is either when the sun is above the horizon, or else just below it. If it is easily visible with the naked eye, then the sky is too dark.

Eyepieces must be carefully selected. The nonachromatic types are quite useless, and a Barlow amplifier is bound to be suspect because of the extra color it may introduce. Any "haunted" oculars will also prove troublesome; the legend of Venus' ghost satellite undoubtedly arose from this cause. The choice probably lies between a Monocentric and a Tolles. High magnifications are not necessary; for routine work, powers of between × 150 and × 200 are satisfactory, although a powerful eyepiece may be of use when the planet is near superior conjunction and so appears exceptionally small. This part of the orbit is grossly underobserved, an extra pity because during these months a large proportion of the illuminated hemisphere is presented.

Refractors and reflectors each offer their particular benefits. A reflector is perfectly achromatic, provided the eyepiece is good, but, as we know, its open tube is much more sensitive than the refractor's enclosed tube to changing air temperature; this happens quite quickly at sunset, when most of the work on Venus is done. For this reason, a refractor may give somewhat steadier images.

When observing Venus, it is most important to be consistent. If a 3-inch refractor is used on one occasion and a 12-inch reflector on the next, one cannot expect the observations to fit into a comparable sequence. Since the eyepieces also have an effect on the view, it is wise to use the same ocular on all occasions. Furthermore, the planet's aspect differs slightly by daylight and by twilight; so, if the time of observation can also be standardized, so much the better. Of course, this is not always practicable, but any departure from routine should be noted in the observing book so that allowance can be made. It is most instructive to compare the views obtained under various conditions of sky brightness. At night, the horns of the crescent appear in brilliant contrast against the sky, while in daylight the terminator merges almost indefinably into the blue sky, making the phase seem reduced. (One obvious by-product of this is that Schröter's effect is even more marked for daylight observations.)

Finding a planet by daylight

Venus appears brightest when about 27 per cent of the disk is illuminated. Therefore, when near elongation, and as a thick crescent, it can easily be spotted during the day, using an altazimuth stand. Once the approximate position is known—and this can be worked out from the positions of Venus and the sun as given in the almanac—the region is searched with the finder or with a low-power eyepiece. Indeed, if the sky is a deep blue, it should be visible with the naked eye. An equatorial telescope, if its axes are graduated, can be set at once on the position. However, when Venus is near the sun in the sky, around the time of inferior or superior conjunction, the glare makes it difficult for an altazimuth telescope to sweep it up at random, and so a rather more scientific approach is desirable.

Let us consider first a simpler case, when a planet is east of the sun and follows it across the sky as the day progresses. First of all, the difference between their right ascensions is found from the almanac. This is always reckoned in hours rather than degrees, since the sky appears to revolve once in twenty-four hours. One hour of right ascension (R.A.) is equal to 15° at the celestial equator, so if we point a telescope to a star that lies on or near the equator, and leave it untouched for an hour, another star will be found in the field of view that lies 15° of R.A. to the east of the first one.

Let us now suppose that an observation of Venus was being made on May 20, 1964. At that time, the planet's phase was only 22 per cent, and it was moving in toward the sun, with inferior conjunction due on June 19. The sun's R.A. on the day in question was 3 hours 49 minutes; that of Venus, 6 hours $23\frac{1}{2}$ minutes. The difference between these values, which is all that matters, is 2 hours $35\frac{1}{2}$ minutes.

Like any other celestial object, the sun is highest when on the *meridian*, an imaginary line running from the north to the south horizon and passing directly overhead. Since observing conditions improve as altitude increases, it is clearly advantageous to catch Venus at this point. Accordingly, the telescope is pointed to the sun at around the time of local noon, the exact time being noted. Just over 2½ hours later (assuming that no clouds have materialized) Venus should be seen in the vicinity.

Now, if Venus and the sun had exactly the same declination, the planet would obviously drift right through the field after the interval of 2 hours 35½ minutes. But this is rarely the case. On the day in question, the sun's declination was 20° north of the celestial equator (written + 20°, south declinations being negative), while that of Venus was + 27°. It therefore follows that the telescope must be raised 7° before Venus will appear in the field. Great care should be taken not to disturb the azimuth axis as the telescope is slowly swept up and down. The planet should be picked up without any difficulty.

In this particular case, Venus was so far from the sun that it could certainly have been found by random sweeping. A difference of R.A. of 2½ hours suggests an angular separation of about 37°, and if we remember that an outstretched hand seen at arm's length subtends an angle of roughly 20°, the approximate vicinity of the planet can easily be estimated. However, the principle is clear enough, and this drifting method can be used very successfully when Mercury or Venus is east of the sun. Unfortunately, it is quite useless when the planet is a morning star, since it crosses the meridian before the sun; under these conditions, the best way is to time the moment at which it will cross the meridian. To do this we must consider the concept of *sidereal time*.

For most civil purposes, the sun is an effective clock. In the northern hemisphere, it is due south at approximately noon, and it takes just 24 hours to make one circuit of the sky—an apparent motion provided by the earth's rotation on its axis. It is thus convenient to suppose that the earth actually rotates on its axis in 24 hours; and so it does, relative to the sun. But its actual, or *sidereal*, rotation, relative to a fixed object such as a star, is accomplished in only 23 hours 56 minutes. This is called the *sidereal day*, because it is a rotation of the earth with respect to the stars rather than the sun (*solar day*).

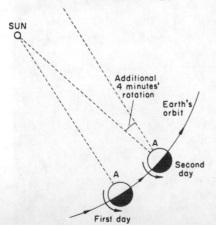

SUN

Additional 4 minutes' rotation

Earth's orbit

A

A

Second day

First day

Figure 44. *The sidereal day. The effect of the earth's orbital motion is exaggerated.*

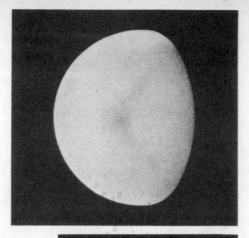

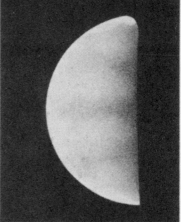

Venus. *Three drawings: Top, April 3, 1951 (76mm OG × 100); middle, June 20, 1975 (115mm OG × 186); bottom, April 1, 1977 (115mm OG × 186) (R. M. Baum, Chester, England.)*

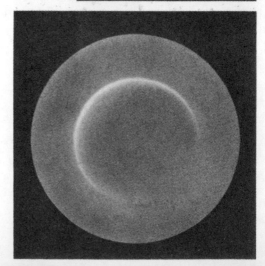

Figure 44 explains why we have the discrepancy of four minutes. On the first day, point *A* on the earth's surface is turned toward the sun at exactly 12 o'clock noon. By the time the second day is reached, the earth has traveled some way along its orbit (the effect is greatly exaggerated for clarity's sake). Consequently, after the elapse of one sidereal day, when point *A* is facing the same direction in space, the sun is no longer there! The angle has changed, and the earth has to spin on for another four minutes before noon occurs again.

The important thing to realize is that distant objects, such as planets and stars, return to the same position in the sky not after one solar day, but after one sidereal day. It is here that the concept of right ascension makes life easy for the astronomer. The longitudes on the celestial sphere are divided into 24 hours, and sidereal time is so arranged that *the R.A. of an object on the meridian is equal to the sidereal time*. The R.A. of Venus on May 20, 1964, was 6 hours $23\frac{1}{2}$ minutes. If we had a sidereal clock, we could know the moment at which Venus was due south without having to refer to the sun at all. Fortunately, there is no need for one, since Appendix VIII gives the conversion from ordinary Universal Time (U.T.) to sidereal time. On the other hand, it is not difficult to make a clock by giving an ordinary clock a gaining rate of about four minutes per day.

This method is universally used by professional astronomers for finding celestial objects. A telescope on an altazimuth mount can be used only for objects on the meridian. An equatorial telescope, however, can find objects anywhere in the sky, provided its axes are graduated accurately. For amateur purposes, however, the equatorial is something of a two-edged weapon. It is undoubtedly convenient for finding objects by daylight, when there are no reference points except the sun; but one learns nothing of the night sky by always setting the telescope by circles. Like an engine in a sailing yacht, it is useful in emergencies, but it should not develop into a habit.

With an altazimuth telescope, the only necessary adjunct is a meridian mark, to insure that the telescope is pointing due south. Some distant and easily defined object, such as a tree or a house (television aerials are particularly useful), are excellent for the purpose, provided the telescope is always used from the same position. Care must also be taken to insure that the telescope's vertical axis really is vertical; otherwise, the telescope will veer off the meridian during sweeping.

Filter observation

A number of amateurs have recently been experimenting with colored filters for planetary observation. The results, particularly in the case of Venus, have caused tremendous controversy; and three groups of observers have emerged. There are those who find that filters can bring out otherwise invisible features; the second, more extensive group has found that they improve contrast but show nothing new; those in Group III have merely found that Venus appears blue through a blue filter and red through a red filter—one of the few unanimous opinions ever voiced by observers of the planet!

Since it has now been proved that some eyes are sensitive to wave lengths of light not responded to by others, it is to be expected that results when using colored filters will vary from observer to observer, just as drawings made in white, or "integrated," light rarely agree with one another. However, some facts are generally proved: Red increases the planet's contrast against the sky, since it absorbs the blue light (for this reason, a red filter can be useful in spotting a planet during the day); whereas blue reduces the contrast, tending to blunt the horns of the crescent and to make the phase appear rather less than it really is. Not surprisingly, "red" and "blue" dichotomy dates differ by several days.

The effect of filters on planet/sky contrast is predictable enough. What is more interesting is that the Group I observers, and to a lesser extent those in Group II, have also found that the dusky shadings appear rather more definite in blue light, and that the cusp caps are also emphasized. There have even been reports that the Ashen Light is best seen with a red filter, though it is hard to say whether this is because the light really is reddish, or merely because of the improved contrast with the sky.

Color filters have not yet led to any conclusive results, and we must await further experiments before anything definite emerges. Venus is undoubtedly the most tricky planet of all to observe, and eyestraining can produce quite remarkable effects, as can the observer's own expectations of what he ought to see. Some of the success claimed for filters may be due to the reduction of glare, so a reflector with a heavily tarnished mirror may be found to give a much improved view, although its performance on dimmer objects may not be as satisfactory. At all events, any experimenting is welcome—provided it is performed with an open mind.

11

Mars

Mars is the only planet in the solar system generous enough to give us even a reasonable view of its true surface. Although, from time to time, its thin atmosphere is polluted by clouds that hide the surface features, the observer is usually treated to the sight of real and permanent markings. It was once thought that these dark regions of the planet, which often look bluish green against the ocher desert, might represent vegetation, but the Mariner probes have told us that Mars is really a volcanic wilderness, dotted with meteoritic impact craters.

In compensation, Mars puts other obstacles in our way. It is only 4,220 miles across, and since it can never approach the earth as closely as can Venus (its minimum distance is 34½ million miles, as against 26 million in the case of Venus), the disk appears relatively small. Moreover, this minimum distance is attained only once every fifteen years or so, when the planet comes into opposition at perihelion. At other times, the gap between the planets is even greater.

Mars is the first of the superior planets; so, instead of confining itself to the twilight sky, it roams right around the zodiac. It takes more than two years to pass from one opposition to the next; and since observations can be usefully made for only two or three months on either side of opposition, opportunities to study it are only fleeting. The distance increases enormously when Mars moves toward conjunction. At its farthest point from the earth, it appears even smaller than the remote planet Uranus, and observations are in any case impossible because it lies in a bright sky. Only Mercury, Venus, and Jupiter can be observed satisfactorily during the day with small or moderate instruments.

At opposition, Mars appears as a small reddish-ocher disk, slightly lighter at the perimeter than at the center due to its atmospheric haze; usually, it bears a white cap at its north or south pole, depending on which hemisphere is tilted toward Earth. On the disk itself, the eye first suspects, then sees with certainty, dark markings that appear of about the same intensity as the lunar maria when seen with the naked eye. As with all planetary work, the first view is a disappointment; the disk appears so small, and the markings so elusive, that it seems a hopeless task to identify even the main features. But perception comes with practice, and with acclimatizing the eye to the object under observation. A 3-inch refractor is quite capable of revealing the better-known dark areas, although to make useful observations even at a favorable opposition an 8-inch reflector is about the minimum aperture required, and a 12-inch is much better.

In Martian observation, perhaps more than in other departments of planetary work, atmospheric steadiness is of paramount importance, and under good conditions a small telescope is capable of performing surprisingly effectively. The explanation seems to be that the Martian markings are of relatively high contrast when its atmosphere is clear, and the disc is intrinsically bright, so that light-gathering power is less important than might be expected. It is widely believed that the superior planetary performance of large apertures is due to their greater resolving power, but, except in conditions of superlative seeing, this is not the case, and their advantage lies in the brilliance of their images, which boost fine, low-contrast features into visibility. It should be remembered, in this context, that the Dawes' Limit of the telescope applies only to the separation of discrete stellar images, and not to the resolution of thin lines, which is how a planet's markings may be considered to be composed. For example, a 3-inch telescope is easily capable of revealing Cassini's division in Saturn's ring, even though it is only about 0.5″ across (about three times narrower than the "resolving limit"), because it is a high-contrast feature. Therefore, planetary observation requires light, but it must be good light, and the small, sharp images of modest apertures can reveal much to a trained eye.

In view of the sensational results of the *Mariner* and *Viking* spacecraft, which have mapped much of the planet with a resolution of a kilometer or less, the amateur may well find himself seriously wondering whether observation of this strange world is of any use. But Mars is not in a lunar state, for it has a thin atmosphere, albeit only equivalent to 6–7 mm of mercury, and this atmosphere is capable of producing, through clouds, apparent or real changes on the surface. The need for documentation of the appearance of Mars is, therefore, as important as it ever was.

What else do we know of our neighbor planet? Its surface is a scarred, volcanic ruin, probably washed by a sea of ocher dust. Some valleys provide evidence that they were formed by water, but we see little direct evidence of it. The polar caps are now known to consist of frozen carbon dioxide, which may evaporate almost directly into the gaseous state in the spring; the thin air, too, seems to contain only carbon dioxide. Biologically, Mars may well be a dead world, but it has intrigued astronomers more than any other planet, ever since Christian Huygens produced the first drawing, in 1659. The feature he saw then, which is shown on the chart in figure 45, is still identifiable: It is a distinctive wedge-shaped marking called the Syrtis Major, following the nomenclature of Schiaparelli, who started observing Mars at the very favorable opposition of 1877 and began the "canal" furore. It is ironic that these features, which do not exist in any shape or form, should, perhaps, have done more to promote Martian observation than anything else.

The dark areas

The sketches made by Huygens and others, who depended on crude aerial telescopes, could not be expected to show much, and the first really revealing observations were made by Schröter. Unfortunately, he held the curious opinion that the Martian markings are only temporary and due to cloud formations. Clouds certainly occur on Mars from time to time, but they are light, not dark; and it is hard to understand how the German observer was misled. His contemporary, William Herschel, was under no such delusion, and by watching the regular progress of the features he was able to announce, in 1784, a rotation period of 24 hours 37 minutes 27 seconds (the modern accepted value is 24 hours 37 minutes 22·7 seconds). The markings are so far from being cloudlike that it is a relatively easy matter to time the return of any particular feature to the planet's central meridian, and if the observations are maintained over a period of weeks, the error is cut down to a few seconds.

So far as is known, Herschel never produced a formal chart of the planet; this task was left to the lunar observers Mädler and Beer, who in 1840 published a map based on observations made with their 3¾-inch refractor. Although necessarily crude by comparison with more modern charts, it is a remarkable piece of work considering the aperture used. The detail shown, which proves how much can be seen by a keen eye, was amplified by later observers, all of whom believed that the dark areas were seas and the ocher expanses dry land, although by the end of the nineteenth century it was generally admitted that there could be no large sheets of water on Mars—as we now know, the planet is virtually or quite waterless.

In the past, it was widely believed that the dark markings go through seasonal color changes—from greenish blue to brown—suggesting the development and decay of vegetation. Although this explanation cannot be correct, the Martian dark areas certainly do, on occasion, appear more colorful than usual. With small apertures they are unlikely to appear anything but brown; but telescopes of 8 inches and upwards will frequently

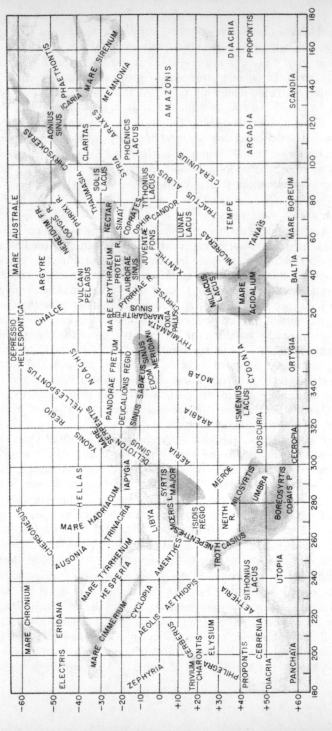

Figure 45. *Map of Mars.*

show the green-blue tint. Contrast with the ocher areas must play a part, but this does not explain any variation, if indeed it exists. The varying transparency of the earth's atmosphere must be borne in mind as well: Jupiter is often referred to as cream-colored, but under conditions of exceptional transparency the disc is almost as silvery as that of Venus; similarly, the common pinkish tint of Mercury belongs to the twilight glow, and not to the planet itself. In the same way, the Martian markings may appear a dull gray brown one night and an unmistakable green the next, of a tint sometimes approaching turquoise. Seen under such conditions, Mars is a beautiful sight.

The undisputed fact about the dark areas is that they change in form. Some transformations seem to be seasonal, while others occur from month to month. For instance, the elliptical area known as the Solis Lacus, which lies in the southern hemisphere, in the Argyre Desert, completely changed its shape between 1926 and 1930, and underwent further transformations in 1939. Some features are totally absent at one opposition and prominent the next. Another widely observed event of recent interest was the sudden darkening of the Aethiopis–Mare Cimmerium region in 1958; by 1963 it had again returned to its normal intensity. There is no doubt at all that these effects are real; the main task of the Mars observer is to keep track of such changes and note whether any pattern of behavior emerges.

The short-term changes are of equal interest, for some are fairly predictable. A good example is the distinctive Syrtis Major, which seems to spread eastward during local spring, encroaching on Libya; to the west, Pandorae Fretum is well known for its seasonal darkening, a darkening that affects many of the temperate areas. Other local darkenings are not predictable, and should be carefully monitored. Conversely, the huge region of Hellas, south of Syrtis Major, regularly undergoes a brightening during the Martian summer, and a dark spot, Zea Lacus, may appear inside it. A feature of the 1975–1976 apparition was the appearance of a distinctive new dark area in the Claritas region, while Nepenthes-Thoth was unusually faint.

The polar caps

On the earth, the seasons are caused by the tilt of the planet's axis. In March and September, the equator appears edge-on to an observer on the sun, neither hemisphere receiving preference. In June and December, however, the north and south poles, respectively, are inclined at the maximum angle of $23\frac{1}{2}°$ toward the sun, bringing extra sunlight and heat to the appropriate hemisphere.

The same state of affairs holds true for Mars. The tilt of its axis, or the angle between orbit plane and equator, is 25° 12′, hardly different from our own, so that the sun's annual rise and fall in the Martian sky is about the same. There are, however, two other differences: Its year is 687 earth-days long, and its orbit is much more eccentric than the earth's. This means that Mars is appreciably hotter when the planet is near perihelion than when at aphelion (on the earth the difference is negligible); and, since southern midsummer occurs near the perihelion point and southern midwinter occurs near aphelion, this hemisphere suffers much greater extremes of temperature. To earthbound observers, a more important consequence is that the south pole, with its glittering cap, is always presented at the favorable perihelic oppositions. As a result, we know the southern hemisphere of Mars in considerably greater detail than the northern.

At their maximum extent, the caps are visible in a very small telescope; but for critical work during the interesting stages of their shrinking, a moderate instrument is necessary. The southern cap, which can extend down to latitude 55° in midwinter and has been known to disappear completely during Martian July, has a large fragment called the Novissima Thyle, which breaks off and leads a short independent existence. The explanation seems to be that Novissima Thyle is a huge plateau elevated several thousand feet above the surrounding surface, so that the frozen material takes longer to melt in the cooler air. The northern cap has a similar offshoot called the Rima Borealis, and sometimes other deformities appear along the edge of the caps. These must be watched for carefully.

It is clear that the caps cannot be thick. They are probably no more than a few inches deep, for the weekly shrinkage during the melting season can be very obvious. Sometimes the gleaming white seems framed by a dark border, known as *Lowell's band*, after the famous American amateur observer of Mars. This is often too distinct to be dismissed as a mere optical illusion, as in the case of the Venusian "caps," and must be due to the temporary darkening of the surface by the melted ice.

The re-formation of the caps cannot, unfortunately, be seen as clearly. As winter approaches, Mars swings away from the sun; and the southern polar region becomes obscured by a hazy white cloud at the very time that its northern counterpart melts and brings spring to the other hemisphere.

Clouds

The carbon dioxide shroud of Mars has only about 1 percent of the ground pressure of our own air, but this is sufficient for it to support cloudy veils, and some form of atmospheric obscuration is observed at every apparition. There are two distinct species of cloud phenomena on Mars. The true "clouds," which are white, are probably similar to our own cirrus clouds, except in being composed of crystals of carbon dioxide rather than of water. Much more obstructive, when they occur, are the yellow clouds, which consist of windborne dust that can obliterate the surface features for weeks. No observer of the 1956 and 1971 apparitions is likely to forget these yellow clouds.

Several white clouds were seen during the 1964–1965 apparition, and in 1967 a cloud formed over Hellas that, when seen near the limb, was bright enough to be taken for a polar cap. Normally, white clouds are most easily seen when they lie near the limb, often irradiating against the sky, and they may be seen more easily on the disc by viewing through a light blue filter, such as the Kodak Wratten 47B, which darkens the surface features. White clouds usually last for a few days only, and any drift should be followed closely, for here we have a clue to the direction of the upper atmospheric winds.

The yellow clouds, the Martian dust storms, being composed of the same material as the ocher regions, are less visible in themselves than through what they obscure. It is curious that the last two perihelic oppositions of Mars, in 1956 and 1971, have been accompanied by planet-wide dust storms. Both of these great storms began in the Hellespontus-Noachis region, spreading mainly westwards and obscuring the south polar cap in a matter of a week or two. The same phenomena were observed at the apparitions of 1909, 1911, and 1924. It has been suggested that the southern hemisphere of the planet always suffers such storms around its summer solstice, although the evidence does not seem to be overwhelming. In fact, the possibility of an unexpected outbreak of storm activity should always be borne in mind, and the observer's first task is to look for any telltale spots that may indicate a new disturbance. These phenomena characteristically break out quickly, becoming visible in a medium-size telescope within hours of their commencement. In 1973 the planet's surface was hidden for three weeks or so in the autumn by another violent storm, but it had cleared by the following January. A minor dust storm was also observed in July and August 1975.

Observing Mars

All planetary observation demands an open mind; and in the case of Mars, after one has studied charts and become intimate with the major surface markings, it is extremely difficult to avoid prejudice. The tendency to sketch in features because they *should* be there has contributed largely to the "canal" controversy. It is significant that no observer ever recorded a thin straight line on Mars until Schiaparelli announced his observations in 1877!

Mars rotates slightly more slowly than the earth. If the Syrtis Major is seen on the planet's meridian on one night, it will reach the same point 37 minutes later on the following night. Conversely, since the planet seems to rotate from right to left, it will be slightly to the right of the meridian at the same hour of observation. As the days pass, the region is carried farther and farther out of view until it remains entirely on the far side during the night's work, returning to the meridian at the former time after an interval of about five weeks. This means that different areas of the planet gradually come under scrutiny, and, depending on the inclination of the weather, one can hope to observe three or four complete cycles of the planet during the time in which it presents a reasonably large disk. Since these opportunities come at intervals of more than two years, every favorable night must be exploited to the full—though this is superfluous advice to the keen observer who has awaited the return of the elusive visitor for many months.

Mars requires the use of high magnifications, since its disk can never be more than 25″ in diameter (at an aphelic opposition it is only 14″), and for guiding under these powers efficient slow-motions are necessary. The earth's rotation quickly carries it out of the field of view, and unless the tube is easily adjustable the observer will consume most of his time and temper in simply keeping the planet in view. With an altazimuth telescope, the best way is to adjust the position so that the planet drifts right across the view. By the time it has reached the central region, where definition is best, any residual tremors will have died down and the observer can concentrate on the scraps of detail that appear and fade away. For viewing the general features, a power as low as × 200 is often useful, since the tube has to be shifted less frequently, but the finer details demand × 300 or × 350 if they are to be seen well. A single-lens eyepiece, if it is well made, can give excellent results, but the field of good definition is so small that an equatorial mounting is almost a necessity.

The focusing requires great attention. This may sound self-evident, but Mars is an extremely difficult object to focus properly, because of the glare at the limb and because not all of the surface markings are well defined. The polar cap is useful, should it be visible; if not, a nearby star offers the best means of achieving a perfect focus. It pays dividends to refocus carefully every five minutes or so until the eye has been taught to remain relaxed.

Some observers adopt a definite scale for their sketches, varying the size of the drawing according to the planet's apparent diameter. There are arguments for and against this practice, but at any rate the disk should never be less than an inch across, and around the time of opposition a 2-inch disk is the most suitable size. The advantage of a standard diameter is that a local printer can run off a supply of blank disks with a blacked-in back-

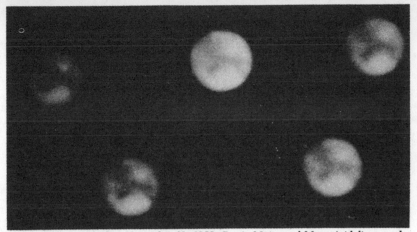

Mars. *Photographed on November 13, 1958. Syrtis Major and Mars Acidalium can be seen quite clearly. The photographer used a 12·5-inch reflector with eyepiece projection and orthoscopic oculars; focal length, 141 feet; exposure time, 1·5 seconds. (Jack Eastman, Jr., Manhattan Beach, California.)*

ground, and these can be pasted in the observing book. They are useful for Mars and Jupiter, but not so convenient for Venus because of the phase effect and the fact that the planet is usually observed against a light sky.

Mars, too, can show a slight phase. This is most noticeable when it forms nearly a right angle with the sun, a position known as *quadrature*, occurring about three months before and after opposition. At this time, the phase is about 88 per cent, and allowance must be made for this when sketching the disk. It is remarkable that Galileo was able to make out the phase with his tiny telescope, although the polar cap and dark markings eluded him.

The rotation of Mars, while considerably slower than that of the giant planets, is nevertheless sufficient to be noticeable after a period of ten minutes. Since it can take easily half an hour to make a drawing and note down the various details—such as the relative intensities of different regions, any unusual development of the dark areas, the state of the polar cap, and the possible presence of clouds—allowance must be made for the spin. The best way of doing this is to sketch only those features fairly near the central meridian, since the western border will be carried out of view, and the right-hand side will be considerably displaced by fresh detail arriving from beyond the limb. A 2B pencil, if used skillfully, will allow both dark and diffuse shadings to be rendered. Some amateurs color their drawings, but it is difficult to achieve really accurate representations, and the best

*September 18, 1971, 20.30 UT.
Note the two bright regions (possibly clouds) at the preceding limb.
Sinus Meridiani is close to the center of the disk. (Central meridian
355°.)*

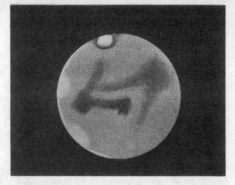

*September 21, 1971, 20.30 UT.
Syrtis Major well shown, contrasting with the much brighter Hellas.
The light diffuse region over Iapigia is the beginning of the great
dust storm of the 1971 apparition,
which was widely observed on the
following night. (Central meridian
327°.)*

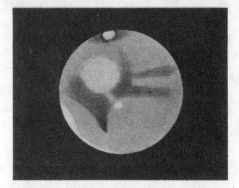

*March 9, 1980, 19.15 UT. Mare
Acidalium is shown, and Tempe is
bright. Note Sinus Meridiani, appearing very foreshortened near the
south-eastern limb. (Central meridian 28°.)*

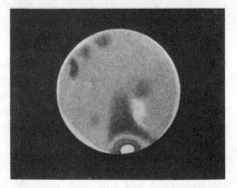

Three drawings of Mars made by Alan W. Heath of Nottingham, England, using a
30cm Newtonian, × 318.

answer is to record all the finer details in the written notes so that there can be no subsequent error in interpretation.

The satellites

Mars has two moons, but they both appear very faint. The larger one, Phobos, is probably about 12 miles across, and circles Mars in a sidereal period of only 7 hours 39 minutes, at a distance of 3,700 miles from the surface. Deimos, probably half the size of Phobos, is 12,500 miles away and has a sidereal period of 30 hours 21 minutes. Both moons are brighter than the 13th magnitude, so that technically they are visible with a 6-inch telescope; however, the glare around Mars is so great that they are hard to see with a 12-inch, even under the most favorable conditions.

The "canals"

The story of the canals, fascinating though it is, belongs more properly to a history of astronomy than a handbook of observation; yet, as with so many other telescopic myths, it is of value in proving just how easily the eye is deceived, and how effortlessly the subconscious takes over the task of recording detail. Schiaparelli, observing in Milan with an 8¼-inch refractor at the opposition of 1877, observed a number of streaky markings that had, in fact, been seen by previous observers of the caliber of Mädler, Lockyer, and Dawes. They had given no hint of artificiality and aroused no special comment. But at the return of 1879, and at succeeding oppositions, the streaks became narrower and straighter, and the mistranslation of the Italian word *canali* (channels) into "canals" simply hastened the unfolding deception. Lowell took up the work at Flagstaff, Arizona, when the Italian retired after the opposition of 1890, not only extending the network of "canals," which he really thought they were, but publishing the evidence for a highly-organized, Mars-wide civilization in his famous book *Mars and Its Canals* in 1906.

We can never know just how Schiaparelli and Lowell were deceived, although there are some suggestive facts. Schiaparelli had to give up observing through failing sight, which ended in blindness; and this handicap, coupled with the strain of observing with a relatively small telescope, might well be expected to produce false effects. Lowell, observing with his fine 24-inch refractor, often worked with it diaphragmed down to only 18 inches. Under conditions of poor seeing, it sometimes helps to reduce the aperture; but Lowell Observatory enjoys conditions as good as anywhere in the world, and in any case it is most unwise to make observations when the seeing is poor. Furthermore, he also drew sharply-defined streaks on Venus, which is notorious for its ghostly shadings!

Whatever the cause, the "canal" observations had a predictable effect: Some amateur astronomers using 6-inch telescopes started drawing linear streaks where their similarly equipped predecessors had merely seen detached markings, and only stopped doing so when the *Mariner* photographs proved beyond any dispute that such features do not exist. The sinuous channels that have been photographed by *Mariner* and *Viking* bear no resemblance, either in size or form, to the objects mapped by Lowell.

We come again to the question: What is the point of observing Mars with a 6-inch; is there anything left to be discovered? Certainly there are things to be *documented,* and much of amateur astronomy is a matter of documentation. It is in the night-to-night patrols, whether of planets, stars, or the moon, that the amateur's strength lies. The sparetime enthusiast is not dependent on the whims of committees and the funds of remote authorities. For some years, leading up to the great age of interplanetary unmanned exploration that is now coming to a pause, if not an end, an international photographic patrol was taking superb pictures of Mars, Jupiter, and Saturn, using large telescopes in favored locations. This has now ended, and as far as earth-based observation is concerned, both the moon and the planets have been handed back to the amateur's tending.

12

Jupiter

The first and largest of the giant planets, Jupiter, is perhaps the finest object on the solar system for observation with a small telescope. Like Venus, it shows a large disk; but it never shrinks inconveniently small, is always fully illuminated, and exhibits far more in the way of definite detail. Moreover, this detail is changing all the time, so that there is always something new and unexpected happening.

Jupiter, Saturn, Uranus, and Neptune are formed on an entirely different pattern from the terrestrial planets. They are nowhere near as dense, since they consist mainly of gas (principally ammonia and methane) that has been frozen into crystalline or liquid form by the intense cold, swathing their surfaces in dense clouds. It seems doubtful that any of them contains a true rocky core; at all events, our observation is confined to the upper layers of their atmospheres. Luckily, Jupiter's cloud features are so definite that there is no serious disagreement over the main details to be seen.

The belts and zones

The basic hue of the disk is silvery-cream, darkening a little at the limbs, where we see the "sunrise" and "sunset" regions. Across the disk, dividing it into various zones, are a number of dark belts lying parallel to the equator. The wider and more prominent of the belts often display a reddish-brown tint, so that Jupiter, despite its surface temperature of $-220°F$,* has a distinctly warm appearance when seen through a telescope of moderate aperture.

*The word "surface" refers to the top of the cloud layer.

The belts, which form the framework against which the finer features appear, develop and fade from year to year, and each apparition of the planet has its special characteristics. Usually the North and South Equatorial belts are very conspicuous; so is the North Temperate Belt, with the South Temperate Belt and the high-latitude features somewhat fainter. However, the appearance of the planet can change quite considerably from one year to the next, in an unpredictable way. The Equatorial Zone, which normally appears as a bright band across the center of the disk, turned yellowish for a time in 1959 and in 1960 began to darken, an appearance that was maintained for the next four years, and a similar darkening occurred in 1980. In fact, it can safely be said that Jupiter *never* looks the same from night to night.

The diagram in figure 46 gives a very schematic view of the main belts and zones, and it must be emphasized that Jupiter never looks anything like this. Quite apart from individual fluctuations, some belts often appear double (the two equatorial belts especially), while the regions near the equator are replete with complex detail. Dark wisps, bright ovals, gaps—all these features pay tribute to the immense turbulence that must be going on in Jupiter's atmosphere. When we remember that the planet's equatorial diameter is 88,700 miles, it is obvious that some of the larger "clouds" are considerably bigger than the earth!

Jupiter's belts and zones are referred to by the following standard abbreviations:

South Polar Region	S.P.R.	South Temperate Belt	S.T.B.
South South Temperate Zone	S.S.T.Z.	South Tropical Zone	S.Tr.Z.
South South Temperate Belt	S.S.T.B.	South Equatorial Belt	S.E.B.
South Temperate Zone	S.T.Z.	Equatorial Zone	E.Z.

Figure 46. *The principal features of Jupiter.*

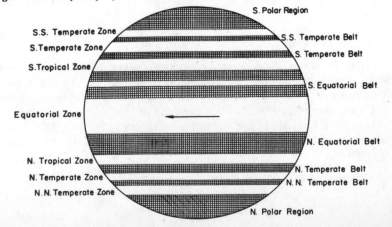

There are corresponding abbreviations for the northern hemisphere. The belts have no defined latitudes; sometimes they also appear in even higher latitudes than the normal position of the second temperate belt, in which case another directional letter would be added; an N.N.N.T.B. is fairly common. When a belt appears double, the components are given north and south subscripts (e.g., S.T.B.$_s$ and S.T.B.$_n$). Two other useful abbreviations—used when describing the longitude of one feature relative to another —are p (preceding) and f (following). Jupiter spins so quickly from right to left that an experienced eye can notice the difference in just two or three minutes, and the use of the p and f terms is obvious.

Rotation periods

The main object of Jovian observation is to find out as much as possible about the planet's atmospheric currents, which cause the various features to drift in longitude by different amounts. Most of these are impermanent, lasting for a few weeks or even less, but while they survive they give us valuable clues about the forces at work. Some regions rotate faster than others, while many individual spots have their own periods, and it is by timing these that the amateur can make his greatest contribution to the study of the planet. This is carried out by noting the instant at which they cross the planet's central meridian; and while the taking of *transits*, as they are called, may sound a dull business, the fascination increases as skill and experience are gained.

Jupiter has two main rotation periods. The main portion of the disk rotates in about 9 hours 55 minutes, but the Equatorial Zone, bounded by the north edge of the S.E.B. and the south edge of the N.E.B., takes about five minutes less. As a result, features in the Equatorial Zone seem to advance steadily relative to the rest of the globe, taking about six weeks to achieve one "lap"—if they survive long enough. Jupiter's rapid spin is demonstrated by the flattening of the disk, which is caused by the extension of the equatorial regions by centrifugal force. Jupiter's polar diameter is only 82,800 miles, and this flattening must be allowed for when making a sketch. The best procedure is to have a blank disk prepared, then have a stock run off by the local printer.

Because the planet rotates so quickly, it is not difficult to time the instant at which a feature crosses the meridian. To begin with, there may be an error of five minutes, but with experience a transit can be timed to the nearest minute, and the longitude of the feature can be worked out from tables. If it then survives for several rotations, its drift can be followed. Occasionally, spots are observed with rotation periods two or three minutes different from that of the surrounding region, so that they move steadily past adjacent features. If a spot gains in longitude, it is said to be *advancing*; if it moves backward, it *retrogrades*.

The *Astronomical Almanac* and B.A.A. *Handbook* give longitude tables for the central meridian of the two main zones: System I is for the equatorial region; System II is for the rest of the disk. These work for synodic rotation periods of 9 hours 50 minutes 30·003 seconds and 9 hours 55 minutes 40·632 seconds, respectively. These periods are quite arbitrary, but are close to the average periods of revolution of features in the two regions. However, when we examine the records more closely we find that different latitudes have their own rotation periods, differing by several seconds from those nearby. Altogether, some twenty different currents have been found to exist, carrying along at their own individual rate any features that happen to lie in them. On top of these, there are occasional short-lived spots with special drifts of their own. Few features show much drifting in latitude.

Much of our knowledge of the behavior of the features and currents of Jupiter is owed to the work of amateur observers. Systematic observation began with A. Stanley Williams, who devised the simple transit system of longitude determination and announced the discovery of nine individual currents in a paper published in 1896. This work coincided with the foundation of the British Astronomical Association, whose Jupiter Section has now amassed records extending back over eighty years; and the work of the Association of Lunar and Planetary Observers in New Mexico has reinforced Jovian observation since its foundation after World War II. Study of the planet with small and medium telescopes is as important now as it ever has been.

The Great Red Spot

Among the intricate but evanescent features that decorate the silvery disk of this great world, one object in particular catches the imagination through its great size, and, still more startling, its persistence. If we can trust a drawing made in 1664, by the British physicist Robert Hooke, it has survived throughout the era of telescopic observation of the planet. This is the Great Red Spot, which Hooke represented as a large elliptical feature in the southern hemisphere. Cassini is said to have drawn a similar feature in 1665. It also figured in a drawing made in 1713 by J. P. Maraldi, and on September 5, 1831, it was drawn by Schwabe, of sunspot fame. But the redness for which it has become so famous did not appear until 1877, and during the period 1879–82, its color was often referred to as "brick-red," a vague term, but certainly descriptive enough to suggest definite activity. Thereafter it faded, becoming gray and occasionally invisible among the other detail of the S.Tr.Z., in which it lies; but occasional revivals have occurred: in 1957, when it assumed an obvious pink hue and could be seen with a small telescope; in 1961; and prominently again in 1966. The Spot measures some 20,000 miles in longitude, by about 8,000 miles in latitude, and it sometimes overlaps the S.E.B.ₛ and produces an indentation in the belt known as the Red Spot Hollow.

The Spot may be a solid body (though a recent theory suggests that it may be the top of a "standing current" in Jupiter's atmosphere, produced by the planet's rotation); but it cannot be fixed to the core. This is proved by its erratic rotation period, which has varied from as little as 9 hours 55 minutes 32 seconds in some years to as much as 9 hours 55 minutes 44 seconds in others. A variation of 12 seconds may seem a small amount, but the mind reels at the thought of the colossal forces required to perturb this vast object.

The Spot is so large that it is not too easy to judge just when the center is on the planet's meridian. The best method is to take three transits, for the p and f ends as well as the middle.

Periodic disturbances

Fresh spots and detail may break out in any region of the disk; but the S.E.B. is the seat of two well-known disturbances. One has not appeared for some years, and may well be periodic; the other is a frequent phenomenon of the planet.

THE SOUTH TROPICAL DISTURBANCE. This was first recorded as such in 1901, although earlier records suggest that it was not a new feature. It took the form of turbulence occurring on the southern edge of the S.E.B. and so extending into the S.Tr.Z. The main characteristic was its relatively short rotation period of about 9 hours 55 minutes 20 seconds, considerably shorter than the mean System II period. The site of turbulence became extended in longitude, and because of its shorter rotation period it began to catch up with the Red Spot, which it did in 1902. When this occurred, there was apparently some mutual attraction, for the Spot was pulled forward from its resting place and suffered a distinct acceleration of several seconds before the Disturbance had passed by!

The same pattern of events was repeated whenever the Disturbance passed the Spot, as it did no fewer than nine times between 1901 and about 1935. After that it suffered occasional darkenings, as in 1940, 1966, and more strongly in 1979–1980, but by 1981 the characteristic activity had shortened and faded once more. A repeat of the 1901–1935 eruption may begin at any time.

S.E.B. ERUPTIONS.   The S.E.B. is the most active region on the planet. In some years it is almost invisible; in other years one can observe almost cataclysmic revivals. The first recorded eruption of this sort was in 1919–1920; the second was in 1928–1929. More recent disturbances, lasting for several months, occurred in 1943, 1949, 1952, and 1958, with a major outbreak as recently as 1971. Clearly, some sort of periodicity is involved, and another eruption may occur at any time. Characteristically, a number of light and dark spots develop, which organize themselves into two zones. Those on the south edge of the S.E.B. have a long rotation period that may be two or three minutes more than the average for System II. The others, on the north edge of the Belt, may have a period as short as 9 hours 52 minutes, and occasionally even less. The active regions therefore drift steadily apart in longitude, and during a severe outbreak they may completely girdle the planet. The Red Spot, which lies near the S.E.B., may disappear for a time during one of these eruptions, as happened in 1958, just after its recent revival to prominence. A further dramatic S.E.B. eruption commenced in July 1975, reaching a peak of activity in September through November. It is noticeable that the longitude of the Red Spot always increases (i.e., its rotation period lengthens) when one of these eruptions occurs.

Other disturbances are seen from time to time. Between 1930 and 1934 a "circulating current" was observed in the S.Tr.Z., where a number of dark spots were seen to retrograde along one line of latitude and then, apparently, to jump a few degrees southward and advance in the direction from which they had come. It is these apparently inexplicable currents, combined with periodic outbreaks of light and dark markings, that make Jupiter a source of such constant fascination. Some long-lived white S.Tr.Z. ovals, which have a slightly longer rotation period than the Red Spot, periodically come into conjunction with it; when this happens, its longitude is temporarily increased, as if it had been dragged back by the attraction of the oval.

Observing Jupiter

Even when Jupiter is near conjunction with the sun, its image has a diameter of about 30″. This is larger than the disk of Mars at a perihelic opposition, so the planet can be observed profitably whenever it is visible in the sky. In temperate latitudes, it is lost from view about six weeks before conjunction, reappearing in the morning sky after the same interval, so that an enthusiastic observer can follow it for about three-quarters of the apparition. Low-latitude observers are even more favored, for it sets at a greater angle to the horizon and so stays visible for even longer. Because of the difficulty of observation, any views obtained near conjunction are of special value, since they act as a link between the bulk of the work done around the time of opposition.

A photograph taken on November 1, 1964, at 01.34 UT, with a 15½-inch Cassegrain reflector at f/90, exposure 1 second on FP3. The Red Spot is on the meridian, and satellite II is casting its shadow on the disc. (H. E. Dall, Luton, England.)

Jupiter does not have as sharp a limb as the other planets. This may well be due to the falling-off of illumination at the edge of the disk, which is usually the easiest part of a planet on which to focus properly. Consequently, medium powers usually work best, which is no disadvantage since the disk appears so large. A good view is given by × 150 on a 3-inch or × 200 on a 6-inch reflector. Even with a 12-inch reflector, powers between × 200 and × 300 give excellent results. A 3-inch refractor will show sufficient detail to allow transits to be taken, but a larger aperture naturally reveals more delicate features and increases the scope of the work.

Disk drawings are of secondary importance to transit work. If some unusual features are seen, it is worth making a "finished" sketch, but normally the main importance of a drawing is as a guide to the features observed in transit. The best way to start is to spend a couple of minutes surveying the planet, looking for any unusual markings and taking note of the spots and streaks that are likely to transit during the session. It is not advisable to make a sketch until the planet has been under observation for an hour or so; in this time, the first features seen to transit will have moved on to the *p* part of the disk, and room is now left on the sketch for the fresh features that have come into view on the *f* part. Once the outlines are drawn in, the finer details can be inserted as their transits are taken, using a watch correct to the nearest minute. If the session is a long one, it will be necessary to make another sketch to cover the new area of surface that the planet's rapid spin brings into view.

The Galilean satellites

Jupiter possesses numerous small moons as well as a very faint ring, but only the four bright satellites detected by Galileo can be seen with normal amateur equipment, and they form one of the best-known sights in the sky. Indeed, they are so bright that some sharp-eyed observers claim to have seen them without optical aid, and binoculars will certainly give a good view. Because of their interest, it is worth giving their details in full.

February 23, 1980, 23.15 UT. Following activity in the S.E.B. and S.Tr.Z., the Red Spot faded during the early part of the 1979-1980 apparition. However, the Red Spot Hollow (on the meridian) was particularly prominent. Note also the disturbed S.E.B. following the Hollow, and the huge Equatorial Zone plumes. (System I: 189°; System II: 54°.)

March 11, 1980, 00.46 UT. During the past few years there has been a revival of the South Tropical Disturbance. This drawing shows the following end of the dark Disturbance on the central meridian. The dark filament marking this end is itself followed by a large white cloud. (System I: 252°; System II: 355°.)

Two drawings of Jupiter made by P. B. Doherty of Stoke-on-Trent, England, using a 419mm Newtonian, × 248.

TABLE IV. The Galilean Satellites

Name	Mean Distance from Jupiter (Miles)	Angular Distance at Elongation	Diameter (Miles)	Orbital Period			Mean Mag.
Io	262,000	2¼′	2,250	1ᵈ	18ʰ	28ᵐ	5·5
Europa	417,000	3¾	1,800	3	13	14	6·1
Ganymede	666,000	6	3,100	7	3	43	5·1
Callisto	1,170,000	10¼	2,800	16	16	32	6·2

The satellites being of the 6th magnitude or above, they would be naked-eye objects were they not masked by the planet's glare. A low-power eyepiece gives a wonderful view of the four moons, their positions changing perceptibly from hour to hour as they revolve around the planet. While observing this charming sight, the colossal size and mass of Jupiter are brought home to the observer by the realization that Ganymede is probably slightly larger than the planet Mercury!

Jupiter's axial tilt is only 3° 05′; thus, unlike the case of Mars and Saturn, we always see the equator virtually edge-on. What is more, the Galilean satellites revolve almost exactly in the plane of Jupiter's equator,

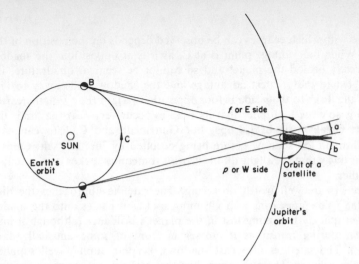

Figure 47. *Eclipses of Jupiter's satellites: Before opposition, when the earth is near* A *and Jupiter is a morning star, a satellite vanishes in its shadow before it reaches the* W *side, or limb, and reappears at the bright limb, being invisible through the range a. After opposition, when Jupiter is seen from* B *and is an evening star, a satellite vanishes at the* W *limb and passes into the shadow behind the planet, reappearing from eclipse at some distance from the* E *limb. At opposition* (C), *the shadow is directly behind the planet and both disappearance and reappearance take place at the limb. Due to the slight inclination of Jupiter's axis, it is sometimes possible for the outer satellite, Callisto, to pass north or south of Jupiter's shadow and avoid eclipse altogether.*

so that to our view they are strung out more or less in a straight line. Their nightly positions around the planet are listed in the almanacs, but with a little practice it is easy to distinguish them. Io moves the fastest, and is never far away from Jupiter. Europa has a rather dull, grayish tint. Ganymede is easily identified because of its brilliance, and even a 6-inch reflector will reveal a definite disk, although surface details cannot be made out except with very large telescopes. Callisto is the slowest moving and faintest of the four. It is worth keeping an eye on the satellites' relative magnitudes, for Callisto has sometimes been recorded as having almost the same brightness as Ganymede, and Io also shows fluctuations. These variations may be due to irregular light and dark patches on their surfaces, but they are not well understood.

These moons are often eclipsed in Jupiter's shadow. Their eclipses occur more frequently than those of our own moon because Jupiter's shadow is much larger than the earth's. At other times, they pass in front of the disk, a phenomenon known by that adaptable word *transit.* As figure 47 shows, the ease with which eclipses can be observed depends on the position of the earth. When our vantage point is at C, as it is at opposition, the shadow lies directly behind the planet and so cannot be seen; at quadrature, we are angled to the greatest advantage and the shadow leads away to one side of the disk: to the p side before opposition (A), to the f side thereafter (B). Io is so close to Jupiter that its eclipses occur very near the limb; the same is usually true of Europa, but Ganymede and Callisto can pass right through the shadow before being occulted by Jupiter itself. Callisto is so far away that the slight tilt of its orbit sometimes makes it miss eclipse altogether.

Transits are well worth observing. The satellite approaches the disk from the f side, appearing as a gleaming spot as it passes onto the shaded limb but quickly becoming lost in the planet's brilliance (although it may reappear during transit as it crosses in front of some unusually dark feature). The satellites also cast shadows, keeping Jupiter well supplied with solar eclipses. These shadows, like the satellites, move across the disk from east to west; they can be seen very easily with a 3-inch telescope, appearing as sharply defined black spots. Just occasionally, one satellite may pass through the shadow cast by another, while mutual occultations can also take place. Unfortunately, the adequate observation of such phenomena requires large apertures.

13

Saturn

Saturn reveals its loveliness in a very small telescope. When the rings are well presented and lie at their maximum inclination towards the earth, as happens every fifteen years or so, good binoculars can show a distinctly oval outline. A 2-inch telescope and a magnification of × 20 are enough to show the rings, and a 3-inch refractor gives a splendid and unforgettable view. On the other hand, though Saturn is a noble showpiece of the heavens, it is a much less convenient object for study than Jupiter. For really useful work, a telescope of at least 10 inches aperture is desirable, although occasional spots have been discovered with smaller instruments.

Galileo's simple telescope revealed to him something protruding from Saturn's outline, but its definition was too poor for him to solve the problem. Succeeding efforts were not much more successful, and it was not until 1659 that Huygens solved the mystery. As a result of patient efforts with a 2½-inch refractor of 23 feet focal length and a magnification of × 100, he was able to announce the planet's amazing ring system. He also managed to see some markings on the disk itself, which, although fainter than those on Jupiter, seem to be essentially similar. This indicates that the two worlds are constructed on the same pattern, with thick, freezing ammonia and methane clouds. However, Saturn's temperature of −250°F has doubtless produced a more quiescent world, and fewer than a dozen really conspicuous features have been observed during the present century, although minor irregularities are common enough. Saturn has its equivalents of the Jovian equatorial and temperate belts, with the accompanying bright zones, but certainly has no permanent features to rival the Red Spot.

Rotation and markings

Saturn's equatorial regions rotate in about 10 hours 14 minutes. This is slightly longer than Jupiter's period, but despite its smaller diameter (75,100 miles), Saturn is actually more flattened at the poles. Clearly, its make-up must be even less substantial, since the more solid a body, the less its tendency to deform; and we arrive at the surprising conclusion that Saturn is actually less dense than water. The image of a planet floating in some colossal ocean is enough to show that its solid core, if any, must be very small. It is a world composed principally of freezing, evil-smelling atmosphere.

As with Jupiter, different parts of the globe rotate at different rates. Exact knowledge of the different regions is frustrated by the paucity of distinctive markings in high latitudes, which only goes to show how vital it is for amateurs with adequate optical means to keep a watch on the planet; a few transits of an identifiable feature would add immensely to our knowledge of these regions. The evidence so far amassed suggests that instead of the planet's being divided into two main rotational zones, the period of rotation steadily increases with latitude, being about 30 minutes longer at a latitude of 60°. Any prominent spot appearing on Saturn counts as a major astronomical event, and must be observed to the full.

As seen with the naked eye, the planet has a characteristic yellowish hue. Telescopically, the disk is seen to be scored by a bright, almost white equatorial zone, with the temperate regions a somewhat darker tint and the belts themselves darker again, often with rather vague edges that seem to melt into the adjacent zone. The polar regions are usually dark, but undergo occasional brightenings. In 1963, for example, the south polar region became almost as light as the equatorial zone, which itself turned unusually dark in 1964. With moderate and large instruments, fine belts can often be seen in quite high latitudes; but they are almost always featureless. Most of the activity occurs in the two equatorial belts, which frequently exhibit a somewhat fluted border with the equatorial zone.

Easily recognizable spots are rare. Cassini is said to have seen a couple of streaks in 1683, but even William Herschel, who paid more attention to Saturn than to any other planet, recorded a conspicuous spot only once, in 1780; this took the form of a large dusky spot on the equator. Nearer the present time, there was a major outbreak of small white spots close to the equator between 1891 and 1894, and again in 1932; all of these gave a rotation period of about 10 hours 14 minutes for the equator. However, the most prominent spot ever recorded was seen a year later, in 1933, being discovered on August 3 from England by an amateur who was rather better known to the public in another sphere—the British stage and screen comedian Will Hay. Hay's white spot was easily visible with a 3-inch refractor, but it quickly extended in longitude and faded away, having been visible for only a few weeks.

Saturn turned active once again during the appearance of 1960, this time in the north temperate regions. Several white spots were seen by amateur observers between March and September, and one in particular, followed from March 31 until May 14, was almost as bright as Hay's. After it had disappeared, several other spots were sighted, the whole zone being in a state of unusual turbulence. They all lay at a latitude of about 60°S, and had a rotation period of about 10 hours 39 minutes, so that they added considerably to our slender knowledge of these regions. These isolated outbreaks, which usually uncover features that are visible in quite a small telescope, prove that persistent observation is essential if we are to make the most of these scraps of evidence. White markings were seen in the Equatorial Zone in 1971 and 1979, and a particularly bright spot was recorded in 1978, but perhaps the most unusual marking of recent years was the brilliant white spot observed in the S.S.T.Z. in November 1976. It is very unusual for markings to be seen in as high a latitude as this.

The rings

Saturn's ring system is one of the wonders of the visible universe. With a diameter of 169,000 miles, and thickness less than one, the rings have been called "the thinnest things in existence"; and they are so slender that when the earth passes through their plane they are invisible, for a day or so, even in large observatory instruments. The rings look solid enough, but they are really very insubstantial, being composed of millions of tiny moonlets just a few inches across. We do not know how they came to be formed, but Saturn's gravitational influence must in some way be responsible. It could be that a former satellite approached injudiciously close and achieved an unexpected immortality, but it is more likely that the rings form the protoparticles of a moon that never coalesced into a solid body.

The system is divided into three major rings, termed A, B, and C as one

Figure 48. *The rings of Saturn.*

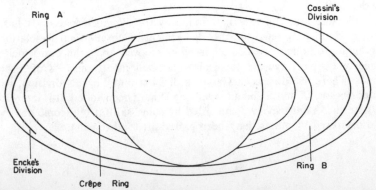

progresses inward; they are shown diagrammatically in figure 48. Ring A is 10,000 miles wide; it has a yellowish tint, rather like the temperate regions of the planet, and is separated from the brighter Ring B by a dark line known as *Cassini's division,* named in honour of J. D. Cassini, who discovered it in 1675. Cassini's division is obvious with a 3-inch refractor when the rings are at their maximum presentation, but is naturally invisible through foreshortening when they are almost edge-on.

Ring B is 16,500 miles wide, of a clear cream color that is usually brighter than the planet's equatorial zone, so that it acts as a useful comparison for estimating the zone's occasional brightenings. The interior part of the ring is somewhat shaded, and between its inner border and the planet lies the much fainter Ring C, or *Crêpe Ring.* This is so dim that it was not discovered until 1850, by W. C. and G. P. Bond, using the 15-inch refractor of the Harvard Observatory, and, independently two weeks later, by the Reverend W. R. Dawes in England, using a refractor of only $6\frac{1}{3}$ inches aperture. It is worth remembering that observers of the caliber of Herschel, Schröter, and Struve all studied Saturn with instruments of considerable power, and yet *missed the Crêpe Ring,* not because it was invisible with the equipment at their command, but because they were not aware of its existence—a proof of Sir John Herschel's dictum:

When an object is once discovered by a superior power, an inferior one will suffice to see it afterwards.

Even so, when we consider that the Crêpe Ring requires an aperture of about 8 inches if it is to be well seen, Dawes' discovery with a 6⅓-inch refractor is certainly a remarkable feat. The ring is about 10,000 miles wide, leaving a gap of 9,000 miles between the inner edge and the outer reaches of Saturn's disk. The different intensities of the rings are explained by the particle density in different regions; Ring B is very closely packed, with the particles perhaps more pulverized and so better reflectors of light, whereas the Crêpe Ring is tenuous and Cassini's division is almost devoid of moonlets. The faint Encke's division, which lies about 4/5 of the way from Cassini's division to the outer edge of Ring A, requires a moderate aperture to be detected.

The close study of the rings made by the *Voyager* spacecraft has, of course, completely rewritten this department of observation. Resolution to less than a mile has revealed a corded or grooved appearance in which the three telescopic rings are themselves resolved into thousands of discrete bands. There are dozens in Cassini's division and Encke's division, while outer rings F, G, and E and an inner ring D have been added. It is interesting to note that faint "spokes," recorded by some skillful telescopic observers and dismissed by others, have been proved after all to exist.

Presentation of the rings

Luckily, Saturn's axis is not as erect as that of Jupiter. If this were the case, our view of the rings would always be very oblique, since they revolve exactly in the plane of the planet's equator. But the axis is tilted from the vertical to the orbit plane at an angle of 26° 45′, so that when one of the poles is turned to its fullest extent toward the earth, we have a reasonably good view of the rings' surface. This phase happens twice during Saturn's sidereal period of 29½ years; the north pole was presented in 1958, while the southern regions were well displayed in 1973. In the intermediate position (1966), the earth passed exactly through the plane of the rings (see figure 49). In 1979–1980 the earth passed through the plane of the rings, and by 1988 the north pole will once again be seen at its maximum presentation.

The edge-on stage is most interesting. To an observer in the position of the sun, there would be only one phase—when he passed through the plane of the rings, and the opposite side came into view. But since the earth and sun make slightly different angles with Saturn, there are actually three definite phases.

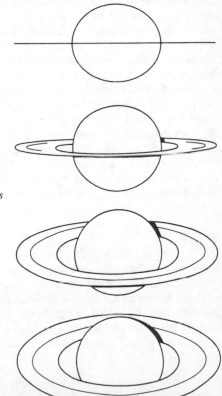

Figure 49. *Presentation of Saturn's rings.*

EDGE-ON TO THE EARTH.   Depending on the position of the earth relative to Saturn, it can pass either once or three times through the plane of the rings. If it is only once, as happened in 1951, the planet is always near conjunction and is unobservable; when it occurs three times, as in 1979–1980, one occasion is near conjunction but the other two occur around the time of opposition. The rings are so thin that they are lost from view, even in a large telescope, for a few days. For instance, when the earth passed through the ring plane in 1920, Barnard, using the 40-inch Yerkes refractor, could see no sign of the rings between November 8 and 13; amateurs had to wait another week before the rings returned to visibility.

EDGE-ON TO THE SUN.   This means that although the surface of the ring system is slightly inclined to our view, the sunlight is striking it horizontally. When this happens it appears very dim, but it is visible in apertures greater than about 4 inches.

EARTH AND SUN ON OPPOSITE SIDES.   This is the most interesting stage of all. It may be that the sun is shining on the north surface, while we view them from the south; in this case, they are seen by light transmitted *through* the rings, instead of being reflected off the surface. Consequently, regions such as the Crêpe Ring and Cassini's division, which contain few particles, transmit the most light and appear bright, while the denser Ring B is almost invisible. These observations are of great delicacy and require at least moderate apertures. During this phase in 1966, which lasted from October 29 until December 18, the rings were glimpsed by some observers using 12-inch reflectors, but there were few reports of their visibility with smaller apertures. They were easily seen with a 3-inch refractor only two days after the earth passed through the ring plane for the third time, on December 18, bringing the sunlit surface once more into view.

The years around the period of ring-plane passage are also the only time when we have a good view of the whole of Saturn's disk. During the rest of its appearance, one hemisphere is more or less obscured by the rings.

Shadows

The ball of Saturn casts a shadow on the part of the rings that is directly away from the sun. At opposition, when the earth is exactly in line with the planet, the shadow is hidden by the disk; but on either side of opposition it is partly visible as a dark line between the rear portion of the rings and the planet's limb. This shadow lies on the *p* side of the rings before opposition, and on the *f* side afterwards. Furthermore, the part of the rings which is in front of the planet casts a shadow on the disk, which may appear on either

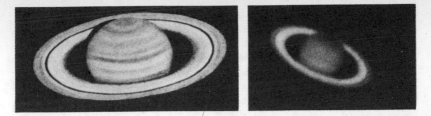

Saturn. Left, *A drawing made on October 28, 1973, at 01.30 UT, using a 10-inch reflector x250. Note the "polar cap" and the high-latitude belts. (D. Gray, Co. Durham, England.)* Right, *A photograph taken on February 20, 1973, with a 15½-inch Cassegrain reflector at f/140, exposure 3 seconds on Plus-X. (H. E. Dall, Luton, England.)*

April 11, 1979, 21.20 UT. In this drawing, the ring is tilted at an angle of 7° to our line of sight. The main division, Cassini's, is well shown, and Encke's division can also be seen. Other features of note are the shadow of the globe on the ring and the ring's shadow on the globe. The S.E.B. is double, and part of the N.E.B. is hidden by the ring.

October 24, 1979, 06.15 UT. This drawing shows the ring as it appeared three days before edgewise presentation. The fully illuminated south face is seen, and this face will not be sunlit again until November 1995. Note the prominent ring shadow across the globe and the two dark condensations on the N.E.B.

Two drawings of Saturn made by P. B. Doherty of Stoke-on-Trent, England, using a 419mm Newtonian, × 248.

the polar or equatorial side of the rings, as illustrated here. When it appears on the equatorial side, it appears rather like a very dark belt on the planet, and the shadow has often been reported as such by unwary observers. The Crêpe Ring sometimes gives the illusion of another belt where it crosses in front of the planet, appearing as a faint shading where the light shines through it. This effect is heightened if the Crêpe Ring is invisible against the black sky beyond the disk.

Occultations by the ring system

Some of the most important observations ever made of Saturn's ring system were secured by two amateurs, J. Knight and M. A. Ainslie, half a century ago. Observing independently from southern England, on February 9, 1917, they watched Saturn pass so near a 7th-magnitude star that the star was occulted by the outer portion of the rings. The star was considerably dimmed when seen through Ring A, but it could be made out clearly enough with a 9-inch reflector and a 5-inch refractor; and twice it brightened spontaneously for a few seconds. One brightening was evidently its visibility through the faint Encke's division, while the other must be taken as evidence of another gap in—or at least a thinning of—the ring particles. When the star reached Cassini's division, it shone out normally. It did not pass behind Ring B, but another occultation occurred on March 14, 1920, when observers in South Africa saw the star shining distinctly through the bright ring.

Saturn quite frequently occults a star—about once every other year—but usually these are too faint to shine distinctly through the rings, and in recent years no stars have been as obliging as those of 1917 and 1920. Occultation predictions are given in the B.A.A. *Handbook*, however, and every effort should be made to observe those that do occur.

Observing Saturn

Although spasmodic white spots can be seen with small telescopes, routine work on Saturn requires at least a 6-inch refractor or an 8-inch reflector, and larger apertures are of course preferable. Nothing less will show the minor irregularities of the belts and such fine features as Encke's division; neither will it allow sufficiently high magnifying powers to be used, for the disk appears almost as small as that of Mars when seen at an aphelic opposition.

TABLE V.   Saturn's Satellites

NAME	MEAN DISTANCE FROM SATURN (MILES)	ANGULAR DISTANCE AT ELONGATION	DIAMETER (MILES)	ORBITAL PERIOD			MEAN MAG.
Mimas	113,300	30″	350		22^h	37^m	12·1
Enceladus	148,700	38	450	1^d	8	53	11·8
Tethys	183,200	48	750	1	21	18	10·3
Dione	234,600	1′	900	2	17	41	10·4
Rhea	327,600	1½	1,100	4	12	25	9·8
Titan	759,500	3¼	3,000	15	22	41	8·4
Hyperion	920,100	4	200	21	6	38	13·0
Iapetus	2,213,200	9½	1,000	79	7	56	9–12
Phoebe	8,053,400	35	100	550½			16·5

(The diameters of the smaller satellites are most uncertain.)

Saturn is an infuriating object to draw, and it is not worth making regular "finished" sketches unless unusual markings are seen. Written notes are normally adequate, especially if they concentrate on the color and intensity of different regions. Some observers have gone so far as to draw up a brightness scale, ranging from 0 (brilliant white) to 10 (the blackness of the sky); but it is extremely difficult to make estimates of such accuracy. A better method is to use Rings A and B as standards for estimating the zones, belts, and polar regions. The great drawback of numerical scales is that they *appear* to be highly accurate, whereas eye-estimates are not very stringent, particularly where color is concerned. An entry such as "Equatorial zone perceptibly brighter than Ring B" really means much more than its evaluation on a 0–10 scale.

The satellites

The nine major satellites are of great interest. Four or five are visible with a 6-inch reflector, and seven with a 12-inch. Their details are given in Table V.

With the exception of the two outer satellites, all the moons revolve almost exactly in the plane of the equator and the rings. This means that when we have an equatorial view, as in 1966, the moons are strung out in a line—as with Jupiter's family—while at other times they are distributed much more irregularly. When the rings are almost edge-on, a large telescope will show the satellites apparently threading their way along the sliver of light, like pearls on a wire: a sight that delighted William Herschel when he viewed the planet in August and September, 1789, with his immense new 48-inch reflector. It was on these occasions that he discovered two inner satellites, Mimas and Enceladus. Mimas is extremely hard to see, not simply because it is faint, but because it is always very near the planet. At its greatest elongations, it has been glimpsed with a 10-inch refractor, but this is an exceptional feat, and many first-class observers have failed with apertures of less than 18 inches. It is actually slightly brighter than Enceladus, but the latter appears more distinct because it is further from the planet, and has been detected with a 6-inch reflector, although an unpracticed eye will naturally require greater optical assistance.

Titan, which is probably the largest satellite in the solar system, although some authorities consider Ganymede to be larger, is so bright that it can be seen with a very small telescope; it was first recognised as a satellite by Huygens in 1655, although there is a curious legend that Sir Christopher Wren, whom one associates more with cathedrals than observatories, saw it independently at the same time. The other bright four (Tethys, Dione, Rhea, and Iapetus) were all discovered by Cassini; Hyperion was found in

1848, and Phoebe was discovered photographically in 1898. The rough order of visibility is: Titan, Rhea, Tethys, Dione, Hyperion, Enceladus, Mimas, Phoebe; Iapetus varies so in brightness that it cannot be included in a list. Rhea is usually visible with a 3-inch, while a 4-inch should pick up Tethys and Dione; Hyperion requires at least a 6-inch refractor. However, so much depends on the state of the air, the condition of the instrument, the nearness of the satellite to Saturn, the presentation of the rings (since when they are open, the glare around the planet is increased), and the observer's keenness of vision, that it is misleading to be dogmatic. Just one of many examples of how eyes differ is revealed in this account by the Reverend T. E. Espin, a well-known amateur observer of past years, of an observing session with the author of the classic *Celestial Objects for Common Telescopes*:

> *A curious instance of difference of vision was well illustrated one superb evening, when Mr. Webb and the writer were observing Saturn with the 9½-inch reflector at Hardwick. Mr. Webb saw distinctly the division in the outer ring [Encke's division] which the writer could not see a trace of, while the writer picked up a faint point of light, which afterwards turned out to be Enceladus, which Mr. Webb could not see.*

Iapetus is exceptional in this family of satellites, because it varies greatly in brightness. When near its eastern elongation, it is about as difficult to see as Enceladus, but at western elongation it is actually brighter than Rhea, and can easily be seen in a 3-inch. *Voyager* photographs have proved that one hemisphere of this satellite has about ten times the reflecting power of the other.

Neither are the other moons of constant magnitude. Rhea has been found to vary by about half a magnitude, and Titan too may show slight fluctuations. Estimates are extremely difficult to make, since they appear fainter when near the planet and so suffer an apparent variation on top of the real one; even so, it is a problem that can be tackled with modest equipment.

Numerous faint satellites have been detected by spacecraft, varying in size from about 130 miles across (1980S1, period 16.7^d), down to about 40 miles for the innermost known moon, tiny 1980S28 (period 14^h 25^m). A total of 17 satellites have had their orbits determined, but others have also been suspected by both *Voyager* probes.

14

The Outer Planets:
Uranus, Neptune, and Pluto

The three remotest planets were all discovered telescopically. Uranus was found more or less accidentally by William Herschel in 1781; and Neptune and Pluto were tracked down in 1846 and 1930, their existence being established by their gravitational pulling on Uranus. Uranus and Neptune are both built on the "giant planet" pattern, although they are considerably smaller than Jupiter and Saturn; but they are too far away to show much of a disk, even in a large telescope, and amateur study of their surface markings is out of the question.

Uranus

Uranus is far bigger than the earth, with a diameter of 29,000 miles, and its freezing gas-clouds seem to reflect light well. But, at 1,775 million miles from the sun, it is so far away that it appears very dim. The naked eye can follow its ambling along the ecliptic without a telescope at all, but the ancient astronomers can hardly be blamed for not noticing it. At the time of writing (1981) it lies in Libra, into which it passed in 1975. The movement is very slow, since Uranus requires 84 years to circle the celestial sphere once.

When its position is known, Uranus can be picked up telescopically as a bright, bluish "star"; careful attention reveals a disk with a power as low as × 40, and anything higher expands it very obviously. However, using a powerful eyepiece also dims the planet so seriously that no detail can be seen. This explains why a large telescope that collects plenty of light is necessary for examination. The only well-established feature of the planet is a bright equatorial zone, with somewhat dusky poles, and there is little chance of anyone adding to our knowledge of its surface features without using a gigantic telescope.

On the other hand, interesting variations seem to occur in its magnitude, and amateurs have been tackling the problem on and off for the past seventy years. The first task is to identify what we might call "natural" effects. Thus, the planet will appear brighter at opposition than near conjunction (a variation of about 0·4 of a magnitude); it will appear brighter when at perihelion than at aphelion (about 0·2 of a magnitude); while the brightness also fluctuates in a period corresponding to its rotation time of 10¾ hours, due to the passage of lighter and darker features across the disk (about 0·1 of a magnitude). We have also to consider the curious tilt of the axis—98°. This means that we sometimes see the pole of the planet in the center of the disk, as happened in 1943, whereas in 1965 the planet presented the normal equatorial view. Since Uranus is appreciably flattened at the poles, its disk shows a slightly larger area when pole-on, an increase that might be expected to augment the magnitude by about 0·3. Clearly, these effects must be weeded out; but even so, discrepancies seem to remain. A German authority, W. Becker of Münster, claimed in 1933 to have discovered a roughly eight-year period extending over about 0·3 of a magnitude, although this was denied by Kuiper and D. L. Harris in 1961. Nevertheless, B.A.A. members investigated the brightness of Uranus between 1952 and 1955, and discovered that it was appreciably brighter than the "official" magnitude of 5·8. This work, done with the simplest of equipment, revised the magnitude of Uranus to about 5·5. Further research in 1969 supported this revision, which has now been incorporated in the B.A.A. *Handbook*.

Fluctuations probably do occur, since the outbreak of bright or dark features will affect the reflectivity, and regular estimates of the brightness would be of value. The method is the same as that described for variable stars (in Chap. 19), where the planet's brightness is compared with that of stars of known magnitude, known as *comparison stars*. In the case of variable stars, the comparison objects are always the same; but Uranus is on the move all the time, so fresh comparison stars must be found and their magnitudes obtained from a reliable catalogue. A very low power must be used, so that Uranus will look like a star rather than a disk, and it will probably be found that binoculars are of far more service than a regular astronomical telescope. It is hopeless to expect quick results, but at least there is nothing complex about the method or equipment.

Uranus has five satellites. The two inner ones, Ariel and Umbriel, are so close to the disk that they are lost in the glare, and it takes a very large telescope to show them. The next two, Titania and Oberon, both of about the 14th magnitude, have been glimpsed with a 6-inch refractor when

Uranus itself was hidden by a field bar; but most observers would do well to catch them with a 12-inch. The fifth moon, Miranda, is so close and faint that it has never been seen visually, although it can be photographed. The very faint rings, discovered in 1977, cannot be seen with any earth-based telescope.

Neptune

Found in 1846 by Johann Galle, after predictions issued independently by two mathematicians, J. C. Adams and U. Le Verrier, Neptune is slightly smaller than Uranus, and a thousand million miles farther away from the sun. It can be seen as a "star" of about magnitude 7.8 with a pair of binoculars. At the time of writing (1981), it lies in the constellation Ophiuchus, and since its orbital period is 165 years it seems to move across the sky at only half the speed of Uranus. The B.A.A. *Handbook* publishes annual charts showing the positions of Uranus and Neptune.

Neptune's somewhat greenish disk can just be distinguished with a 4-inch refractor, and its satellite Triton, which is probably a little smaller than Saturn's Titan, is somewhat easier to see than the outer satellites of Uranus. The other moon, Nereid, appears very minute. It may be that Neptune shows magnitude fluctuations of the same kind as Uranus; the matter has not been fully investigated, and it may provide a promising field of investigation.

Pluto

Pluto was found from the Lowell Observatory, Flagstaff, by Clyde Tombaugh in 1930, after independent calculations had been made by Lowell himself and W. H. Pickering. It then lay in Gemini, and so slowly does it move across the sky that even now it has only reached Virgo, three "doors" along the zodiac.

It is of no interest to the amateur, since a telescope of about 8 inches aperture is needed to show it at all, and it simply looks like a faint star. However, it is satisfying to glimpse the point of light that marks this frozen, almost forgotten world and its large satellite Chiron.

15

The Minor Planets

The small, rocky bodies circling more or less between the orbits of Mars and Jupiter are often referred to as *asteroids*, a word meaning "starlike," although, like all the other members of the sun's family, they shine by its reflected light. But the term is reasonable enough, for only the greatest telescopes can show even the largest and nearest minor planets as definite disks, and the amateur's task of identification might appear hopeless. Luckily, there is a very easy way of distinguishing one from the other, for the stars have retained the same patterns in the sky for thousands of years, whereas an asteroid moves appreciably from night to night.

All the planets, in their orbital movement around the sun, appear from the earth to creep along the zodiac. When Mars, for instance, passes close to a naked-eye star, its movement over the course of a day can be detectable without optical aid. Jupiter, more remote and therefore slower-moving, takes longer to reveal its drift, since it takes 12 years to circle the zodiac, whereas Saturn takes 29½ and the outer planets even longer. Nevertheless, the earliest astronomers noticed this movement and termed them "planets," or "wanderers." So, to identify an asteroid once its approximate position is known, the observer simply draws the star field and reexamines it a couple of nights later. One of the "stars" will have shifted slightly, proving its planetary nature.

Discovery of the asteroids

This process is all very well once it is known which region of the sky contains the asteroid; but the original asteroid-hunters, affectionately known at the time as the "celestial police," were faced with an immense task. In the early nineteenth century, when interest was at its height, there were no accurate maps of the fainter stars; each observer had to plot and "patrol" his own allotted fields along the zodiac, watching for any suspicious shifting of a "star." The vagaries of weather and moonlight made this process alone difficult enough; but the detection of a wanderer was only the beginning. Its movement then had to be measured so that an orbit could be worked out, and it had to be done immediately and accurately, for the tiny body could easily be lost again among the myriads of stars. But the job was done, and done well. By 1847, eight asteroids were known; by 1891, 323. After that, photography provided a quicker, more convenient way of recording faint stars, and the discoveries snowballed into the thousands.

One asteroid, Vesta, sometimes reaches magnitude $5\frac{1}{2}$ and is readily visible with the naked eye. Another, Pallas, can reach magnitude 6·3 at a favorable opposition. Both have diameters of between 250 and 300 miles. The largest asteroid, Ceres, first to be discovered, is about 480 miles across but can never exceed the 7th magnitude, which suggests that it is composed of rather darker substance than the highly reflective Vesta. Altogether, about forty asteroids can exceed magnitude 9·0 at opposition, and are therefore visible with 2-inch binoculars, while hundreds are brighter than the limiting magnitude of a 3-inch refractor, which usually lies somewhere between the 10th and 11th magnitudes.

Location and plotting

Since the B.A.A. *Handbook* and *Astronomical Almanac* give ephemerides for the brighter minor planets visible during each year, it may sound a simple matter to find them. If the observer is lucky enough to possess a copy of the *Atlas Eclipticalis,* or Webb's *Star Atlas,* or one of the other maps showing stars as faint as the 9th magnitude, the asteroid's position can be plotted directly, and the star field compared with the low-power telescopic view. If the asteroid is of the 7th or 8th magnitude, it should immediately be obvious as an extra "star" shining where none is marked in the chart.

One precaution is necessary when adopting this line of attack. The stars' positions, in terms of right ascension and declination, are changing steadily as the years pass, and it is necessary to adjust the asteroid's published position to correspond with the epoch of the chart. This can be done by using the precessional table in Appendix VI. If, for instance, an asteroid's position was being plotted in 1980, it would have been necessary to subtract the precessional difference for 30 years to fit it accurately into the coordinates of the *Atlas Eclipticalis* drawn up for the year 1950.

Most amateurs, however, depend on *Norton's,* which shows stars down to the 6th magnitude only. In this case, a good deal more work is involved, but the challenge is greater and the final identification more satisfying. The method here is to plot the minor planet's position on the atlas, and to draw a map of all the telescopic stars visible within an area of perhaps two degrees square around the critical position. We can be reasonably sure that one of these "stars" is, in fact, the minor planet; but there is no direct way of identifying it except by its motion. On the following night, therefore, either the field is redrawn, or the chart is compared with the telescopic view to see if the visitor has betrayed itself by shifting its position. This procedure is not as easy as it sounds, for a slight inaccuracy in the original map may conceal the motion, or the planet may lie outside the region plotted.

A better way of identifying the object is to turn one's telescope into a transit instrument. Stretch a hair across the field stop of a low-power eyepiece, so that it appears in sharp focus, and orient it so that a star in the field of view trails along its length; it must then lie in an east-west direction. If the sky is so black that the hair cannot be seen, diffuse a little light into the field of view by judicious use of a flashlight shining obliquely onto the objective (bright-field illumination); this is easier to arrange than a light shining on the hair itself, and the loss of the faintest stars will not matter when the minor planet is above, say, the 9th magnitude, which it probably will be if this relatively crude method is being used.

Now rotate the drawtube through 90°, so that the hair is north-south. Having made a sketch of the stars that include the minor planet, set the telescope a little ahead of the field and make a note of the time that each star transits the hair, writing the times down on the drawing. On the following night repeat the timings. Typically, a minor planet near opposition is moving along the ecliptic at a rate of about 1′ per day, so it should transit the hair about four seconds earlier, compared with the other stars. The method cannot be used if the body is at a stationary point, which it will reach about two months before and after opposition, since its daily motion at this time is negligible.

Even when the asteroid is found, there is no guarantee that it will remain within reach. A sequence of cloudy nights can mask its motion into a new and unknown region of the sky, so that fresh charting is necessary before it can be identified again. But there is always the challenge of following it over a long period, watching it brighten as it approaches opposition and dim again as it swings away from the earth and toward conjunction.

It must be admitted that study of the minor planets is hardly a profitable branch of amateur astronomy. Physical observation of their surfaces is clearly out of the question; even Ceres shows only the minutest disk when viewed with the largest telescopes in the world. Moreover, study of their color and of their light variation, which can be taken as evidence of rotation, has been undertaken by professional astronomers using more accurate equipment than the amateur usually commands; Appendix V lists some details of the more important minor planets. But, because of the rather large eccentricities of many of the orbits, we sometimes find a normally dim and inaccessible asteroid approaching the earth sufficiently close to be visible with modest equipment. Under these conditions useful observations of magnitude and light variation can be made. The technique, comparing their brightness with that of nearby stars whose magnitudes are known, is the same as that used in variable star work (see Chap. 20). It must be remembered, however, that these little planets are moving across the sky all the time, so the star field is changing and fresh comparison stars must be found.

Occultations of stars

The greatest chance for an amateur to make a really important contribution comes when a minor planet passes in front of a star. These occultations are, however, excessively rare. The B.A.A. *Handbook* publishes predictions for "appulses," or close approaches to stars; but the diameters of even the larger planets are so small that the chances of a perfect line-up are remote. There is, nevertheless, an air of excitement about awaiting a very close approach, since our knowledge of their motions is not perfect, and an error of even a second of arc in the ephemeris may make all the difference between a positive and a negative observation. Nevertheless, some interesting observations have been made in recent years, and forthcoming events are predicted well in advance. Surprisingly, the first widely observed occultation of a star by a minor planet was as recent as 1975, when Eros occulted the naked-eye star κ Geminorum, making it appear to disappear momentarily! Several other events have been seen since. The asteroid 6 Hebe occulted δ Ceti on March 4–5, 1977, while in 1978 both 532 Herculina and 18 Melpomene passed in front of faint stars. On December 11, 1979, 9 Metis also occulted a star. This apparent increase in the number of phenomena is at least partly due to the greatly improved precision with which the positions of stars and minor planets can be determined.

It is clear that the track along which an occultation will be observable must be relatively narrow, since even the larger minor planets are only a few hundred miles across. Timing the duration of the occultation allows the size to be calculated with considerable accuracy, and useful work has been done by amateurs. In addition to the expected occultation, brief secondary disappearances of the star were recorded in the case of Herculina, and suspected for Hebe and Melpomene. The only reasonable explanation seems to be that these bodies are accompanied by one or more satellites, or possibly that they possess a cloud or ring of material held within their very weak gravitational field.

16

Comets and Comet-hunting

A bright comet is one of the finest of all natural phenomena. Like a total eclipse of the sun, it is as impressive to the layman as to the fully trained astronomer. No scientific knowledge is required to appreciate the spectacle of a milky tail stretched across the sky, and it is perhaps for this reason that more misconception surrounds the subject of comets than any other branch of astronomy.

A comet is a mass of gas, dust, and icy particles, that revolves, like a planet, under the sun's gravitational influence. There is, however, one important difference: its orbit is very eccentric, so that whereas at perihelion it may be only a few million miles away from the sun, at aphelion it recedes far out into the depths of interplanetary space. Only around the time of perihelion do comets glow brightly; throughout the rest of their orbits, almost all the known specimens are so faint that no telescope can pick them up, although mathematicians can calculate approximately where they are.

The size of the cloud composing a comet may be anything from a few hundred thousand miles across to a million or more, but in terms of actual mass there is less matter in the greatest comet than in a small asteroid. The *nucleus*, which is the central collection of solid particles, is rarely more than a few miles across, and the surrounding aura of gas and dust is inconceivably tenuous. On several occasions during recorded history, the earth has passed through these shadowy regions with no noticeable effect; although a collision with the nucleus itself might have serious consequences, the chances of such an encounter are extremely small. For all their fire and fury, comets are ethereal bodies.

Life cycle and movements

A comet usually brightens up into telescopic visibility at about the distance of Mars. If it is a known body, certain observatories that specialize in this work (e.g., the United States Naval Observatory, Flagstaff station) will be carefully photographing the region of the sky where it is expected to appear, until finally a tiny smudge of light signals the return. Although these predictions, known as ephemerides, are usually remarkably accurate, there are always slight errors, or *residuals*. One of the first tasks of the astronomer is to measure accurately its position relative to the stars, so that the ephemeris can be corrected. At the same time, a circular is issued announcing the recovery. The clearinghouse for cometary information is the I.A.U. Telegram Bureau at Cambridge, Massachusetts (Western Union Telex/TWX Number 710–320–6842); most national societies, however, issue special circulars. Ephemerides are listed annually in the B.A.A. *Handbook,* and notes on current returns are included in the monthly *Sky & Telescope.*

The comet continues to brighten as it approaches the sun. This occurs partly because of the increased illumination, but also because, as the temperature rises, the material composing the head of the comet is vaporized and starts to glow, so that it emits light as well as shining by reflected sunlight. The fuzzy *coma,* the matter surrounding the nucleus, begins to expand, while the force of the atomic particles that are flying away from the sun—known as the *solar wind*—brushes some of the coma backwards to form the *tail.* Generally speaking, the closer the comet approaches to the sun, the longer and more spectacular the tail becomes; the coma and nucleus also glow more brightly. Greatest brilliance is reached at perihelion, but at this time the comet is often so close to the sun in the sky that it is quite invisible; so it is best observed just before and after perihelion, when it is still bright but can be seen again a dark sky.

It must not be thought that all comets reach the "sword-in-the-sky" stage. Relatively few do, but these are the ones likely to be acclaimed in the popular press. Of the six or eight comets that are observed every year (which includes new ones discovered as well as known ones recovered), by far the greatest number are telescopic objects, approaching the sun no closer than about a hundred million miles, and perhaps not even growing a tail. About one a year brightens sufficiently to be visible with the unaided eye from some part of the world; once every five years or so, the sun is visited by a comet bright enough to attract general attention; and perhaps five really brilliant ones may be seen in a century. The brightest of recent years was Comet West of 1976, the nucleus of which could be seen with the naked eye when the sun was still above the horizon. This may have been the brightest since the famous daylight comet of 1910, which in turn was outshone by the brilliant comet of 1882, one of the "sungrazing" group of great comets.

More than 500 comets, bright and faint, have been discovered since the invention of the telescope. There are records of naked-eye sightings going back to ancient times, so they are clearly plentiful; the solar system probably contains hundreds of thousands. Every year, two or three new ones are discovered (no fewer than 13 were found in 1975), and many faint ones doubtless slip past perihelion unseen. Comets do not keep to the zodiac as do the planets, since their orbits can lie in any plane; because there is no way of telling just where a new comet is likely to appear, comet-hunters need both immense perseverance and luck. It is therefore not surprising that a good proportion of discoveries have been made by a line of dedicated amateurs.

The popular idea of comets "flashing across the sky" is wildly mistaken. When they are far away from the sun they seem to move among the stars at planetary speed; even when they are near perihelion, and therefore moving fastest, they stay in the same region of the sky from night to night. Halley's comet, at its last return, was visible with the naked eye for about two months, and was followed telescopically from September, 1909, until April, 1911. The great comet of 1882 was a naked-eye object for five months and could actually be seen in broad daylight when it was only a few degrees away from the sun. Of course, few comets are as brazen as these, but even the less spectacular examples, such as the two naked-eye comets of April and August, 1957, could be followed with a small telescope for several weeks.

Known comets are generally sighted when they are far away from the sun and several months from perihelion; but new discoveries are usually made when they are close to the sun and so appear much brighter. When a new comet is found and its movement investigated, its image can often be made out on routine sky photographs taken several weeks previously by professional observatories. It would be far too great a task to examine every photograph for such tiny features; for this reason, many new comets slip through the initial net and avoid discovery until they are quite bright. The second of the two conspicuous comets of 1957 was discovered when it was a naked-eye object, and many have been visible with binoculars when first picked up.

Comet-hunters

Historical literature contains many references to bright comets. Halley's comet, which returns to the sun every 76 years, has been periodically recorded, back to 239 B.C., and other bright comets have appeared from time to time. Naturally, the recent records are more complete, especially since the invention of the telescope; indeed, the age of modern observation can be taken as dating from 1758, the year in which it was proved that comets revolve around the sun, and are therefore true members of the solar system.

This was a posthumous triumph for Edmund Halley, the British Astronomer Royal who in September, 1682, observed a bright comet and carefully tracked its path across the sky. Its general similarity to the comets of 1531 and 1607 made him suspect that they were actually the same body; accordingly, he predicted a return for 1758, expressing the hope that others would search for it. Halley died in 1742, but in 1758 his hope was fulfilled: Charles Messier, the French "ferret of comets," and Johann Palitzsch, an amateur astronomer living near Dresden, both swept the sky at the appointed time, Palitzsch first spotting it on Christmas night, 1758. This was a sensational achievement, for it not only proved the nature of a comet's movements, but also established the telescope as a powerful weapon of discovery. The comet was named after Halley not because he discovered it, which is the usual method of nomenclature, but because of his remarkable prediction.

Palitzsch made no further discoveries, but Messier, no doubt both inflamed by his own failure and inspired by his rival's success, undertook regular comet-hunting with a 2-inch refractor, using a power of only × 5 and therefore having a very wide field of view. By his death in 1817 he had discovered 13 comets in the period 1759–1801; but he was closely challenged by the French observer Pierre Méchain, who discovered 10 between 1781 and 1802, while five were found between 1786 and 1797 by Caroline Herschel, William Herschel's sister and assistant. Her brother had made her a special "comet-sweeper"—a 4-inch Newtonian reflector of 27 inches focal length, with a magnification of × 20 and a field of just over 2°—with which she made her discoveries.

However, of all these comets, not one is likely to return to the sun's vicinity for many generations; their orbits are so extensive, probably reaching out to Pluto and farther, that one course will take hundreds or even thousands of years. The majority of comets, including all the really brilliant ones, belong to this "once-only" class, which explains why a new one may appear unannounced at any time.

Another notable comet-astronomer of the time was the German physician Heinrich Olbers, who for forty years observed and swept the sky for comets from his house in Bremen; he found one with a period of 73 years, which last returned in 1956. He also used his considerable mathematical talents to devise a system of orbit computation known as *Olbers' method.* This period also saw the king of comet-hunters, Jean Louis Pons, of Marseille, at work. He began his career as doorkeeper at the local observatory, receiving casual instruction from the director, but he discovered 27 comets between 1801 and 1827—a record for a single observer. One of these, discovered in 1812, returns to the sun every 72 years. In 1884, at its next return, it was found quite independently by the well-known American observer W. R. Brooks, and is now known as Comet Pons-Brooks. Brooks himself, observing from near New York, discovered 22 comets between 1883 and 1911, and his contemporary, E. E. Barnard, who worked at the Lick and Yerkes observatories and spent his spare time observing and photographing the night sky, found 16 between 1881 and 1892.

Brooks and Barnard were both associated with large observatories, and it might be argued that they had equipment superior to that of most amateurs, and so had a better chance' of success; but no matter what the equipment, it must be used with patience and enthusiasm. As we have noted, many comets are quite bright at discovery, and it does not require a large telescope to show them. Indeed, a small and maneuverable instrument is more or less essential, both from the point of view of easy manipulation and because of the wide field. Great telescopes have very restricted fields of view, and an aperture of between about 4 and 10 inches is best for cometary work.

During the past thirty years, comet-hunting has passed through a number of phases. A Czechoslovakian team made many discoveries during the 1950s, and Japanese amateurs swept successfully in the following decade. Occasional discoveries were made from the United States and Great Britain (where G. E. D. Alcock discovered four comets between 1959 and 1965), but it is curious that more work has not been done in these countries, particularly the United States, which has produced no observer to rival Leslie Peltier, who discovered eleven new comets between 1925 and 1954. The southwestern states, in particular, enjoy almost ideal conditions for comet-hunting, far superior to those in most of Europe.

The 1970s have belonged to one man: William Bradfield, who lives near Adelaide, Australia. Using a home-constructed telescope with an old plate-camera portrait lens of 6-inch aperture for the objective, he has to date (1981) discovered 11 new comets since starting sweeping in 1971. It is significant that not one of his finds has been a co-discovery, an indication of how poorly the southern skies are being covered.

Nomenclature and ephemerides

When a comet is first sighted, whether it be new or periodic, it is given a provisional designation consisting of the year followed by an italic letter signifying the order of detection. It is also known popularly after its discoverer, or, in the case of independent detection, co-discoverers. In November, 1956, for instance, two Belgian astronomers named Arend and Roland found the eighth comet of that year. Since it was new and not periodic, it was titled Comet Arend-Roland; it was also known as 1956h.

Comet Pons-Brooks, at its third return since discovery, was the third comet to be sighted in 1953; it was therefore designated 1953c. This provisional identification is used during the apparition of the comet, but is subsequently replaced by the permanent designation: the year and order of perihelion passage, given in Roman numerals. Since Arend-Roland reached perihelion in April, 1957, the third comet to do so in that year, it has become known as 1957 III. The name itself is not always definitive, since one observer may discover a number of different comets.

Comet ephemerides give the position of the comet, in terms of right ascension and declination, at intervals of a week or ten days, so that its path may be plotted on a star chart. Terms used in ephemerides include these:

T—Date of perihelion passage, usually given to decimals of a day.

i—Angle, or *inclination*, of its orbit with respect to that of the earth.

q—Distance of the comet from the sun at perihelion.

e—Eccentricity of the orbit. If e is less than 1, the orbit is elliptical. If it is 1·0 or greater, the comet will probably never return to the sun, since its path is parabolic or hyperbolic and does not close on itself.

P—Period between perihelion passages. This refers only to comets with elliptical orbits. If the period is less than about 200 years, the comet is referred to by the symbol $P/$; thus, $P/$Pons-Brooks, $P/$Halley, and so on.

r—Distance of the comet from the sun on any given date.

p—Distance of the comet from the earth on any given date. All distances are given, not in millions of miles, but in *astronomical units* (A.U.). One A.U., which is the mean distance of the earth from the sun, is 93,000,000 miles.

Equipment

Most visual comet discoveries, as against photographic ones, have been made with telescopes of between 4 and 6 inches aperture, with magnifications varying from × 25 to × 40. Low magnifications are essential, for comets are diffuse bodies, very difficult to detect under a powerful magnifier; moreover, a low magnification provides a large field of view, so that a given area of sky can be covered in fewer sweeps. These requirements are very convenient, for low-power work is relaxing to the eye and carries incidental advantages: the mounting need not be perfect, since vibration is less noticeable than with high-power work, and the objective can possess appreciable flaws and still be serviceable, since perfect definition is not necessary when it comes to distinguishing a fuzzy patch of light from a star. The important characteristics of a comet-sweeper are good light-gathering power and a wide field of view—between 1° and 3°.

Refractors are definitely in vogue for this kind of work. Caroline Herschel and Denning were almost alone in using reflectors; the most successful American observers, Brooks, Barnard, Swift, and Peltier, all used refractors, while 4- or 5-inch aperture binoculars are used by some modern observers, including Alcock. They are also in vogue at the Skalnaté Pleso Observatory, Czechoslovakia, from where an observing team discovered 19 comets between 1948 and 1960. However, many of the Japanese observers use reflecting telescopes.

Of course, giant binoculars are expensive and difficult to obtain, and a suitable refractor will cost as much or even more. An ordinary instrument is not very suitable, since it is difficult enough to get a field of view of even 1° with a 4-inch refractor. This is because the focal length of such an object glass is commonly about 60 inches, giving an image scale of about

Armchair mounting. *Dr. Henry E. Paul demonstrates his armchair mounting for a pair of binoculars and short-focus refractor. Mounting is controlled by convenient hand cranks. This "Sky Sweeper" chair was originally designed and built by Dr. Edgar Everhart, who used it in his discovery of the comet 1964h, which now bears his name. (Sky Map Publications, St. Louis, Missouri.)*

1°/inch. Since the internal diameter of the drawtube is only 1¼ inches, the preferable field of 1½° or even 2° is quite out of the question.

To overcome this difficulty, some enterprising telescope manufacturers have designed special telescopes for comet-hunting, often called *rich-field* telescopes. By having an object glass with a relatively short focal length, they embrace a very wide field of view. For instance, if the 4-inch lens mentioned above had a focal length of only 30 inches, the image scale would be halved; using the same eyepieces would give a field of view twice as big. Such an object-glass would not be convenient for high-power work, since eyepieces of a very short focal length would be required; but they are ideal for comet-sweeping. Unfortunately, it is difficult to correct object glasses of short relative focal length, since errors of figuring are enhanced, and they are rather expensive.

In the light of these facts, it seems curious that short-focus reflectors have not come into vogue. This involves using a telescope with a mirror not of the usual f/8 variety, but of f/4 or f/5, giving a smaller focal image and a wider field of view with standard eyepieces. Anyone who has ground himself an orthodox mirror will be able to make a satisfactory short-focus type; the finish need not be perfect, since the instrument will be used with low powers. A mirror of 6 or 8 inches diameter, with a focal length of between two and three feet, could form the basis of an inexpensive and most powerful astronomical tool. For instance, a 6-inch mirror with a focal length of 30 inches (f/5), used in conjunction with a low-power eyepiece, could give a field of view of over 1½° and show stars down to the 12th magnitude. Since most comets are of 8th or 10th magnitude when discovered, the opportunities afforded by such an instrument are clear enough.

But, however desirable a special instrument may be, an ordinary reflector of moderate aperture should not be looked down on. Denning discovered his five comets with the 10-inch telescope he used for planetary work, using a power of × 32 or × 40; and he rightly instanced a point that is often overlooked, namely, that many telescopic comets are so small, almost starlike, that too low a magnification will result in some being missed. He usually swept with × 40, keeping a × 60 eyepiece at hand with which to examine any small, suspicious objects.

Predicted comets

Let us now come down to earth and examine the business of sweeping up and following known comets. This can be almost as exciting as finding a new one, since comets rarely behave exactly as predicted by the mathematicians.

In addition to a suitable telescope, the other necessity is a good star atlas, such as *Norton's*, which is clearly marked with lines of right ascension and declination, so that the path of the comet can be drawn in lightly in pencil. It is then a matter of sweeping the appropriate region of the sky, which is commenced when the comet is judged bright enough to be visible with the equipment available. If it is a well-known and regular visitor to the sun, its ephemeris will have been published well in advance. Ephemerides usually give its likely magnitude, but the comet may differ from the prediction by several magnitudes. There is also the difficulty of assigning an accurate value to a diffuse object, so that brightness classification is inevitably somewhat vague.

Three factors are likely to hinder the observer as he searches: If the comet is low in the sky, it will appear faint through horizon haze; if too close to the sun, it will be drowned in twilight; if the moon is up, the sky may be too light. Generally, only a very bright comet can be seen within about 15° of the sun, since this is the limit of the main twilight arc; and remnants of twilight linger on for another 10° or so. In short, to be seen against a dark sky a comet should be about 30° away from the sun; even then, it will be low in the sky and dimmed by haze. If it is likely to be faint, therefore, it must be sought some time before and after perihelion, unless its perihelion distance happens to be unusually large. This point is often not appreciated, for a light sky will drown a quite considerable comet, and many beginners search hopefully in the bright twilight and are disappointed. As for moonlight conditions, these change from night to night and opportunities must be seized as they occur.

Sweeping must be done with scrupulous care. To allow for possible errors in position, it is best to cover an area 2° or 3° square around the predicted point. Move the telescope very slowly in horizontal sweeps; at the end of each traverse, slightly raise or lower the instrument and make an *overlapping* return sweep, so that every part of the region is covered two or even three times. For this work, an altazimuth telescope is much more convenient than an equatorial. If no suspicious hazy object is seen in the region, then the comet is too faint for the aperture and another sweep must be made under more favorable conditions.

An invaluable technique when trying to glimpse faint objects is the trick of *averted vision*, in which the observer directs his gaze at one part of the field while concentrating his attention elsewhere. Often a very dim star or nebulous object that has been invisible to the direct gaze becomes relatively conspicuous when sighted out of the corner of the eye! Also, since a moving object attracts the eye more powerfully than a stationary one, a gentle side-to-side motion of the telescope may amplify a half-suspected hazy patch into clear visibility.

Efficient dark-adaptation of the eye is obviously of the greatest importance. About half an hour is required to bring the retina to full sensitivity; once this is gained, it must not be destroyed by shining a bright flashlight onto the star chart. A dim red light, just sufficient to allow reference without straining the eye, is an essential adjunct. It is also a good idea to protect the idle eye from stray light. Closing it while observing through the other puts a strain on the ocular system; a better answer is to wear an old pair of spectacle frames, one rim holding a patch over the idle eye, and the other left empty to admit the eyepiece.

When the comet has been sighted, the most important observations concern its magnitude, the appearance of the coma and nucleus, and whether or not it has a tail. One good way to estimate the magnitude is to memorize the appearance of the coma—which will probably be a faint, spherical haze 2′ or 3′ (minutes of arc) across, then to rack out the eyepiece until the extrafocal images of the stars have the same diameter, noting the magnitude of the one whose brightness most closely matches that of the comet. It is quite possible to estimate the magnitude of a comet to within about 0·5, once the observer is used to the appearance of stars of different brightness in his instrument. If more accurate results are required—and they are probably not justified by the crudeness of the method—two or three different comparison stars can be used, their exact magnitudes being found from a catalogue. Alternatively, should a known variable star field lie nearby, its own comparison stars may be pressed into service.

By this technique, the total, or *integrated*, magnitude of the comet is found (i.e., the magnitude of the starlike point that would be produced if its coma were compressed), and this value can be of use to computers in correcting the predicted magnitude. At the same time, notes should be made of the diameter of the coma (in minutes of arc), and the appearance of the nucleus. Some comets have an almost stellar center, while others are either weakly condensed or have no noticeable nucleus at all; the latter is frequently the case when they are far from the sun. The tail, if any is visible, will be best seen with the lowest available power; and, because of the better light-concentration provided by instruments of very low power, it may be seen better in the finder or with a pair of binoculars.

Some comets have changed markedly in appearance from night to night. Fluctuations sometimes occur in their total light, while the tail may develop or decay. Neither must it be assumed that because a comet is conspicuous before perihelion, it will be equally noteworthy afterward. Alcock's second discovery, 1959 VI, was never recovered after it disappeared into the sun's rays; and the hopes raised by Comet Ikeya-Seki, discovered by Japanese observers on September 18, 1965, as it was moving in towards perihelion, were dashed when it faded out. This was in spite of confident predictions of a brilliant display.

A comet, if near the earth, shows perceptible motion relative to the stars in just a few minutes. Few amateurs have the equipment necessary for accurate measurement of cometary positions, but interesting observations can sometimes be made if the comet passes in front of a star. Donati's comet of 1858 passed in front of the bright star Arcturus in the constellation Boötes, but the star's light was undimmed, proving the tenuous nature of even a great comet. If the comet is seen to be approaching a star, a rough estimate of its motion will indicate when the transit is due to occur. If it has a well-defined nucleus, and it passes directly over or very near the star, an accurate timing may be of great value in calculating its orbit—provided, of course, that the star in question can be identified afterward and its position obtained.

Any peculiarities should also be noted. Comet Arend-Roland, for instance, developed a sunward-pointing spike, or "beard," and sometimes the tail is curved instead of straight and shows considerable structure.

Comet-hunting

Although an average sweeping time of 300 hours per comet has often been quoted, a value such as this means little in practice. Comet Candy, 1960, was found by an observer who was testing an eyepiece on the star κ Cephei; Meier's first comet was discovered in 1978 after only 50 hours of sweeping. Pons discovered comets at the rate of about one a year, while Brooks, with a more powerful instrument, was twice as successful. So much depends upon observing conditions, and, it must be admitted, luck. From England, Alcock swept for 1,418 hours before making his first discovery, although he used superior equipment to find his four comets in 735 hours of work. Roy Panther, who made his first discovery on Christmas evening, 1980, had been searching for just over 600 hours. Bradfield's first six comets were found in a total of 873 hours, although he had to sweep for 260 hours to find the first one. His technique is to sweep horizontally an arc of about 90°, starting near the horizon and working upward, studying the western sky after dusk and the eastern sky before dawn. Bennett, who found the bright comet of 1970, also sweeps horizontally, whereas Alcock, with his large mounted binoculars, prefers vertical sweeps. No two observers have the same technique, but all share the desire to cover the region of dark sky near the sun as comprehensively as possible. It is interesting to note, however, that the successful Czech team at Skalnate Pleso found 19 comets in 11 years, with an average time of only 100 hours for each discovery; and they were confined to the morning sky because of high mountains in the west. Their success, together with other evidence, suggests that new comets are more likely to be discovered in this region than in the western sky after sunset.

It is worth investigating this more fully; most books simply mention

Comet Seki-Lines. *Photographed at Gates Oasis in the Tucson mountains on April 10, 1962. Film used was Royal Pan. Focal length was 7 inches; exposure time was 40 seconds.* (*Donald J. Strittmatter, Tucson, Arizona.*)

that comet-sweeping should be carried out in the region of the sun (where they are most likely to be bright), and leave it at that. But if several years are to be spent sweeping for comets, it is sensible to study the economics of the situation. A few years ago, M. J. Hendrie, an amateur comet-hunter, raised the following interesting points in the B.A.A. *Journal*:*

Comet discoveries can be divided into two distinct classes: those made intentionally, by amateur or professional observers searching the region near the sun; and those achieved quite by accident, on photographic plates exposed for some other purpose. At the moment, three large observatories are leading the "accidental" discovery field: Palomar, Lick, and Lowell. All these take routine "patrol" plates, intended to spot any stars that may suddenly flare up, and it is not surprising that they have discovered the majority of new comets. Of the 61 comets discovered between 1948 and 1960, only 26 were found visually, most of these being sighted at Skalnaté Pleso.

* Vol. 72 (1962), 384–396.

If we then accept the fact that most comets brighten up into telescopic visibility somewhere between the orbits of Mars and the earth, it is obvious, from figure 50, that when we look toward the region of the sun, we are looking through a much greater volume of comet-holding space than when we look away from it. The interesting point is that the professional patrol cameras tend to scrutinize just this region away from the sun, as it is most favorably placed for stellar work, though least likely to hold comets. Far from being discouraged by the professionals' success, amateurs should take heart; the fact that relatively few are being sighted in the more favorable regions indicates that some are surely being missed, and that more observers are required to maintain a watch of the regions nearer the sun!

There is another, more suggestive, point. Considering the random appearance of comets, it might be expected that the number discovered to the east of the sun (i.e., in the evening sky) would about equal the number found to the west (in the morning sky). But this is not so, at least for the more conspicuous visitors. About three times as many discoveries of bright comets have been made in the morning sky, and this may be due to the more thorough evening searches, which reveal comets while still rather faint. The morning sky, for obvious reasons, is less well patrolled. Clearly, then, the best place and time to discover a bright comet is in the region of the eastern horizon, before dawn.

Comet-hunting is carried out in much the same way as sweeping for a known object, except that a much larger area of sky is covered. The instrument is moved slowly and regularly in azimuth, and stopped whenever a hazy gleam is spotted. It may be a comet, but it is far more likely to be a nebula. And this is where the tedious work begins.

When first seen, a comet usually appears to be a faint grayish stain on the sky. It is certainly unlikely to have developed a tail, the classical clue, and unfortunately the sky is full of dim objects that can easily be mistaken for the coma of a small comet. These are the *nebulae*, clouds of glowing gas that lie among the stars, and the *galaxies*, star systems similar to the Milky Way, which are so remote that they appear as mere blurs of light. Part III of this book deals with the observation of these objects; for the moment, we are concerned only with their nuisance value. For this purpose, they will be lumped together, very inaccurately, under the term "nebulae,"

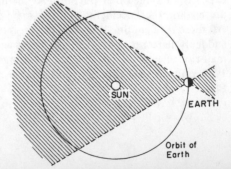

Figure 50. *The distribution of comets.*

since both types of object appear very similar when viewed through a small or moderate telescope.

So when a nebulous object comes into view, the first task is to note its position among the stars and to consult a star map. Like the stars, nebulae keep their positions from century to century, so it is a fairly straightforward matter to check up. *Norton's* marks the positions of more than a thousand of the brighter ones, and the chances are that the "comet" will be identified as a nebula that was catalogued many years ago, probably by Herschel himself.

If the object is not marked in the atlas, it does not necessarily indicate a comet discovery. There are several thousand nebulae in the sky that are within the range of an amateur's telescope, and *Norton's* lists only the brightest. Becvar's more comprehensive *Atlas Coeli*, which is now easily obtainable, goes considerably fainter than *Norton's* and aids identification. The final step is to consult the standard work on the subject, Dreyer's *New General Catalogue*, issued in 1888 and reprinted in 1953 by the Royal Astronomical Society, London. This lists over 7,000 nebulae and star clusters, giving brief notes on each and their exact positions in the sky. If the position of the suspicious object is noted, the N.G.C. will certainly provide the answer. If it is not listed as a nebula, it is a good idea to examine it with a more powerful eyepiece, since two or three faint stars lying close together can occasionally give the impression of something fuzzy and extended.

Two final checks must now be made. The first is to consult the year's ephemerides and to make sure that it is not the return of a known comet— although, if its recovery has not yet been announced, it must certainly be communicated right away. If this search is negative, the final vindication rests with the object itself. If it is a comet, it will be moving relative to the stars, and this motion should manifest itself in an hour or two at the most.

As soon as its cometary nature has been established, a telegram should be sent to I.A.U. Telegram Bureau, Cambridge, Massachusetts (if observing from the United States); or, in any other country, to the national observatory. All that is required is its position, measured as accurately as possible; the date, hour, and minute at which this position was taken; its rate and direction of motion; and its apparent size and brightness. It is also a good idea to mention the instrument used. This should give the professional observers all the information they need to turn their powerful telescopes onto the region and take over the task of following this new visitor to the sun.

This sequence of events is in the mind of every comet observer, but of course *Norton's* and the N.G.C. between them have a sad habit of dashing hopes. Nebula after nebula will be picked up, and most of the observer's time will be spent in consulting reference lists; but the situation improves as the same area is swept over night after night. Soon, out of initial disorder and frustration, comes a wonderfully intimate knowledge of the region. Every comet-hunter testifies to the compensations afforded by this type of work; these lonely vigils with the stars have an edifying effect on the mind, and the discovery of the wonders of the night sky will prove in the end to be as rewarding as any comets that may be picked up.

It is far better to work a small region thoroughly than a large area vaguely. At the same time, the work should be done as rapidly as is consistent with thoroughness, for the fertile region of dark sky near the sun is visible for only a short time; all too quickly, during morning work, the east begins to lighten until it is ruinously bright. It is, however, worth glancing at the twilight region to make sure that no really prominent visitor has appeared out of the sun's rays.

The following account was written by Roy Panther soon after the discovery of his first comet, 1980u, from his home near Northampton, England:

The evening of Christmas Day 1980 was fine and clear. I remembered reading an account by J. H. Elgie, a North-country Edwardian amateur astronomer, that he could see the Great Bear at lower culmination on this day at 4:50 P.M. I ventured out at this time and did just that, as he did 73 years ago.

After dark the sky was very clear and I could not resist this rare English event for telescopic work. Firstly I was to make an observation of Comet Meier with the 20-cm reflector (f/4), but the altitude was getting low and I could not see it. I thought of using the 31-cm (f/5) reflector to locate it, but it was too low for this, so that saved me the trouble. With the 20-cm f/4 reflector I considered obtaining a set of observations of Comet Stephan-Oterma, now in southern Auriga, but this could wait as there was plenty of time.

Using this telescope with × 35 I settled down to a comet sweep at 17.55 U.T., starting at 3 Vulpeculae to ι Draconis, sweeping northwards. I was troubled by a distant mercury vapour street lamp shining through the boughs of several leafless trees. A sheet attached to two poles pushed into the ground put a stop to that. But this was not the end of it. A strong wind was tearing at the sheet and keeling over the poles. So once again I had to leave the eyepiece and push the poles into the ground further. Peace returned at last and sweeping continued. The star fields towards the Galaxy were gorgeous. M.56 [in Lyra] went by in the field bright and clear. Sometime later a dim "M.56" was located in a field of faint stars. This immediately aroused my suspicion as a stranger, as I had a rough idea where it was by the length of the sweep.

On looking through the finder and seeing ε Lyrae near the centre of the field, I knew I had a suspected comet. This was at 18.50 U.T.

I fixed the position and checked all my charts for nebulae and clusters at this spot. Finding none of these objects charted I looked up all the ephemerides of known comets around, and then the N.G.C.

I could not detect any motion after 15 minutes, but was not unduly worried. After [I gave] the information to Michael Hendrie, Director of the British Astronomical Association's Comet Section, he immediately took a photograph of the region when the altitude was becoming very low and almost hopeless and confirmed my discovery . . .

I began my first comet sweep on the evening of July 22, 1947, using a 3-inch refractor, × 22. Since then, and using a variety of instruments, I have spent a total of 601½ hours searching on 699 nights. Three more new comets have been missed by only a few degrees of sky.

Weather conditions and moonlight

To have much hope of success, a comet-hunter must live in an area favored with truly dark skies; this means rural or at least suburban conditions. This, combined with clear east or west horizons (preferably both), is a *sine qua non*.

Weather conditions can also affect matters, but here there is scope for initiative, and it is a matter of taking chances when they come. Few sites can rival Lick Observatory, which claims to enjoy 300 clear nights in the year, but it is amazing what can be done if a close watch is kept on the sky. Frequently, large breaks occur in the densest cloud; and if it has recently been raining heavily the atmosphere is likely to be of exceptional clarity, and hence extremely favorable for the detection of faint comets. Looking through the records, it is found that July and August have claimed more comet discoveries than any other months, with a noticeable preponderance of discoveries in the second half of the year. Denning, in his book *Telescopic Work*, analyzes the 289 cometary discoveries made between 1782 and 1890, noting that 38 comets were found in August, and that 123 were found in the first six months as against 166 in the remainder of the year. Hendrie's analysis of the discoveries between 1948 and 1960 shows the same tendency, with only 25 of the 61 being discovered in the first six months; there can be little doubt that weather conditions are responsible. Denning himself, who observed meteors as well as planets, managed to average about 36 hours of clear sky every month; and he not only observed from "cloud-laden" England, but also confined his meteor watches to periods when the moon was out of the sky. Sky conditions vary enormously in different regions of the United States, the West Coast being generally more favorable than the East; but probably no place is less favorable than the British Isles, so that an enthusiastic observer should be able to log 200 or 300 hours in a year—especially if he has the necessary tendency toward insomnia.

Naturally enough, far fewer comets have been discovered around the time of full moon than when it is a crescent. Halley's Comet, at its 1910 return, reached its greatest brightness during the moonlit evenings, and so was seen at a disadvantage. But opportunities present themselves even at these unfavorable times. The moon, just before full, sets a couple of hours before dawn and so leaves a precious interval of dark sky; and after full the evening sky is available for some time before it rises. These moments should be profited by, since the majority of comet-hunters leave off work at this time, and the sky is less well patrolled, giving the persistent watcher an extra chance of success.

"False" comets

In 1781, Charles Messier issued a catalogue of 103 star clusters and nebulae that has remained a useful standard until the present day—even though thousands of fainter objects have since been recorded. The story goes that he was so frustrated by continually alighting on these objects and at first taking them for comets that he drew up the list as a warning of "objects to avoid." This is probably untrue; it is far more likely that he was interested in them for their own sake, but his catalogue is of value in listing some objects (prefixed by the initial M.) that do appear distinctly cometlike. For example, M.13 and M.92 in Hercules are globular star clusters (see page 260) that appear very much like the nucleus and coma of a bright telescopic comet when they are viewed with a low power. Among other bright false comets are M.2 (Aquarius), M.3 (Canes Venatici), M.15 (Pegasus), and M.49 (Virgo). The observer should begin by looking at these objects, and at the others in Messier's list, with his comet-sweeper.

Of more interest still are the dimmer and more treacherous objects— more treacherous because there are far more of them. Swift described a nebula in Draco as looking just like a small comet; this is N.G.C. 6654 (the letters refer to Dreyer's *New General Catalogue*). Its position is R.A. 18^h 26^m, Dec. $+73°$ $6'$, so that it lies near the north celestial pole. It is actually very near the bright star Chi Draconis, and if its position is marked on the chart it can be sighted. Nearby is a brighter nebula, N.G.C. 6643, which should be found without difficulty with a 4-inch refractor. This one is marked in *Norton's*. If these nebulae prove difficult or impossible to find, then the chances of finding a faint comet are low, although a bright one may still be obvious. Alcock has mentioned a nebula in Leo Minor, N.G.C. 3344, as looking very much like a comet. It lies between stars 40 and 41, and he comments: "What makes it look so much like a comet in a small, low-powered instrument is the fact that it has a star superimposed near the p rim, so that it appears to have a small sharp nucleus with a fainter following tail." The constellations Draco, Leo, Ursa Major, and Virgo are so full of nebulae that it requires great determination to sweep them and considerable courage to announce a discovery!

Some bright periodic comets

The regularly returning comets have periods ranging from $3\frac{1}{3}$ years (Encke's) to 76 years (Halley's), and some of the brighter members are mentioned below.

P/ENCKE.   The second periodic comet to be "discovered" by analysis, this time by Johann Encke, later director of the Berlin Observatory. Before that, he had been a gifted mathematician, and it was this aptitude that induced him to compare the comets observed in 1786, 1795, and 1818, and to realize that they were the same body. The period of Encke's comet is much shorter than that of any other, and since its predicted appearance in 1822 it has been observed at every return. It was last seen in the autumn of 1967, when it approached the sun to within 30 million miles, and it is due back in 1973–1974. Near perihelion, it is usually visible with binoculars.

P/TUTTLE-GIACOBINI-KRÉSÁK.   The cumbersome name of this comet is due to its independent discovery in 1858 by three observers. It has a period of about $5\frac{1}{2}$ years, but due to computing errors it was not picked up again until 1907, after which it evaded discovery again until 1951. It was once more missed in 1956, but came within the range of moderate telescopes at its 1962 return. This comet is subject to large fluctuations in brightness, and on two occasions during its 1973 return it flared, in a matter of a few days, by an unprecedented factor of up to 10,000 times, reaching naked-eye visibility. The reason for such behavior is not understood. At future apparitions its position can be profitably examined with small instruments, even though its predicted magnitude is below their threshold.

P/GIACOBINI-ZINNER.   This comet has a period of $6\frac{1}{2}$ years. It was bright enough to be well seen at its 1959 return, and it appeared again in 1966. The material of this comet was associated with prominent showers of meteors (the Giacobinids or Draconids) seen in 1933 and 1946; the relationship between comets and meteors is mentioned in the following chapter.

P/FINLAY.   This is another short-period comet which is sometimes seen with small instruments. It last returned in 1960, and has a period of nearly 7 years.

P/SCHWASSMANN-WACHMANN I.   This curious body moves in an orbit of very slight eccentricity, its distance from the sun ranging from that of Jupiter to that of Saturn! Although normally extremely faint, it has occasionally exhibited bursts of activity, becoming as bright as the 10th magnitude.

P/PONS-BROOKS.   With a period of more than 70 years, this comet has been observed at only three returns. At the last, in 1954, it was a naked-eye object.

P/OLBERS.   At its 1956 return, this comet just reached the 6th magnitude, but was never striking. Period, 73 years.

P/HALLEY.   The brightest of the short-period comets, but apparently not as striking now as it was two thousand years ago, due to the steady destruction of its material by successive returns to the sun.

The repeated loss of material that occurs at each perihelion passage explains why the short-period comets are always relatively small and faint: Their volatile constituents are becoming exhausted. Only the unexpected comets, which have never been recorded before, have a chance of being really bright. Even then, it is a matter of luck whether the earth is in such a position in its orbit that the comet can be seen well against a dark sky. Furthermore, with the growth of town lighting, no comet, however brilliant, may ever again amaze whole populations as did the great comets of 1811, 1843, and 1882 that shone in relatively unpolluted skies. Probably the most widely acclaimed comet of the twentieth century was the Daylight Comet of 1910, which far outshone Halley's of the same year. Comet Ikeya-Seki of 1966 was a brilliant morning object, but faded unexpectedly quickly after perihelion. Comet Bennett was a superb sight from temperate latitudes in the dawn sky at Easter, 1970, looking like a white scimitar against the stars.

However, Comet West of 1976 was probably the brightest and most active comet of recent decades. The following description by the well-known comet observer John E. Bortle of Stormville, New York, describes the impression he had of it as it passed perihelion:

Somewhat more than an hour before sunset of February 25, I readied my equipment and the comet was found immediately in the 32-cm reflector working at full aperture and with the Sun occulted by my house. . . . After perhaps 20 minutes' observation with the telescope, the 15 × 80 binoculars were set up. In these the comet was astonishingly bright and the tail was quite apparent even though the Sun was still nearly half an hour from setting. Unquestionably, the finest view was through a pair of 10 × 50 binoculars 15 minutes before sunset. With these, the condensation looked like Venus and a tail 15' long trailed behind it, brighter at the edge than in the centre. At this time the Sun was still several degrees above the horizon and in no way blocked from the observer's sight. At the same time the comet was also visible to the naked eye!

This observation was made within a few hours of perihelion. When it reappeared in the morning sky in early March and could be observed against a dark sky, a tail length of 30° or more was reported, and the nucleus was seen to disintegrate into separate fragments. Comet West has now vanished back into the chill of space, and may not return to the sun's vicinity for a million years. But another visitor to match or even outshine it must even now be approaching the region of the inner planets; and a patient amateur may well give the first warning of its coming.

17

Meteors and Meteor Showers

On some clear and moonless nights, particularly in the fall, a casual sky-watcher will notice spasmodic streaks of light flying among the stars. Initially, they appear to be moving quite at random; but if a watch is maintained for an hour or two, it will become apparent that many of these streaks are radiating from certain regions of the sky. Evidently meteors, or "shooting stars," are not the haphazard bodies they seem to be.

Meteors have been known from the earliest times, but not until the early part of the nineteenth century did astronomers take them seriously enough to investigate their nature. They had previously been regarded as purely atmospheric phenomena, with no astronomical connection. But the night of November 12/13, 1833, precipitated a radical transformation, when, as the astronomical historian Agnes Clerke related in her *History of Astronomy in the Nineteenth Century*, ". . . a tempest of falling stars broke over the earth."

> *North America bore the brunt of the display* [*she wrote*]. *From the Gulf of Mexico to Halifax, until daylight with some difficulty put an end to the display, the sky was scored in every direction with shining tracks and illuminated with majestic fireballs. At Boston the frequency of meteors was estimated to be about half that of flakes of snow in an average snowstorm. Their numbers, while the first fury of their coming lasted, were quite beyond counting; but as it waned, a reckoning was attempted, from which it was computed, on the basis of that much diminished rate, that 240,000 must have been visible during the nine hours they continued to fall.*

Numbers were what impressed the layman, and rightly too; but there was more to the matter than that. Perceptive observers noticed that the meteors did not appear to move in random directions, but seemed to radiate from a definite region—a point in the constellation Leo. Moreover, as the night wore on and the earth's turning took Leo high in the south and began to drop him on his nose in the west, the meteors still streamed from the same area. This proved conclusively that they were independent of the earth; in fact, they were interplanetary bodies. Thus, overnight, meteors were shifted from the meteorological to the astronomical sphere.

It was recalled that meteors had been seen in great numbers in November, 1799; and a further shower of these objects, now known as *Leonids* because of their place of origin in the sky, occurred in November, 1866. Europe had a better view this time, and, although the meteors' performance was inferior to the display of thirty-three years earlier, it was by far the densest falling-star shower in European memory. And by this time, astronomers had arrived at some conclusions. Observers such as Heis in Germany and the lunar astronomer Schmidt at Athens had discovered that many of the ordinary meteors which flash from night to night have similar regions of emanation, known as *radiants*. The explanation is that the meteors are really traveling in parallel paths; when they enter the earth's atmosphere and are consumed in a streak of fire (as in fig. 51),

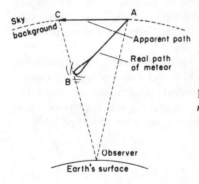

Figure 51. *Real and apparent paths of a meteor.*

the observer sees their apparent motion as *AC,* whereas the real path is *AB.* This is because everything in the night sky, from nearby meteors to the remotest stars, appears to be at infinity, so that the observer is not aware that most meteors are much closer to him at the end of their path than at its beginning. A meteoric *shower* occurs when the earth passes through a *swarm* of particles, although "swarm" is perhaps a misnomer when we remember that actually the individual meteors may be hundreds of miles apart. Similarly, the term "shower" evokes visions of streaks in great abundance, whereas only on rare occasions are meteors to be seen at rates higher than about one a minute.

It now remained to find out where they went and from whence they came; and Schiaparelli, in the days before he turned his attention to Mercury and Mars, came up with results that at first amazed the scientific world, but were soon confirmed. First, he showed that meteors are true members of the solar system, revolving around the sun in closed orbits. Second, he pointed out that in some cases these orbits more or less coincide with those of known comets.

The meteor/comet relation noted by Schiaparelli was clinched by the 1866 Leonid display, which was well observed by astronomers. Schiaparelli now found that they moved in the same orbit as Tempel's comet, which had been seen early that same year and which had a period of 33 years, coinciding with the interval between the great showers. Any remaining doubts were dispelled six years later by Biela's comet, a short-period object which had been seen to divide into two parts at its 1845 return, and which reappeared in 1852 as two separate comets! Severe disintegration had obviously set in; this was confirmed when it escaped detection at the next two returns. The comet seemed to have gone to the grave; and then, in 1872, when it should have reappeared, there occurred a great meteor shower. This marked the remains of Biela's comet; its last solid fragments were being spread along its orbit, and it would never be seen again. Nowadays, the shower itself has become very feeble, only a few, slow meteors radiating from the constellation Andromeda every year during the month of November.

Yet it would be wrong to think that all meteor showers have established cometary relations. More than a thousand detectable showers are known, and of these only a few comets are officially recognized as "parents." Although there may be many associations, the comets themselves have either long since disintegrated, or have simply not been observed.

Some well-known showers

Under good conditions, between six and a dozen meteors will be seen every hour on most nights of the year. Some of these are *sporadic*, which means that they are unrelated to any particular swarm and move around the sun in independent orbits. Others belong to some of the many faint showers that may number only a single meteor in several hours of watching. These form the great majority of established radiants, and extensive watches are needed to form any firm deductions about these streams. But at certain times of the year Earth passes through much denser swarms, producing obvious displays. The principal ones are listed by name in Table VI.

TABLE VI. Most Prominent Night-time Meteor Showers[a]

NAME OF SHOWER	PERIOD OF DETECTABLE METEORS	DATE OF PEAK ACTIVITY	MAXIMUM VISUAL HOURLY RATES[b]	RADIANT COORDINATES (DEG.)[c] RIGHT ASCENSION	DECLI-NATION
Quadrantids[d]	Jan. 1–4	Jan. 3	50	231	+50
Corona Australids	Mar. 14–18	Mar. 16	(5)[e]	245	−48
Virginids	Mar. 5–Apr. 2	Mar. 20	(less than 5)	190	0
Lyrids	Apr. 19–24	Apr. 21	8	272	+32
Eta Aquarids[d]	Apr. 21–May 12	May 4	12	336	0
Ophiuchids	June 17–26	June 20	(20)	260	−20
Capricornids	July 10–Aug. 5	July 25	(20)	315	−15
Southern Delta Aquarids[d]	July 21–Aug. 15	July 30	30	339	−17
Northern Delta Aquarids[d]	July 15–Aug. 18	July 29	15	339	0
Pisces Australids	July 15–Aug. 20	July 30	(20)	340	−30
Alpha Capricornids	July 15–Aug. 20	Aug. 1	5	309	−10
Southern Iota Aquarids	July 15–Aug. 25	Aug. 5	(10)	338	−15
Northern Iota Aquarids	July 15–Aug. 25	Aug. 5	(10)	331	− 6
Perseids[d]	July 25–Aug. 17	Aug. 12	50	46	+58
Kappa Cygnids	Aug. 18–22	Aug. 20	(5)	290	+55
Orionids	Oct. 18–26	Oct. 21	20	95	+15
Southern Taurids	Sept. 15–Dec. 15	Nov. 19	(5)	52	+14
Northern Taurids	Oct. 15–Dec. 1	Nov. 19	(less than 5)	54	+21
Leonids	Nov. 14–20	Nov. 17	(5)	152	+22
Geminids[d]	Dec. 7–15	Dec. 13	50	113	+32
Ursids	Dec. 17–24	Dec. 22	15	217	+80

[a] Adapted from D. W. R. McKinley, *Meteor Science and Engineering* (New York, McGraw-Hill Book Company, Inc., 1961) and Joseph H. Jackson, *Pictorial Guide to the Planets* (New York, Harper & Row, 3rd edition, 1981). [b] Number of meteors observable visually by a single observer at maximum shower activity. [c] +, North of the celestial equator, south of it, —. [d] Among the stronger and more consistent meteor showers. [e] Figures in parentheses less reliable than other figures for visual hourly rates and period of detectable meteors.

It must be remembered that the dates given in Table VI are only approximate. Some meteors may be visible outside the limits, and the time of maximum may differ by a day or two from year to year, due to leap-year adjustments. The R.A. of meteor radiants is commonly given in degrees instead of hours and minutes, but the conversion is not difficult when one remembers that 1^h of R.A. is equivalent to 15°. The radiant point of the Taurids cannot be given with any accuracy, for this is a long-lived, diffuse shower with a number of radiants scattered around Taurus and Aries.

A meteor is a small body that may be anything from the size of a grain of sand to that of a pebble; anything smaller will give too fugitive a streak to be noticeable; a larger object will light up the whole sky and be termed a *fireball*, or *bolide*. Since all the various swarms seem to consist of particles of much the same size, it might be imagined that all showers will produce similar meteors. This is very far from the case, however, since it all depends on the angle at which they strike the earth.

Meteors, like comets, are traveling around the sun in large and eccentric orbits. This means that at the earth's distance from the sun, which is when contact occurs, they are traveling faster than the earth. This is not difficult to understand. If the earth suddenly slowed down from its speed of $18\frac{1}{2}$ miles per second, it would sweep in toward the sun; if it accelerated to about 25 miles per second, it would fly off at a tangent and move into a more or less cometic orbit. This is, in fact, the velocity at which meteors are traveling when they reach the earth's vicinity, although this speed varies slightly from shower to shower.

Because meteor swarms do not keep to the counterclockwise, uniform-plane restriction that binds the major planets, it is evident that they can meet the earth at various angles. The extreme cases are shown in figure 43, where swarm A is moving in a direction more or less opposite to the earth's orbital motion, while swarm B is traveling in the same general direction as the earth. It follows that a meteor in swarm A will pass through the earth's atmosphere at a velocity consisting of the sum of the orbital speeds— $25 + 18\frac{1}{2}$ miles per second. A B meteor, having to overtake the earth at its rear, will achieve a relative velocity of only $25 - 18\frac{1}{2}$ miles per second. It will therefore appear in the sky as a much slower object, perhaps leaving a faint luminous trail, or *train*, whereas an A meteor will be seen as a swiftly moving streak. Of the showers listed in Table VI, the Lyrids, η Aquarids,

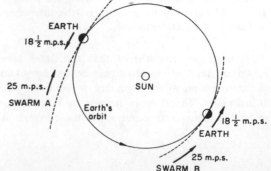

Figure 52. *Meteor velocities.*

Perseids, Orionids, and Leonids are swift; the Quadrantids, δ Aquarids, Geminids, and Ursids are of medium speed, because they are hitting the earth more or less "beam on"; the Taurids are slow, lasting perhaps a second or two.

QUADRANTIDS. This is a most awkward shower to observe. It is short, the radiant is rather near the sun, and it has a very brief maximum that occurs between 6^h U.T. on January 3 and the same time on January 4, the exact hour depending on the year. Sometimes, when this period occurs in daylight, only a few meteors can be seen during the hours of darkness. The most favorable conditions for observation occur every few years, when maximum activity takes place a couple of hours before dawn, and the radiant is reasonably high in the sky. The shower earns its name from the now-forgotten constellation of Quadrans Muralis, which lay in the Ursa Major–Boötes region.

This shower has been largely neglected by amateurs, partly, no doubt, because of the difficulties of observation. Nevertheless, even though conditions at the 1965 display were rather unfavorable, one amateur recorded 49 meteors in 33 minutes, while another wrote: "A bright Quadrantid is a characteristic electric blue color, with an expanding, mottled, silvery train." This elusive shower seems worthy of wider attention.

LYRIDS. One of the few spring showers appearing April 19–24. The radiant can be well observed in the morning sky, reaching its highest altitude, or *culmination*, at around dawn, and the shower is well seen from northern temperate latitudes. At maximum activity, 6 or 8 swift meteors should be seen within an hour.

η AQUARIDS. To be seen at their peak on May 4. These are two distinct showers with radiant points in Aquarius, so they are distinguished by citing the bright star nearest to the radiant. This is only 70° away from the sun and it rises not long before dawn; therefore, observation is difficult. It also lies on the celestial equator, so not many meteors are likely to be seen by observers in high northern latitudes. They are interesting, nevertheless, from the suggested association with Halley's comet.

δ AQUARIDS. These meteors are of medium speed, and so are easily distinguished from the early Perseids that appear at about the same time in mid-July; because of their lower velocity, they appear yellowish rather than blue-white. The radiant lies south of the celestial equator, and so is poorly seen from latitudes higher than about 40°N; but observers in other regions may expect to see from 30 to 60 meteors an hour on the night

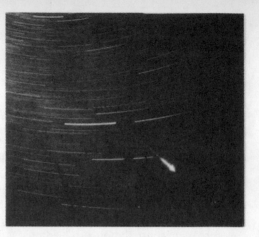

Perseid meteor. *T. Lloyd Evans used a stationary camera with an f/4 Ross Xpres lens of 5-inch focus for this photograph. Since the exposure time was 32 minutes, the stars trailed into arcs, and this brilliant exploding meteor was caught near one corner of the plate. The dark object is the outline of a weather vane.*

of maximum, while they show considerable strength on succeeding and preceding nights. This is the only major shower whose radiant lies well south of the equator; the meteors are of interest in passing very close to the sun at perihelion and in not being associated with any known comet. It is curious that the southern hemisphere should be so devoid of strong showers. A great deal of work has been carried out by both amateur and professional observers, and the deficiency is undoubtedly real.

PERSEIDS. Probably the best known and most reliable of all the annual showers. The meteors usually reach their maximum on the night of August 11/12; up to 60 swift meteors an hour may be seen. Early Perseids are seen in mid-July, and persist for about a month, with the radiant shifting northeastward from night to night, beginning in Andromeda and finishing up in Camelopardus. This is an effect of parallax, caused by the earth's motion through the swarm. These "August meteors" have been recorded for at least a thousand years.

ORIONIDS. A prominent display that reaches an intensity of 10 or 20 meteors per hour on October 21 depending on the altitude of the radiant; low-latitude observers have the best view, since Orion is on the celestial equator.

TAURIDS. This is a complex, long-enduring system of slow meteors. The two most prominent radiants lie in R.A. 54°, Dec. +21°; and R.A. 52°, Dec. +14°; but they have about the same duration, and it is not easy to distinguish between them. However, the first group (the Northern Taurids) is somewhat faster than the other. Rates of about 10 meteors per hour occur at maximum, which comes in early November.

Two-point Leonid meteors. *This 3½-minute exposure was taken from Kitt Peak at about midnight U.T. November 17, 1966. A total of 61 Leonids radiate from the Sickle. A Voigtlander 105-mm f/3·5 lens was used with Tri-X 120 roll film, developed 12 minutes in D19 at 68°F. (Dennis Milon, Tucson, Arizona.)*

LEONIDS.   These interesting meteors are the remnants of the famous showers seen in the past. The meteors, which reach their peak on November 17, are swift, with an unpredictable hourly rate; in recent times it has been low, but in 1961 there was a definite revival, increasing over the succeeding years and culminating in the magnificent 1966 display, during which observers in the United States recorded thousands of meteors per hour. The extraordinary concentration of these meteors is indicated when we find that observers in Europe, who had dark skies a mere nine hours before the Leonids' greatest intensity, recorded rates of only two or three a minute. It will be interesting to see what happens to the Leonids in future years.

GEMINIDS.   A fine December shower, short-lived but prominent. The radiant reaches a considerable altitude for northern temperate observers; under good conditions the rate should match that of the Perseids.

URSIDS. A rather faint, short-lived shower of medium-speed meteors, to be seen the week before Christmas. The radiant lies near the polestar.

Meteor observation

The amateur's reign in detection and analysis of meteor showers lasted from about 1833 until 1945. But since World War II, as a result of the unprecedented development of electronic techniques, radar has to a very great extent supplanted the visual observer. When a meteor hurtles down through the atmosphere, the intense heat of its passage not only causes it to glow, but also chemically disrupts, or *ionizes*, some of the molecules in the surrounding air. The meteor itself is far too tiny to be picked up on a radar set, but this transient streak of ionization can be recorded, and its velocity and direction can be measured very accurately. In addition, clouds, moonlight, even daylight, have no disturbing effect; radar observations can therefore be carried out under all conditions.

On the other hand, since radar observation lies in the professional field it must be involved in definite schedules. This means that the watchful amateur may pick up some unexpected event, such as the intense Leonid display of 1961. It must also be remembered that what is a bright meteor to the visual observer may not be "bright" to a radar set. Since radar only detects the ionized trail, it reacts most strongly to swift meteors; this is because their greater velocity produces a higher temperature, and hence a greater degree of ionization. A bright but slow-moving meteor, on the other hand, being much cooler, may leave little or no detectable trail.

This discrepancy was well brought out on the night of December 5, 1956, when an unexpected meteor shower struck the earth. Amateur observers in New Zealand, Australia, and South Africa were amazed to find a great number of meteors radiating from the constellation Phoenix, a southern group; an hourly rate of up to 100 meteors was recorded. It so happened that the meteors were also picked up by radar workers, but the hourly rate was considerably less, a fact explained by their slow speed. Eyewitnesses recorded them as being exceptionally bright, rivaling Venus or, in a few cases, even the moon, and leaving conspicuous trains. Moreover, since amateurs were distributed throughout the area over which the shower was visible, it could be kept under observation for a long period as the earth spun on its axis, bringing fresh regions of the globe to face the shower. In this case, the amateur results were of definite value.

In the case of the well-known showers, the amateur's most useful duty is simply to count the number of meteors visible from hour to hour, noting any interesting features, such as brilliance, length of path, color, train, and other effects that may seem unusual. A note must also be kept of the number of sporadic meteors seen, as well as those radiating from the active area.

To determine the activity of a shower in definitive terms, it is not enough to record the *observed hourly rate* (OHR) and leave it at that, since the number of meteors visible to an observer, if we neglect meteorological and other factors, depends on the altitude of the radiant. If the radiant is low, some meteors will occur below the horizon while others will be lost in the haze. To produce uniformity, the results must be corrected to *zenithal hourly rate* (ZHR), which is the number of meteors that would be seen were the radiant directly overhead. Table VII gives the correction factors for different altitudes. If a watch of several hours' duration is maintained during the night (shorter spells are of little use), the number of meteors seen every hour or half-hour can be corrected to ZHR.

TABLE VII.   Altitude correction factors

ALTITUDE	FACTOR	ALTITUDE	FACTOR
0°		27·4°	
	0·1		0·6
2·6		34·5	
	0·2		0·7
8·6		42·5	
	0·3		0·8
14·5		52·2	
	0·4		0·9
20·7		65·8	
	0·5		1·0
27·4		90·0	

If the radiant altitude is equal to one of the tabulated values, the upper factor should be taken. For example, the factor for 27·4° is 0·5 and not 0·6.

When recording meteors in this way, one should note their magnitudes. Some showers consist predominantly of bright meteors, while others are more evenly graded. The time of maximum activity is not necessarily the same for all magnitudes, and the Geminids, for example, have a slightly earlier maximum for faint meteors than for bright ones, due to a slow, selective drift of the different-sized bodies into separate orbits. It is the normal practice in meteor work to note the approximate magnitude of the faintest stars visible near the zenith.

Unexpected showers such as the Phoenicids, which lasted for only a few hours and have not been seen since 1956, are very rare indeed; but if unusual meteoric activity is noticed, attempts should certainly be made to find the radiant. This is done by plotting as many trails as possible on a star chart, afterward extending them back until they meet in what should be a small area of the sky—the radiant.

The moment a meteor is seen, and while its path across the stars is still fresh in the memory, a piece of string, or a straight rod known as a "wand," is held up at arm's length so as to coincide with the line of the track. Notes are then made of the positions (relative to the stars) of three points along this track, so that it can later be reproduced in the atlas; and the position of the beginning and end of the meteor's trail is also recorded. Other matters, such as its duration and color, are of relatively minor importance. The success of this method depends greatly on the experience of the observer, but in occasional emergencies even poor results are better than none at all. The same method should be adopted if a brilliant fireball is seen, since other observers may have made a similar observation, and the parallactic shift of the meteor against the stars as seen from different stations enable its true path through the atmosphere to be calculated, although the results are bound to be poor compared with reductions made from photographs (pages 240, 241).

Among the new showers of recent years that have been recorded on a regular basis are the June Lyrids, first reported in 1966 and with a ZHR of about 8 (radiant at $18^h 32^m$, $+35°$, maximum on June 15), and the Cepheids, with a similar ZHR (radiant at $23^h 30^m$, $+63°$, maximum on November 9), first noticed in the early 1970s. The fact that even regular showers can surprise was proved by the unexpectedly brilliant Perseid display in 1980, when the sky seemed to be alive with bright meteors and fireballs on the night of maximum, and a ZHR of 120 was recorded by some observers. In 1973 and 1975, observers in Hungary and the United Kingdom reported a strong shower around March 23, radiating from near η Geminorum.

Just occasionally, curious meteors are seen. If one happens to be traveling directly toward the observer it will show no lateral movement at all, simply shining out and fading away. Sometimes, too, a meteor describes a path that is twisted rather than straight, probably due to some irregularity on its surface.

Moonlight is the bane of the visual meteor watcher. A full moon will drown all but the brightest objects, and conditions vary from year to year, some showers being well seen while others are very unfavorable. The Quadrantids, δ Aquarids, and Orionids always occur within a few days of the same lunar phase, as do the April Lyrids and the Geminids. It is fortunate that the two strongest showers of the year, the Quadrantids and the Perseids, always occur one half-lunation apart, so that the more one is hindered by moonlight, the darker will the sky be for the other.

The well-known scientist J. B. S. Haldane once remarked that "the observer of meteors requires a clear sky, a thick coat, a notebook, a knowledge of the constellations, infinite patience, and a tendency to insomnia." This list is fundamental but not inclusive. To it can be added: a deck chair, allowing the observer to recline at a comfortable angle; a lightweight lapboard to hold the observation log; and a dim red light worked by a convenient switch so that notes can be made immediately a meteor is seen.

Telescopic meteors

Any observer who habitually uses a low magnification and a large field of view, as in comet-hunting and variable-star observation, will occasionally have his attention distracted by a *telescopic meteor*. Sometimes a fast-moving, naked-eye meteor flashes across the field, but the truly telescopic variety are faint, and many seem to move relatively slowly, so that their paths can be followed without difficulty. There is as yet no satisfactory explanation of this anomalous velocity. One might be inclined to attribute it to their great height, if calculations did not show that their altitudes would be above the farthest reaches of the atmosphere, where there is no resisting medium to cause a glow. Another explanation is that they always move in a path directed more or less toward the observer, so that they do not seem to cover a large arc of sky. Whatever the explanation, the effect is undoubtedly real. Some telescopic meteors move so slowly that they can be seen distinctly as starlike points of light, fading out after paths covering a degree or less of the sky.

Since the introduction of radar into the naked-eye meteor field, some amateurs have turned to a study of these objects. As yet, however, the field is hardly touched, and most telescopic meteor reports come from comethunters. For example, Denning noted 95 telescopic meteors between May and November, 1890, while sweeping for comets with his 10-inch reflector. Altogether, he observed 635 during 727 hours' sweeping. Much more recently, Alcock recorded 201 between April 1 and October 11, 1964, using 25 × 105 and 11 × 80 binoculars, which brought his total since the beginning of 1953 to 1,698. Moreover, amateur observers have lately enjoyed a boost in the supply of faint moving objects—the various items of hardware moving in orbits well above the earth's atmosphere. These now exist in such profusion that it is impossible to keep track of them all. Their low velocity and long paths distinguish them from meteors, but they sometimes appear in impressive numbers; during a week's work, an amateur recently counted 48 telescopic meteors and 27 satellites! The great multitude of satellites and rocket fragments now orbiting the earth could

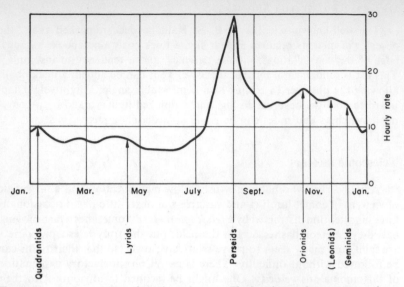

Figure 53. *Meteor activity throughout the year. The hourly rate here is averaged over a fortnight, so that only showers of considerable duration have an effect on the curve.*

perhaps be put to good use by the observer of telescopic meteors, since they offer a useful standard against which to reckon meteor rates. If transparency is poor, the visibility both of satellites and of meteors will be approximately equally affected, so that the relative frequency of one against the other will not change unless there is a real increase or decline of meteor activity.

Very little has as yet been published on the subject of telescopic meteors. The following notes are based on researches carried out by a British amateur, K. B. Hindley, who specializes in the observation of these objects.

FREQUENCY. Naked-eye meteors show an annual activity curve like that in fig. 53. There are two minima, one in February, the other at the beginning of June, while the extended Perseid shower provides a marked peak at the beginning of August. In the autumn the general rate is considerably higher than during the rest of the year; added to this, we find a considerable diurnal variation. During the early morning, the observer will see two or three times as many sporadic meteors as in the evening, because, as the night wears on, the earth's rotation has carried him into the region facing the planet's direction of motion. Consequently, the morning meteors are faster (and therefore brighter) than those seen in the evening, and there are more of them, since, as explained earlier, the evening meteors have a much harder job to do in overtaking the earth at its rear.

Telescopic meteor activity shows the same general change. After the

June minimum, activity increases until the second half of July, when the rates have a broad maximum until the first half of November. A smaller increase then occurs, coinciding with the appearance of the Geminid meteor stream; subsequently, the curve falls to the February minimum. Using a 5-inch short-focus refractor, with a magnification of × 18 and a field of view slightly exceeding 3°, rates of between 5 and 9 meteors per hour are the average on good nights.

NAKED-EYE SHOWERS. It might be supposed that telescopic meteor activity shows an increase on the nights of the major naked-eye showers, but this is not usually so. In the case of the Perseid and Leonid showers, for instance, the majority of meteors are bright; the same is true of the δ Aquarids and the Taurids. All these showers are deficient in faint meteors. The Geminids and Quadrantids, on the other hand, show considerable telescopic activity right down to the 10th and 11th magnitudes. Indeed, the Geminids are active telescopically before they become noticeable with the naked eye.

TELESCOPIC SHOWERS. Of even more interest, perhaps, are the exclusively telescopic showers. Both the end of August and the end of September are active telescopically, with a number of faint streams, many of which have probably not yet been identified. One of the most interesting showers of all has its radiant near the star 11 Canis Minoris, and is active on December 11. It seems that these meteors move in a very elongated orbit that carries them within 3,500,000 miles of the sun at perihelion— ten times closer than sweltering Mercury.

However, telescopic showers are not easy to identify. This is because the sporadic rate is relatively high—there may well be more sporadic meteors than shower meteors, so that prolonged watches may be necessary before any preferential direction of flight becomes obvious.

OBSERVATION. The study of telescopic meteor rates is straightforward enough, but if attempts are being made to find a radiant, considerable preparation is necessary. The field of view must be drawn, preferably tracing it from a large-scale atlas, such as *Atlas Borealis* or *Eclipticalis*, or possibly from the *Bonner Durchmusterung*. The paths of meteors seen are then drawn directly onto the chart, and their magnitudes can be estimated from those of the stars in the field. After a suitable interval of time, the telescope can be shifted to another, nearby field, and the process repeated. By this procedure, any divergence of the trails from a common radiant will become obvious. For the same reason, meteors from a known radiant can best be analyzed by choosing two fields 10° or 15° from the radiant, but forming an angle of 90° with it.

Half-hour watches are best. The strain of viewing an unchanging field is considerable, and the observer tends to find himself hypnotized by one of the stars, so that frequent rests are necessary. The observer must stop when he finds his powers of concentration failing, for the subsequent inaccuracies will prejudice the good work performed while he was fresh. Comfort and alertness are major contributors to the accuracy of the final results; and, provided they do not involve other adverse factors, no measures designed to increase the observer's sense of well-being are to be scorned.

PART III

THE STARS AND NEBULAE

18

The Stars

Apart from eruptive and long-period variable starts, those curious, remote suns that brighten and fade in an inexplicable and unpredictable manner, there is little original work open to the amateur beyond the confines of our own solar system. The stars constitute the professionals' realm. They pose problems that require great telescopes for their investigation, while the mathematics involved is so advanced that no person without specialized, university-level training has the slightest hope of making useful contributions. Times have changed drastically since William Herschel and the other great pioneers swept unexplored skies. In this age of modern astronomy, there is an essential difference between the two branches of the science: Amateur astronomy is mainly *qualitative*, or descriptive, whereas professional astronomical research, involved with explanations of behavior and in deriving theories in terms of mathematics applied to physical laws, is entirely *quantitative*. It depends on the measurement of actual amounts rather than on generalities. On the whole, the amateur simply does not have either the apparatus or the knowledge required for such measurements.

Since a 3-inch refractor can reveal something like a million stars over the whole sky, it would be presumptuous to say they are of no interest. Yet, sadly, this is what many amateurs do. Intent on following lines of useful work, they have no inclination to stand and stare at something that, however beautiful, is of no immediate concern to them. Such people are missing half the inexhaustible delight and fascination of astronomy. Indeed, scientific research has become so hallowed that it takes considerable courage to admit that one derives pleasure from making technically useless observations! The study of the heavens from a purely esthetic point of view is scorned in this technological age.

Except, then, for a discussion of certain variable stars (Chap. 19), this section might be considered a plea for "useless" observations; an invitation to every telescope owner to spend an occasional evening away from Jupiter's cloud belts and the moon's inhospitable crags, to wander instead through the almost limitless tracts of the heavens. Here there are wonders in plenty—double stars, multiple stars, stars of strange colors; here and there one detects gleams of nebulae, and just occasionally a superb, encrusted star cluster masses its members against the black sky. The observer has left the solar system far behind him; he is wandering where in our time no man will ever wander, with only his thoughts and speculations for company.

Stars and the Galaxy

To give a full description of the nature of these objects would require much more space than would be appropriate in a work devoted to practical observation. In any case, I have dealt with them at length in *Stars and Planets*, so that only a brief outline of the main facts will be given here.

The sun is a star, unremarkable on the grand scale but of tremendous importance to us, since it is the controller and benefactor of our solar system. However, it is only one among perhaps 50,000 million stars that together form our particular *galaxy*. Even the nearest and brightest of these stars are, however, so far away that they shine very faintly in the sky, and we cannot see them with the naked eye until nightfall. Even then, only about 6,000 stars are detectable over the whole sky. The rest—millions of them—are so far away that a telescope is necessary to show them. The larger the telescope, and the more light it collects, the fainter the stars that it will reveal; but even the greatest telescopes in the world are unable to show all the galactic stars, so that the value of 50,000 million members is a statistical estimate only.

Some of the stars are far more luminous than the sun; others are much fainter. This depends partly on size, partly on temperature; the size of a star can be measured only indirectly, but it is not difficult to classify a star in terms of temperature by noting its color. Some are white, with a tinge of blue or green; these are the hottest, with surface temperatures of about 25,000°C. Pure white stars are considerably cooler (about 11,000°C), whereas a tinge of yellow suggests a temperature of about 7,500°C. The so-called yellow stars, of which the sun is one, are about 6,000°C and therefore temperate by stellar standards. "Orange" and "red" stars are still cooler, with temperatures down to about 2,600°C. These colors are subtle; there are few glaring stellar tints. To call a star "red" is to exaggerate its hue considerably, but after some practice the regular observer can distinguish different shades or colors quite easily. As an extreme example, there is a clear difference between the blue-white star Vega (overhead in temperate northern latitudes during late summer) and the red Antares (low in the south at the same time); clearly, Antares is the cooler of the two.

Most of the stars in the Galaxy are immensely far apart by inter-planetary standards. If we construct a scale model of the solar system, with the sun about 4 inches across, the earth will be about 25 feet away, the size of a grain of sand, and Pluto will be about 300 yards away. But the nearest star to the sun, represented by another 4-inch globe, would have to be placed almost 1,500 miles away! It is obvious, then, that the whole Galaxy is of enormous extent, and that the mile is an absurdly inadequate unit of distance. Instead, astronomers make use of the *light-year*. A ray of light travels at 186,000 miles per second, so that it takes 8 minutes 20 seconds to reach us from the sun; and a light-year, the distance traveled by light in one terrestrial year, is just under six million million miles. The nearest star to the sun is $4\frac{1}{3}$ light-years away, which means that the light reaching our eyes left that star $4\frac{1}{3}$ years before—and this is just the *nearest* star!

The stars in the Galaxy are grouped together, forming a regular system. At the center is a roughly spherical nucleus with a diameter of about 11,000 light-years, and extending from this nucleus are two immense trains of stars that have been wrapped into a spiral by the system's slow rotation. The sun does not lie near the nucleus, as we might like to think; instead, it occupies a not very impressive position out on one of the arms, about halfway from the center to the perimeter. The Galaxy itself measures about 60,000 light-years across, and since it is very flattened, we naturally see more stars per square degree of sky when we look through the plane of the arms than at right angles to them. This explains the *Milky Way* effect, the Milky Way being our "inside" view of the galactic arms. Even a small telescope resolves it into depthless swarms of stars.

We cannot see the nucleus of the Galaxy. It is hidden from our eyes by clouds of interstellar dust, so much of our knowledge is hypothetical. These dark, obscuring clouds, known as *dark nebulae*, occur in many parts of the Milky Way. In the constellation Cygnus, for example, the naked eye per-ceives apparent irregularities and gaps in the nebulous course of the Milky Way, and these are simply our view of dark nebulae projected against the bright background.

Some, but not all, of the stars are lone wanderers, like the sun. Since, without exception, they are so remote that they do not show real disks in the greatest telescopes, they are not particularly interesting unless they happen to be of some unusual color, or vary in brightness. But other stars form fascinating telescopic objects. For instance, they may be "double," consisting of two or more individual stars lying so close together in the sky that a telescope is required to resolve them; or they may form a spectacular *star cluster*. These two classes of objects, together with the nebulae, form the main diet of the stellar observer.

The constellations and stellar nomenclature

From earliest times, the star patterns have been divided into rough groups, or *constellations*. There is nothing very scientific about a constellation; it is simply an arbitrary region of the sky and until recently, in fact, the boundaries between adjacent constellations were not well-defined. In 1930, however, definite demarcations were established by the International Astronomical Union, and the limits are now properly fixed.

The ancient astronomers were content simply to recognize the brightest stars in each constellation, and they associated their pattern with the outline of some suitable figure: the Dipper, or Great Bear (Ursa Major); the Hunter (Orion); the Scorpion (Scorpio); and so on. This was all very well in the early, rough-and-ready days, but as soon as people took a more critical interest in the stars it became necessary to list them individually. The first "modern" atlas, issued in 1603 by the German astronomer Johann Bayer, allotted to each bright star in every constellation a Greek letter, followed by the genitive form of the constellation's Latin name. In general, too, he classified them in order of brightness, so that the brightest star in Boötes (the Herdsman) is known as Alpha (*a*) Boötis. This system worked so well that it is still in use. The more prominent stars are also endowed with proper names, usually derived from ancient Greek or Arabic apellations: Alpha (*a*) Boötis is known as Arcturus (the Bear Watcher); Alpha (*a*) Scorpionis, as Antares (Mars-like); Alpha (*a*) Lyrae, as Vega (the Falling Vulture); and so on.

Since the Greek alphabet has only twenty-four letters, and most constellations have more than twenty-four naked-eye stars, some other system was required to make the cataloguing really comprehensive. This was achieved in Flamsteed's catalogue of 1725. John Flamsteed, first British Astronomer Royal (1675), observed all the naked-eye stars visible from the latitude of Greenwich ($51\frac{1}{2}°N$), from the very brightest down to those of the 6th magnitude. In his catalogue he classified the stars in each constellation in order of right ascension, starting at the western boundary with 1 and running up to whatever number the constellation contained. In so doing, he reclassified all Bayer's stars, so that by his reckoning Arcturus, for example, is 16 Boötis. In these cases, however, Bayer's Greek letter is usually retained.

Catalogues of telescopic stars (i.e., those below the 6th magnitude) are numerous, but these are of less concern to the average amateur; the only telescopic stars likely to interest him are doubles and variables, and these have their own special classification. Mention must nevertheless be made of Friedrich W. A. Argelander's monumental general catalogue of all stars down to a declination of −2°, subsequently extended by others to cover the southern sky as well. First issued in 1862, and listing some 458,000 stars, Argelander's catalogue is still widely used. It ignores the constellations; instead, it covers the sky in bands of declination just 1° wide. Any star of between the 6th and 9th magnitude is usually referred to by Argelander's reckoning.

Stellar magnitudes

The magnitude of a star, as we have already seen, refers not to its size but to its brightness. Although early astronomers were content to call the brightest stars "1st magnitude" and the faintest "6th magnitude," the system has now been refined to two or three decimal places. The three brightest stars in the sky, Sirius, Canopus, and Arcturus, are actually assigned negative values, and the magnitude ratio has been adjusted so that a ratio of 5 magnitudes is equivalent to a brightness difference of exactly 100 times. A gap of one magnitude corresponds to a brightness ratio of about 2½ (more accurately, 2·512). In this connection, Table VIII may be of interest.

TABLE VIII.   Stellar Magnitude and Brightness

MAGNITUDE DIFFERENCE	BRIGHTNESS DIFFERENCE	MAGNITUDE DIFFERENCE	BRIGHTNESS DIFFERENCE
1·0	2·51	7·0	631·0
1·5	4·0	7·5	1,000·0
2·0	6·3	8·0	1,585
2·5	10·0	8·5	2,512
3·0	15·9	9·0	3,981
3·5	25·1	9·5	6,310
4·0	39·8	10·0	10,000
4·5	63·1	11·0	25,120
5·0	100·0	12·0	63,096
5·5	158·5	13·0	158,490
6·0	251·2	14·0	398,110
6·5	398·1	15·0	1,000,000

From this guide it follows that the brightest star in the sky, Sirius, with a magnitude of −1·44, is almost a thousand times as bright as a star of the

6th magnitude. Yet the faint stars—so faint that they can be seen only with a telescope—are so much in the majority that they send us far more *total* light than do the naked-eye stars. In this connection, it is interesting to examine the total number of stars of each magnitude that are to be found in the sky. An examination of Argelander's catalogue, which covers just over half the sky, gives the numbers of stars for each broad magnitude division as shown in Table IX.

TABLE IX.   Magnitude Distribution of Stars
(numbers of stars in a one-magnitude interval)

1ST	2ND	3RD	4TH	5TH	6TH	7TH	8TH	9TH
20	65	190	425	1100	3200	13,000	40,000	142,000

The proportion of faint stars becomes still greater when we delve into the dimmer regions. Over the whole sky there are probably five million stars between magnitudes 11·0 and 12·0. They add considerable illumination to the background of the night sky—especially along the course of the Milky Way, in which we see probably 90 per cent of all the stars in the Galaxy.

Double stars

On a clear night, a glance at Zeta (ζ) Ursae Majoris (known as Mizar), second star from the end of the Dipper's "handle," will show that it is not alone. Mizar is a 2nd-magnitude star, and very close to it there is a fainter, 5th-magnitude star. If a telescope is turned onto this pair, which together constitute a naked-eye double star, Mizar will be seen to consist of two stars of magnitudes 2·1 and 4·2. These are so close together—the angular separation (known as *distance*) being only 14"·5—that a telescope is required to reveal the two separate stars. When viewed through the telescope, the 5th-magnitude naked-eye companion is of course seen very clearly.

Something like one-fourth of all the stars in the sky are double. Some, like Mizar, are easy; they can be resolved with a small telescope. Many others tax the powers of great instruments. But there are literally thousands within the range of a 3-inch refractor. Some consist of stars of about equal magnitude; in other cases, one star is so much brighter than its companion, or *comes* (pronounced kō-mez), that the second star is difficult to see. Some doubles have components of finely contrasted tints. And since most of these pairs are genuine stellar systems, with the two members slowly revolving around each other under the tie of gravity, we have cases where their appearance gradually changes as the years go by, the components apparently either moving closer together or else opening up. Few sights are more impressive than the steady progress of one of these so-called *binary* systems over the years, the only evidence the amateur has of actual movement in the stellar heavens. Even if the pair is of the *optical* kind, where the stars are not physically related at all, but simply happen to lie in almost the same line of sight, it may still have its interest. The amateur with a 3-inch refractor will find here an inexhaustible fund of interest and pleasure.

A double star such as Mizar is defined by three quantities: first, by the magnitudes of the components, commonly given to an accuracy of 1/10 (e.g., 2·1 and 4·2); second, by the distance, in seconds of arc (e.g., 14″·5); and, third, by the *position angle* (P.A.), which is the orientation of the fainter component relative to the brighter.

Position angle is measured in degrees, counterclockwise, starting from north, through east, south, west, and back to north again. Thus, N = 0°, or 360°; E = 90°; S = 180°; W = 270°. It is important to remember that *directions are reversed* in the field of an astronomical telescope, so that to an observer in the northern hemisphere, north is at the bottom, with east to the right (fig. 54). On the other hand, this is correct only when an object is on the meridian. When newly risen, the P.A. orientation is tilted over to the left; when near setting it is inclined to the right. It is important to bear this in mind when making an estimate of position angle. The P.A. of Mizar is 150°, which means that the comes (companion) is to the southeast. The complete details are written: 2·1, 4·2; 14″·5; 150°. Where a double star consists of three or more components, they are labeled *A*, *B*, *C*, etc., in order of decreasing brightness.

The terms "preceding" and "following" are also very useful in double-star work. They refer to the apparent drift of the object through the field of view, which is always from east to west. Combined with the north and south points, they enable rough positions to be assigned: *north-following* (*nf*) indicates the 45° quadrant; *south-preceding* (*sp*) that at 225°; and so on. By the use of these terms, plus an estimate of distance in either seconds or

Figure 54. *Position angle. This shows the telescopic (inverted) view to an observer in the northern hemisphere.*

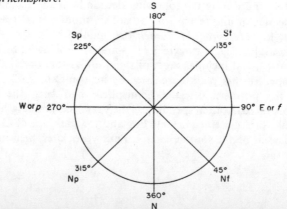

minutes of arc, the position of any star in the field can be assigned quite accurately; although for precise measures a special micrometer is necessary.

We have already seen how the resolving power of a telescope depends on its aperture. Knowing the aperture, it is possible to decide which double stars are likely to be resolved. The resolving limit of a 3-inch is about $1''\cdot5$; therefore it is no use trying to split a double whose components are $1''\cdot0$ apart, although the image may be elongated if conditions are favorable. Other factors are involved, however. The resolving limit applies to stars of the same brightness and of about the 6th magnitude. The brighter the star, the larger its diffraction disk, so that it may prove impossible to resolve a pair of 2nd-magnitude stars, even though they are slightly wider than the "theoretical" limit. On the other hand, if the stars are too faint, they are difficult to glimpse. Even more serious is the effect of one star being much brighter than the other. A case in point is the brilliant Sirius—Alpha (a) Canis Majoris—which has a 7th-magnitude companion with which it forms a binary system, the distance varying from about $2''$ to $11''\cdot5$ in a period of fifty years. When at its elongation the comes is hard to see even with a moderate instrument, and when closest (a position known as periastron), the intense glare of the bright star makes it invisible in the greatest telescopes.

The double stars listed in Chapter 20 have been chosen for their visibility in small telescopes. The brighter ones were listed in Bayer's and Flamsteed's catalogues, and have been nominated accordingly. The fainter ones, and the southern objects, are listed according to the catalogues of the various observers who have specialized in this kind of work. Herschel himself discovered more than 800, all being identified by the letter H; but this designation is rarely used, for his stars were remeasured and re-catalogued by the great Russian astronomer F. G. W. Struve. Struve's work was begun in 1819, three years before William Herschel's death, and in 1824 Fraunhöfer's splendid $9\frac{1}{2}$-inch refractor, then the biggest in the world, was installed at the Dorpat Observatory, where Struve worked. With this instrument he discovered about 2,200 new double stars between 1825 and 1827, and during the following decade he made accurate micrometrical measures of all the known pairs. This monumental research was embodied in his *Mensurae Micrometricae*, published at St. Petersburg (now Leningrad) in 1837. It runs to 3,112 stars, all designated by the Greek capital sigma (Σ), and includes most of the doubles that are visible with a small telescope. Further pairs were listed by his son Otto ($O\Sigma$).

So far, the northern celestial hemisphere had been the center of attention. But in 1833, after spending eight years reexamining his father's stellar objects, Sir John Herschel left England for the Cape of Good Hope to lead the assault on the southern skies. Four years later, he returned with

2,102 new doubles, identified by the letter h. In this way, the Herschels, father and son, and the Struves, father and son, exhausted the entire sky of its more obvious pairs. Even so, an unsuspected number remained, the components so close together and difficult of separation that they had so far been overlooked. Of the stars detected by subsequent observers, the catalogue of most interest to the amateur is that of S. W. Burnham, a Chicago lawyer who later became a professional double-star observer at the Lick Observatory. In 1900, he published a catalogue of 1,290 new doubles, more than one thousand of which he had discovered from his back garden during whatever leisure time was afforded by his legal profession. The Greek letter beta (β) designates these stars, not many of which are within the range of a 3-inch. Burnham himself as an amateur used only a 6-inch refractor, which testifies to his remarkable keenness of vision, since most of his pairs are extremely close and difficult.

The following abbreviations are the ones most likely to be used for doubles visible with a small telescope:

β = S. W. Burnham (1906)
H = W. Herschel (1782–1822)
h = J. Herschel (1847—southern)
Hh = J. Herschel's catalogue of W. Herschel's doubles (1833)
Ho = G. W. Hough
Hu = W. J. Hussey
I = R. T. A. Innes (1927—southern)
L = N. L. de Lacaille (1847—southern)
Σ = F. G. W. Struve (1837)
OΣ & O$\Sigma\Sigma$ = Otto Struve (1850)

Mention should also be made of a standard work, R. G. Aitken's *New General Catalogue of Double Stars* (1932), listing 17,180 pairs (Carnegie Institution, Washington). The most recent and exhaustive work is the Lick Observatory's *Index Catalogue* (1964).

Star clusters

The Galaxy contains two main types of star clusters. In some parts of the sky, particularly along the track of the Milky Way, we find glorious assemblages of stars sprinkled across the sky like glittering crushed glass. Some of these so-called *open* clusters are such superb spectacles that the observer instinctively returns to them again and again. The impression wrought by these crowded suns is overwhelming. In some open clusters the stars are bright, perhaps set in festoons like a diamond brooch; in others, a vast aggregation of faint specks powders the night sky. Presented with these sights, above all others, the observer suffers a feeling of awe that can be matched by nothing else in the sky. Here, immensity and delicacy are fused into a quite unearthly concept of grandeur.

The *globular* clusters, on the other hand, are somewhat disappointing objects in anything less than an 8- or 10-inch telescope. These are colossal balls of stars, massed together in numbers estimated by the hundred thousand, and are impressive objects in photographs; yet the individual stars are packed so closely together that a small telescope shows only a spherical blur, with perhaps a hint of individual stars around the margin. Not many globulars are known—a small telescope will reveal perhaps thirty over the whole sky—but they are interesting enough, even if nowhere near as spectacular as the open clusters.

Nebulae

One of the most notorious points of confusion in the field of astronomy is the difference between *nebulae* and *galaxies*. For there *is* a difference, not withstanding the fact that in the older catalogues they have all been listed as "nebulae"! Let us examine what is now known about these two classes of objects.

True nebulae *belong to the Galaxy*. They are clouds of gas and dust, very thin but of immense extent, in most cases being thousands of times larger than our solar system. The dark nebulae, already mentioned, are visible only because of the stars they hide from view; but of more interest are the *bright nebulae*, which appear as glowing wisps of matter. Their individual appearance varies enormously. The Great Nebula in the constellation Orion is easily seen with the naked eye, but only because it is exceptionally near the solar system. Most of these nebulae are rather faint and frankly unspectacular, but here and there the telescope will pick up well-known examples. The *planetary nebulae*, or *planetaries*, form an interesting subgroup, for they show small but definite disks, rather like ghostly planets. However, not more than a dozen planetaries are accessible with a small telescope.

As well as these galactic nebulae, there are many other dim gleams of light in the sky. A 3-inch refractor will show quite a few in Leo, Virgo, and Coma Berenices, and there are accumulations elsewhere in the sky. These objects, unlike the nebulae, *do not belong to our galaxy*. They are independent galaxies in their own right, each containing perhaps as many stars as there are in our own system. Millions of light-years away, they appear to us as mere stains against the sky. The universe is known to contain millions of galaxies, of which our own star system is an unremarkable member. Most of the *external galaxies* are very remote, but it is not surprising that the nearest ones should be bright enough to be conspicuous in even a small telescope; indeed, one of these can be seen with the naked eye in the constellation Andromeda. Lying more than two million light-years away, it appears to our vision as an elongated blur of light; yet this is a star city like our own galaxy, comprising thousands of millions of stars!

We have no direct means of telling how far away these other galaxies are, because every celestial object, from the moon to a remote star system, appears to be at infinity. To establish deep-space distances involves indirect investigations, and it was not until the present century that the external galaxies were finally identified as such. Before then, since the individual stars could not be seen with the available telescopes, they had simply appeared to be gaseous nebulae; accordingly, they are so listed in the classic catalogues. Modern catalogues, of course, differentiate between the two types of object; but the old lists, of which the N.G.C. is the most comprehensive, are still in general use. To the eye, indeed, there is no perceptible difference; they all look nebulous, and the designation has remained. Hence, an object classed as a "nebula" may, in fact, be a mass of glowing gas inside our own galaxy—but it is more probably an external galaxy.* As a general visual guide, the external galaxies have a regular appearance, whether spherical, elliptical, or extremely elongated, whereas the galactic nebulae are more irregular and the planetaries are quite sharply defined. In the constellation notes in Chapter 20, the distinction will be made wherever possible.

In general, nebulae and clusters have been listed together in the same catalogues. William Herschel, in his pioneer work on the northern sky, divided them into eight different classes:

 I. Bright nebulae
 II. Faint nebulae
 III. Very faint nebulae
 IV. Planetary nebulae
 V. Very large nebulae
 VI. Very compressed clusters
 VII. Compressed clusters of bright and faint stars
 VIII. Coarse clusters

This is a most convenient system, for it offers a preliminary guide to the observer. Thus, a bright planetary nebula in the constellation Draco is listed as H.IV.37—or, more simply, 37^4—while two beautiful open clusters in Perseus are listed as 33^6 and 34^6.

Sir John Herschel's subsequent visit to the Cape produced a fine

---

* Because of their visual similarity—they all look nebulous—they are sometimes referred to as "extragalactic nebulae"; but this term is misleading.

southern-sky sequel to his father's list of 2,500 nebulae and clusters. His total came to 1,708, and subsequent additions brought the number of combined observations up to 5,079. These were published in a great catalogue in 1864, which was afterward extended by Dreyer to more than 7,000 objects and published in 1888 as the *New General Catalogue*, which remains the standard list. Perhaps a thousand of the objects listed are discernible with a small telescope (though most of these are simply gleams of light), so the *N.G.C.* is of value to all observers. Unfortunately, it lacks the elder Herschel's system of classification, so that without the descriptive catalogue it is impossible to tell whether a certain object is a cluster or a nebula. Because of this, most of Herschel's original designations have been retained, while Messier's list of 103 nebulae and clusters, subsequently extended to 107, is also a useful guide to the bright objects.

Magnification and resolution

So far as magnification is concerned, each double star imposes its own conditions. A wide pair, such as Mizar, can be easily divided with × 50; a higher power, which separates the stars even more, loses some of the effect. The same is true of the beautiful double Beta (β) Cygni, where the stars are yellow and blue (partly a contrast effect), at a distance of 35″. Here, a power of × 30 is adequate; it shows Beta's splendid tints and also includes many other stars in the field of view. On the other hand, there are pairs such as Epsilon (ε) Boötis (3·0, 6·3; 3″·0; 340°), whose stars are so close that a high magnification is required to give a clear separation. The regular observer will become accustomed to these idiosyncrasies.

The nebulae, being dim and ill-defined, require a low power to concentrate the light and show them to the best advantage. Many are quite imperceptible with a powerful eyepiece, because of the loss of contrast; and some very large ones are so faint and extended that they are more conspicuous in the finder than with the main telescope. Similarly, open clusters are generally best seen with low powers and wide fields, whereas planetary nebulae and globular clusters require considerable magnification.

The resolution of a double star depends to a great extent on the seeing. Take for example the bright red star Antares (Alpha Scorpionis), which, with respect to color, has been aptly termed "the rival of Mars." This is a 1st-magnitude star with a 7th-magnitude companion at a distance of 3″. On an unsteady night, the image of the bright star flares and expands far beyond its nominal size, sometimes extending beyond the first and second diffraction rings. Under these conditions, it is clearly hopeless to look for the comes. Only when the great star subsides into something like its proper diameter can the faint attendant be seen alongside its rays. Obviously, the telescope cannot be blamed for failing to find it on bad nights. Antares is a notoriously difficult object to resolve from the latitude of the British Isles or extreme northern United States, where it is always low in the sky; but from stations farther south it is much easier, since it appears higher above the horizon, where the seeing conditions are superior.

The above example proves that the statement (encountered in so many books) that double stars are a test of telescopic performance is gravely inaccurate. They are far more truly a test of atmospheric steadiness. By far the best test of a mirror or object glass is the intrafocal and extrafocal test, which can be performed under almost any atmospheric conditions. Stellar resolution may be a by-product of optical excellence, but there is no reason to make it a criterion.

Magnitude limits

The concept of "limiting magnitude" is another widely-fostered misconception. It has been claimed that a 3-inch telescope should show stars down to magnitude 11·4; a 6-inch, down to 12·9; a 12-inch, to 14·4, and so on. However, there are so many influential factors at work that the unwary beginner may be unjustifiably disappointed when his performance fails to match with the tables. On some nights, it may be difficult or impossible to see the 9th-magnitude companion to the polestar—Alpha (a) Ursae Minoris—with a 3-inch. On other nights, it may be obvious, and stars as faint as the 11th magnitude may be glimpsed. By indulging in deep breathing before making the observation, then holding his breath and using averted vision, a friend of mine claims to have glimpsed stars of the 13th magnitude, using a 3-inch refractor. Continued practice and refinement of technique will naturally lead to superior results.

Whatever its capabilities on a particular night, a telescope cannot be expected to approach its limiting magnitude in the case of a star very near a more brilliant one. This is partly due to the effect of glare; there may also be some temporary desensitizing of the retina in the region of the bright image. The satellites of Saturn indicate this anomaly very well. For instance, Rhea can be seen with a 3-inch when near elongation, but in so small a telescope disappears from view when it approaches the planet's limb. Ariel, innermost of the four bright Uranian satellites, is the most difficult to see, though of the four it is the brightest; this paradoxical effect misled many early observers and is responsible for errors even today.* The companion to Sirius, a famous teaser, used to be rated 9th magnitude

* The usually authoritative B.A.A. *Handbook* lists Ariel as being fainter than both Titania and Oberon.

because of the difficulty of observation, but is now generally accepted to be as bright as the 7th. It is advisable to use the highest possible power for observations of this sort, since this increases the apparent separation. A high magnification also has the effect of darkening the sky background, thus augmenting faint points of light.

Atmospheric steadiness is essential for delicate observations of this sort. To the naked eye, the sky may appear black and crowded with faint stars; but these conditions are of no help to the telescope if the seeing is bad, for the atmospheric ripples will blur the pinpoint images of the tiniest stars and efface them as effectively as haze. The nebulae, being extended objects, do not suffer materially from the effects of bad seeing, but faint companions to double stars demand both transparency and steadiness, a combination that occurs only on rare nights.

When preparing to observe faint objects, the eye must be given a chance to become thoroughly dark-adapted. Here again we can take a hint from William Herschel, who knew that the only chance of detecting dim stars and nebulae was to give his eye the greatest possible sensitivity before commencing observations. As Agnes Mary Clerke, the astronomical historian, wrote, in *The Herschels and Modern Astronomy*:

> His [*William Herschel's*] *sense of sight was exceedingly refined, and he took care to keep it so. In order to secure complete "tranquillity of the retina," he used to remain twenty minutes in the dark before attempting to observe faint objects; and his eye became so sensitive after some hours spent in "sweeping," that the approach of a 3rd-magnitude star obliged him to withdraw it from the telescope. A black hood thrown over his head while observing served to heighten this delicacy of vision. Details are "of consequence," he wrote, "when we come to refinements, and want to screw an instrument up to the utmost pitch."*

It is worth pointing out that even Herschel's "bright nebulae" are mostly dim objects in a 3-inch refractor, and that class II and III objects are mostly imperceptible without a moderate aperture. The remarkable fact that he *discovered* many of these with a 6-inch reflector testifies to his "delicacy of vision."

For bright double stars, where transparency is of little account, adequate observations can be made from towns. But the observation of faint stars and nebulae is severely handicapped by urban conditions, and it must be admitted that the night sky of most towns does little to gladden the stargazer's heart. The atmospheric haze, its effect heightened by nocturnal illuminations, utterly blots out a nebula, while the swarms of faint stars in an open cluster will also be lost to view. Just occasionally, conditions improve—I have on a few occasions glimpsed the brightest regions of the

Milky Way from near the center of London—and formerly invisible objects may be detected; but such revelations act more as a frustration than anything else. Facts must be faced: The town-dweller's opportunity for useful work lies with the moon and planets, whose brilliant disks are dimmed and steadied by the haze; in this field, therefore, he may well have an advantage over those observers who enjoy a black sky and a brilliant Milky Way.

The celestial sphere

It may be wise at this point to deal more fully with the concept of right ascension and declination, since the pinpointing of an object on the celestial sphere is of great importance in stellar work.

The stars and nebulae retain the same relative position from night to night and, to all intents and purposes, from century to century. Their only obvious movement is their diurnal passage across the sky, and this comes as a result of the earth's rotation. Since the earth spins from west to east, the stars appear to revolve in the opposite sense, from east to west.

Since all celestial objects appear to be equally far away, the night sky assumes the appearance of a colossal sphere, with the stars attached to its inner surface, revolving around the earth once in a sidereal day. Figure 55 shows an external view of this imaginary sphere; certain fundamental points on the earth's surface are mirrored on the sphere. Projecting the axis makes it touch the sphere at just those two points that appear to be stationary in the sky: These are the north and south celestial poles, around which the rest of the stars seem to revolve. Since the angle of the earth's axis

Figure 55. *The celestial sphere.*

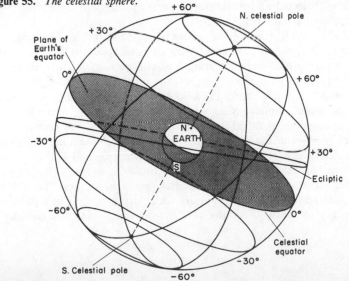

is virtually constant, the celestial poles remain in the same place. The north pole is marked approximately by the polestar, of the 2nd magnitude, which happens to lie nearby; but the south pole occurs in a rather barren patch of sky and does not have a conspicuous marker. An observer on the earth's equator would, theoretically at least, see both poles at his north and south horizons; as one travels northwards, the north celestial pole rises higher and higher in the sky. If one ventured as far as the earth's north pole, the polestar, would be directly overhead.

In a similar way, the celestial equator can be inscribed on this imaginary sphere by extending the plane of the earth's equator. An equatorial observer sees the celestial equator as an imaginary line running from the east to the west horizon, and passing overhead. To an observer moving progressively north or south away from the equator, it is inclined lower and lower in the sky; by the time he reaches the north or south pole, it is level with the horizon. A polar observer will always see just one hemisphere of stars, the other half of the sky being permanently hidden from view by the body of the earth.

Now that the sphere has been inscribed with poles and equator, lines of declination can be added. These are analogous to terrestrial latitude, the equator being 0° and the poles 90°. With respect to longitude, or *right ascension* (R.A.), it is more convenient to divide the sphere into 24 hours than into 360°, since this provides a direct conversion to its rate of rotation.

The zero position for right ascension (the celestial equivalent of the Greenwich meridian), is derived from the apparent motion of the sun. As we have seen, the earth's annual circuit of the sun makes the sun itself appear to revolve around the celestial sphere. One way of demonstrating this is to stand a chair in the middle of a room and walk slowly round it; the chair will be seen projected against successive parts of the room, until it seems to have completed a full circle.

Since the earth's axis is tilted with respect to its orbit at an angle of $23\frac{1}{2}°$, the sun's apparent path around the celestial sphere is inclined to the equator at this critical angle, crossing it at two points. The first, traversed on about March 21 when the sun is traveling northward, is called the *vernal equinox*, and this point marks the 0^h line of right ascension. The other, reached on about September 23 when the sun is sinking into the south celestial hemisphere, is called the *autumnal equinox*. This, of course, is on the opposite point of the equator, and marks the 12^h meridian. It follows from this that the part of the ecliptic lying between 0^h and 12^h is in north declination; the remainder, from R.A. 12^h to 24^h, is south.

This is the way in which the celestial sphere is marked out with its fundamental reference lines. The stars keep virtually the same positions from century to century, but the sun, moon, and planets are constantly shifting, always keeping to the region of the ecliptic, so that they cannot be marked on a star map.

Precession

There is, however, a very slight drifting of the stars. This occurs through a slight wobble, or *precession*, of the earth's axis; it is turning, very slowly, through a small circle, rather like a dying top, and the drifting effectively pulls the grid of right ascension and declination with it. The result is that the stars, relative to this grid, are changing their positions.

Precession is a very leisurely process. It will take about 26,000 years for the celestial poles to make their revolution; nevertheless, the change from century to century is appreciable. Since many useful catalogues are now out of date by this amount, it is necessary to apply a slight correction if an object's position is to be plotted accurately on a modern atlas. The *Bonner Durchmusterung*, for example, is correct for the epoch of 1855; there is therefore a considerable discrepancy between the positions given in this and those given in *Norton's*, which is based on the 1950 epoch. The tables in Appendix VI will enable the observer to allow for precession when working with old atlases.

19

Variable Star Observation

The observation of variable stars deserves a separate chapter, because it is one branch of stellar astronomy in which the amateur can do really useful work. These are stars which change in brightness over periods ranging from a few hours to a year or more, while some fluctuate quite erratically. Generally speaking, it is only the erratic variables that are of interest to the amateur, for the predictable ones have been closely studied at professional observatories.

Variable stars can be divided into two broad classes. There are the *eclipsing* types, which are simply binary systems seen more or less edge-on to the orbit, so that the stars periodically occult each other; and the *intrinsic* type, where the star actually changes in luminosity due to physical instability in its gaseous shell, which in many cases produces fluctuations of pressure which, in turn, affect its temperature and brightness. The light fluctuations of the eclipsing type are, of course, perfectly regular. When one star is occulted by the other, the total brightness is reduced; and since these systems are so close that the individual components cannot be distinguished even with the greatest telescopes, the effect is of a single star appearing first to dim, then to brighten again.

Variable-star nomenclature

While many of the bright variables had been catalogued as fixed stars (albeit unwittingly) by Bayer and Flamsteed before their vagaries were noticed, the telescopic variety have received their own system of nomenclature. One, devised by Argelander, who discovered several variables during his work, identifies the stars in each constellation by the letters R, S, T, and so on, up to Z. This provides for only nine stars, so after that they are designated RR, RS, RT, and so on, followed by SS to SZ, TT to TZ—and so on, to ZZ. This allows for 54 variables in each constellation; if more are present, we return to the beginning of the alphabet with AA to AZ, BB to BZ, and so on, down to the letter Q (but omitting the letter J). This extremely cumbersome system, which allows for a total of 334 variable stars in each constellation, has been established for a century and is still used in standard lists. If the constellation contains more stars than can be accommodated by Argelander's nomenclature, subsequent variables are designated as 335V, 336V, etc. Argelander began his system with the letter R to avoid any possible confusion with the A–Q lettering used, instead of Greek letters, to identify the naked-eye stars in some southern constellations.

A more convenient system of identification, often used in conjunction with the above symbols, was introduced by Harvard Observatory. It consists of six numbers, the first four referring to the right ascension and the last two to the declination of the variable. For example, SW Geminorum, at R.A. 6^h 53^m, Dec. $+ 26°$, is referred to as 065326. If the declination is southern, the last two numbers are underlined.

Eclipsing variables

The most famous eclipsing variable in the sky lies in Perseus. It is known as Algol—Beta (β) Persei—and it can be found very easily, for it lies south of the line joining Alpha (a) Persei and Gamma (γ) Andromedae. For most of the time, Algol shines at a steady magnitude of 2·3, which in brightness is midway between Alpha (a) and Zeta (ζ) Persei; but every 2½ days it fades quite rapidly, dimming to 3·5 (fainter than Gamma) in five hours. In another five hours it regains its original magnitude, and the cycle begins again.

Algol belongs to the subclass of *dark-eclipsing* variables, since one of its components is much larger and dimmer than the other. The main drop in brightness therefore occurs when the dark component occults the bright one. When the reverse happens, the total brightness drops only very slightly. Actually, there is a "secondary minimum" in the middle of the 2½-day spell during which this occurs, but the fall is only about $\frac{1}{20}$ of a magnitude—imperceptible without accurate measuring instruments.

If, however, the two components are about equally luminous, the secondary minimum will be much more noticeable, and there will be two distinct fadings during the stars' revolution around each other. The prototype of this *bright-eclipsing* variety is Beta (β) Lyrae, near the brilliant Vega. The period is 12 days, during which time it drops from its maximum value

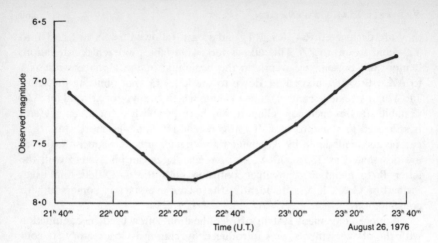

Figure 56. *The eclipsing binary RW Cassiopeiae. This series of observations of a minimum was made by the writers using 12 × 40 binoculars, on August 26, 1976.*

of 3·4 to 4·3 (main minimum) and to 3·8 (secondary minimum). Both these stars can be represented by the light curves shown in figure 57.

There is some scope for amateur work on eclipsing binary stars, and the Variable Star Section of the British Astronomical Association has a number of stars of magnitude 7–10 on its program. Since many eclipsing binary stars have probably not been observed by professional astronomers for years or even decades (since, in fact, their periodicity was first determined), the accumulated errors in their periods may have become large enough to be detectable by simple visual determination of the minima. Anyone with an aperture of six inches or more could usefully turn his hand to checking the fainter eclipsing binaries, say about magnitude 11, that are to be found by the hundred in the *General Catalogue of Variable Stars* (GCVS). A comprehensive photographic atlas such as Vehrenberg's *Atlas Stellarum* will be helpful, or the enterprising amateur could photograph the field and identify the variable by off-setting from brighter stars whose position is known. The predicted time of minima is obtained from the data given in the GCVS. All that has to be done is to compare its brightness with that of nearby stars in the manner described on page 280, to see if it undergoes a fall of brightness around the predicted time. Once its current time of minimum has been "bracketed," so to speak, convenient future minima can be predicted, and a thorough series of estimates made at 10-minute or half-hour intervals, or whatever seems appropriate bearing in mind the rapidity of the magnitude change. If plotted on a graph, as in figure 56, the time of minimum can be determined. It will be necessary for the observer to decide on the magnitude values of his comparison stars, although these need not be true values, since all that is needed is a yardstick against which to measure the variable's change of brightness with time.

Some eclipsing binaries have periods that are an almost exact multiple of the earth's day, such as ZZ Bootis (4.9917^d) and Y Cygni (2.9965^d). Their minima therefore pass through "seasons," lasting weeks or months, of continuous visibility and invisibility.

Intrinsic variables

These can be grouped into the following classes: *Cepheids* and allied types, in which the star has a very regular period, usually of a few days; *Long-period* variables, red giants with fairly predictable periods, usually of the order of a year; *Semiregular* variables, stars with poorly-defined periods; and *Irregular* stars, which, as their name suggests, are unpredictable. The stars in the two latter classes usually have ranges of a magnitude or so, while the long-period stars range up to six magnitudes or even more.

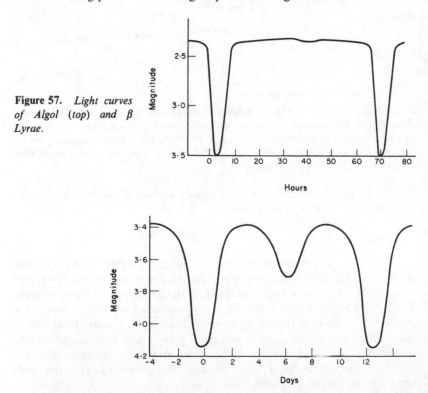

Figure 57. *Light curves of Algol (top) and β Lyrae.*

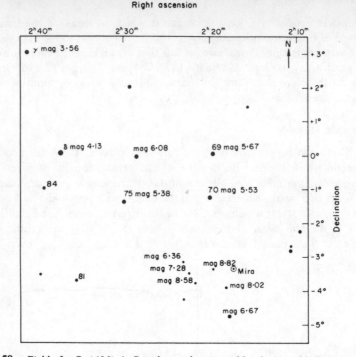

Figure 58. *Field of o Ceti (Mira). Based on a chart issued by the Variable Star Section of the British Astronomical Association, this shows a field of 9° around the variable, and will allow magnitude estimates to be made with binoculars. Numbers against some of the stars refer to Flamsteed's catalogue of 1726. Not all stars are given a magnitude, since some are unsuitable for estimates.*

Finally, there are the *Eruptive* types, which include those spectacular objects the *novae*, or exploding stars.

Cepheids

Cepheids and RR Lyrae stars are two forms of regular, pulsating giant stars, the main difference being that RR Lyrae stars are commonly found in globular clusters. A bright example is the prototype of the Cepheids, Delta (δ) Cephei, which varies from magnitude 3·8 to 4·6 and back in a period of 5½ days. A feature of most of these stars is their sharp rise to maximum (Delta Cephei takes just over a day) and their slower fall to minimum. These variables are much larger and more luminous than the sun, but most are too far away to be seen with the naked eye. They have been closely studied, and offer little scope for amateur work.

Long-period variables

The type-star of this class of variable is Omicron (o) Ceti (Mira). It can be sighted in the rather barren northeast corner of Cetus; it lies slightly south of the celestial equator. Mira is a "giant" star, much bigger than the sun, but it has a lower density and is cooler, which means that its color is redder. In fact, most of the semiregular variables have a definite red hue, which helps in identification.

Mira's caprices were first noticed as far back as 1596, when the Dutch observer David Fabricius discovered that Cetus contained a new 2nd-magnitude star. He first noticed it on August 13, and followed it for some weeks until it had sunk below naked-eye visibility. Some years later, in 1609, he reobserved this curious object, apparently unaware that Johann Bayer, while collecting material for his famous star map "Uranometria," had seen it in 1603, not recognizing it as a variable star, and had recorded it as Omicron Ceti. It was sighted twice again, in 1631 and 1638, but not until 1639 were these sightings established as being of one and the same object, and another thirty years went by before a definite periodicity of about 11 months came to light.

But Mira, while admitting to a general law of fluctuation, has characteristics of its own. Its period can vary from about 300 to 360 days; and while it usually falls to about the 9th magnitude at minimum, detectable with a 2-inch refractor, its maximum brightness is unpredictable. In November, 1868, for instance, it reached only the 5th magnitude and was visible with the naked eye for only a few weeks; in 1799, on the other hand, Herschel recorded it as being almost as bright as Alpha (a) Tauri (Aldebaran), which is of the 1st magnitude. At a very bright maximum it has been seen with the naked eye for several weeks, as in 1969. It is hardly surprising that Omicron was called Mira, the Latin word for "wonderful." Usually it reaches the 3rd or 4th magnitude, rising from minimum somewhat faster than the subsequent fading; but it is dangerous to trust to average form, and Mira is well worth watching for fluctuations in behavior. As an aid to its identification, figure 58 shows a chart of the region.

Mira-type stars, or LPVs, have periods of between 150 and 500 days, and, because of their large ranges and continuous fluctuations, they are popular with amateur observers. One of the most interesting is Chi (χ) Cygni, which can reach the third magnitude at maximum, with an extreme range of 11 magnitudes.

Semiregular variables

The semiregular variables encompass a large class of red giant stars, like Mira, but with much smaller ranges and less well defined periods. Generally speaking, these stars have been neglected by both amateur and professional observers, and more intensive observations are needed. Many bright ones can be studied with binoculars throughout their range.

Irregular variables

Although the term "irregular" applies to a number of types of variable, it is generally used in connection with red giants of no established period, and stars known as "nebular variables," many of which lie in the Orion Nebula and are subject to sudden fluctuations, often occurring in the space of an hour or less.

A number of bright irregular variables form objects for naked-eye study. The most prominent example is Alpha (α) Orionis (Betelgeuse), which, to an observer's view, marks the Hunter's left shoulder. The extreme range of this fine orange star is 0·1 to 1·4, and the bright Aldebaran, in Taurus, whose magnitude is 0·78, makes a convenient comparison star. A period of five years has been quoted, but this is doubtful. A fainter irregular variable is Mu (μ) Cephei, Herschel's "Garnet Star," which shines with a beautiful claret color. Mu Cephei's tint is rivaled by Hind's "Crimson Star," the LPV R Leporis.*

Mu Cephei varies from magnitude 3½ to 5; Alpha (α) Herculis, another bright example, fluctuates between 3 and 4. Another kind of irregular star, Gamma (γ) Cassiopeiae, at the middle of the "W," is white. It is known as a *shell star*, since it is believed to be surrounded by a glowing gaseous shell, the source of variability. In 1936–1937 it brightened to magnitude 1·4, falling to its more normal level of 2·4 in a few months. Small fluctuations still occur, and, since it may brighten markedly at any moment, this is a star that should be glanced at from week to week; Beta (β), magnitude 2·4, is a suitable comparison.

Eruptive variables

As their name suggests, the stars under this heading are subject to bursts of violent activity. Under the general heading of the U Geminorum stars, or "dwarf novae," we find a fascinating family of close binary stars in unstable equilibrium. Periodically a transfer of material from one star to the other takes place, accompanied by an outburst of energy, and the telescopic star brightens or "rises." The type star, U Geminorum, is nor-

---

*Two nineteenth-century observers, J. Birmingham and T. E. Espin, produced extensive catalogues of *red stars;* the final list, drawn up by Espin in 1888 and incorporating the work of both observers, is often referred to. If a star is marked "E–B," it is certain to have a reddish tint.

mally of magnitude 14½, but three or four times a year it can rise to magnitude 9 literally overnight. Other dwarf novae rise more frequently, and the observer must always be prepared for a surprise, which is why observation of these stars is currently so popular.

Most dwarf novae are fainter than magnitude 14 at minimum, but rises can be covered with instruments of 6 inches aperture, or even less. The brightest, SS Cygni, reaches about magnitude 8½ at maximum. A sub-class of these stars, the Z Camelopardalis variables, often suffer a standstill on the decline, where they may remain for months before sinking back to minimum.

Another remarkable family contains the "flare stars," red dwarfs which can rise by several magnitudes in a few minutes, dying down just as quickly and perhaps remaining dormant for weeks or months. Some amateurs have conducted flare-star patrols by pointing a stationary camera at the field and allowing the earth's rotation to make the stars trail; the flare star, which will not normally be visible, will leave a short trail if a flare occurs during the exposure period.

The term "eruptive" can also be applied to the R Coronae Borealis stars, which are like U Geminorum stars in reverse: They spend most of their time at maximum, suddenly dropping precipitously to a brief minimum. The type-star is normally visible as a 6th magnitude object in the Northern Crown, but at intervals of perhaps several years it drops to the 14th magnitude, or even fainter, in a few days.

Corona Borealis also contains another unpredictable star, the variable T, aptly nicknamed "the Blaze Star." Normally of the 9th magnitude, in May, 1866 it shot up more or less overnight to the 2nd magnitude. After a week of naked-eye visibility, it vanished from view once more, returning to its original brightness until February, 1946, when it surged up to the 3rd magnitude. Once more it faded rapidly, and it is now visible as a magnitude 9·5 object about a degree south of Epsilon (ϵ). No doubt it will one day flare up again, although it may not be for years; but whenever it does, the first warning will probably come from an amateur who has noticed "something wrong" with the familiar group of Corona Borealis (the Northern Crown).

The most famous "blaze star" of all lies in the southern hemisphere, but for some time now it has been quiescent. This star, Eta (η) Carinae, was first catalogued by Edmund Halley, who noted it as a 4th-magnitude object while observing from the island of St. Helena in 1677. Since the Greek astronomer Ptolemy, who drew up a star catalogue in the second century A.D., had not noted a star in its place, it seems certain that at that epoch it was beyond naked-eye visibility. Drastic changes soon followed.

In 1687, and again in 1751, it was seen as a 2nd-magnitude object, fading down between times to its more usual luster; while in 1827, it shot up to the 1st magnitude, declined slightly, and finally summoned a burst of energy that brought it up almost to the level of Sirius, the brightest star in the sky! After holding this position for several years, this extraordinary star began to fade until, in 1868, it was lost from naked-eye view. By 1886, it had reached a minimum magnitude of 7·6, at which it has remained to the present time. Whether it will ever repeat this splendid course, only time will tell.

Novae and Supernovae

The most remarkable "variables" of all are those stars which seem literally to explode. Once every three years or so, on an average, one of these inconspicuous 10th- or 11th-magnitude stars suddenly suffers some internal instability, forcing the surface layers outwards in a colossal eruption that blasts it up into naked-eye visibility. It takes just a few hours to increase its light output by perhaps 50,000 times, remaining for a day or two perhaps the most luminous star in the Galaxy. So far as astronomers can tell, there is nothing exceptional about a star that decides to turn itself into a *nova*, and there is no way of forecasting where and when the next one is likely to appear. The only statistical probability of use to nova-hunters is the 90 per cent chance of one occurring within the Milky Way, since this is where we see the majority of stars.

A bright nova was found on February 6, 1963, by a Swedish amateur astronomer, Elis Dahlgren, and, independently, by the American comet-hunter L. C. Peltier, of Delphos, Ohio. On that date it appeared of magnitude 3·9, lying on the border of the constellations Lyra and Hercules.

Still more recently, on July 8, 1967, the British comet- and nova-hunter George Alcock was rewarded with the discovery of a bright new star in the small constellation of Delphinus. When discovered, after only twenty minutes' sweeping with his 11 × 80 binoculars, its magnitude was 5·6. He immediately notified the Royal Greenwich Observatory, and on the following night the discovery was confirmed by professional astronomers. Nova Delphini has proved to be a most unusual star; instead of sinking down quite rapidly after its outburst, it fluctuated between magnitudes 3·5 and 5·5 for several months, attracting the fascinated attention of amateurs all over the world. Less than three months later, on October 27, 1967, the variable star RS Ophiuchi, which had suffered nova-like outbursts in 1898 and 1933, was discovered by amateurs in Europe and the United States to have flared up again.

Further nova discoveries by amateurs have followed. Alcock found his second, in Vulpecula, in 1968, while the veteran Japanese observer M. Honda found Nova Serpentis and Nova Aquilae in the following year. All of these were about magnitude 5 at maximum, and would not have been noticed without careful searching. Alcock made another discovery, of a 7th magnitude object in Scutum, in 1970, and Kuwano, in Japan, found Nova Cephei 1971 at magnitude 7½. A further discovery by Alcock, of a 6th magnitude nova in Vulpecula in 1976, was followed by the detection of Nova Sagittae 1977 by another English amateur, John Hosty. The very bright naked-eye nova in Cygnus, seen in August 1975, was independently discovered by several observers. Honda also discovered a 7th magnitude nova in Aquila in January 1982.

Nova Sagittae was discovered as a result of a cooperative nova patrol operated by amateurs in the United Kingdom, in which both visual and photographic methods are used. Over the years, a large stock of negatives covering the Milky Way has accumulated, and any suspicious star can quickly be checked at the central clearinghouse. The discovery was communicated to the I.A.U., and relayed to both amateur and professional telegram and telex subscribers, within a few hours of its confirmation. It is significant that almost all "live" nova discoveries have been made by amateurs, and it reflects great credit on those observers who are prepared to spend hundreds of hours scanning the night sky.

These suicidal stars, which afterwards dim down as if exhausted and return to their quiescent state, have proved a rich harvest for the casual sky-watcher. Every amateur who spends regular evenings with the stars, whether hunting up doubles or nebulae or watching for meteors in the quiet reaches of the night, soon develops an intimate knowledge of the star groups. The first step in becoming an amateur is to learn the main constellations: the Great Bear, or Dipper; Orion; Cygnus; Cassiopeia; and other distinctive star patterns that can never again be mistaken. For these groups, a very simple star map of the kind published regularly in many newspapers is the best, since it shows only the brightest stars and does not invite the confusion of a more detailed chart. But once these "guide groups" have been found, the smaller constellations fit themselves into place. Vulpecula, Sagitta, and Delphinus emerge in the region south of Cygnus; even the stragglers, like Draco and Pisces, which contain few bright stars in proportion to their great length, suddenly appear quite distinctive in a region of sky that formerly seemed barren and undistinguished.

By this time, the amateur will certainly be more familiar with the sky than are many professionals, who rarely need to look at the stars at all; it is not surprising that many of the naked-eye novae have stuck out like sore thumbs to regular sky-watchers. The lunar observer Julius Schmidt, who just missed discovering the remarkable rise of T Coronae Borealis in

1866 (it rose three hours after he had observed the region, and was found by the red-star observer John Birmingham), was the first to observe Nova Cygni, in 1876—a 3rd-magnitude object near Rho (ρ) Cygni. In 1891 and 1901, the Scottish amateur meteor observer Dr. T. D. Anderson discovered two novae; the second, Nova Persei, was a splendid zero-magnitude star that he noticed shining near Algol. Espin found Nova Lacertae in 1910; several observers saw Nova Aquilae in 1918; and W. F. Denning picked up another nova in Cygnus in 1920. In 1934, a nova appeared in Hercules, not far from the site of the 1963 nova; it was discovered by J. P. M. Prentice, then director of the Meteor Section of the British Astronomical Association. The year 1936 produced no fewer than three naked-eye novae, with two more being picked up by an amateur experimenting in astrophotography. This is certainly an impressive tally, and since nova-hunting essentially requires nothing more than a clear sky, and a good knowledge of the constellations, it is remarkable that more amateurs do not make a five-minute survey of the Milky Way a regular part of their observing program. As with comet-hunting, success will eventually reward the persistent.

Even more drastic than the novae are *supernovae*, stars that explode so violently that they emit as much light as the other galactic stars put together! These are excessively rare; only two have been recorded in our Galaxy in the last five hundred years, in 1572 and 1604. At their peak, these stars seemed as bright as the planet Venus and were visible with the naked eye in broad daylight, so there is little chance of an eruption of this type being overlooked.

Comparison stars

Observing a variable star consists of making estimates of its magnitude. These estimates are made by using certain standard *comparison stars*, afterwards working out the variable's magnitude from the accurate values available for the comparison stars. Societies dealing with the observation of the semiregular variables issue standard guide charts for each star, showing the nearby stars and giving magnitude values. In the case of naked-eye estimates of the bright irregulars, it is up to the observer to choose his own comparisons; the same is true if a nova happens to put in an appearance. Of course, it is necessary to make sure that none of the comparison stars is itself a variable—this has happened more often than one might imagine!

Like all programs in amateur astronomy, variable-star estimates require practice and experience before they can be made of the highest accuracy. Experienced observers can sometimes reckon to $\frac{1}{10}$ of a magnitude, when conditions are favorable; but consistent accuracy to $\frac{1}{2}$ is good enough for most practical work. Indeed, *consistency* is the watchword. It does not matter too much whether one's estimates are slightly bright or slightly faint, provided they all err by the same amount. The *form* of the light curve is the important thing, so that one can follow the rises, falls, and standstills as they occur. To achieve consistency, it is necessary to use the same set, or "sequence," of comparison stars for all estimates, and to make sure that these standard magnitudes all come from the same catalogue. There are sometimes inconsistencies even in the charts and sequences provided by observing societies, and it must not be assumed that just because a value is given to $\frac{1}{100}$ of a magnitude, which is quite usual in professional catalogues, it is necessarily accurate to anything better than $\frac{1}{10}$. A gross error will be introduced if the magnitudes for different stars are taken from different catalogues, since these sometimes differ by as much as half a magnitude.

It is worth examining the reason for these inconsistencies, for it sheds light on one of the problems of the variable-star observer. Inconsistencies arise because stars are of different tints. Most are white, with a hint of blue at one extreme end and of yellow at the other; as we descend the temperature scale, the color deepens into yellow and red. The human eye is most sensitive to yellow-green light; it is also fairly sensitive to blue, but it is rather insensitive to red. Moreover, eyes themselves vary. Some people have a much more delicate appreciation of subtle color differences than others, and it is well known that some eyes can detect rays that are quite beyond the grasp of normal vision. With some observers, too, each eye responds slightly differently. This is bad enough, but, to make matters worse, most standard star catalogues have been prepared photographically, and the usual photographic plate is especially sensitive to the blue and ultraviolet regions of the spectrum. Thus, if a star is very blue, it will appear brighter photographically than it does visually; if it is red, the reverse happens. This gives us the basis for a star's *color index*, which may be loosely defined as the difference between its photographic magnitude (using a blue-sensitive plate) and its visual magnitude. If a star is appreciably red, it might appear of magnitude 5·0 to the eye but only 6·0 on a photograph; the difference (6·0 — 5·0) gives + 1·0 for the color index. If the index is negative, the star must be blue. For the most reliable magnitudes, comparison stars should be as white as possible, with a very small color index. Many charts, to avoid this error, use accurate magnitudes derived by visual means, but this is not always possible.

A permanent difficulty is added to the situation when the variable itself has a definite tint. This is usually the case, for most LPV's have a conspicuous reddish hue, a problem we shall turn to presently.

Finding a variable marked in *Atlas of the Heavens,* or a similar atlas, may sound an easy task—until one remembers that the star in question may be of the 11th or 12th magnitude! It is therefore necessary to have a large-scale guide chart showing the more prominent stars in the region. This region is first identified by using the finder, or a low-power eyepiece. The stars are then used as pointers to the variable itself. The first attempt may take some time, especially if it lies in a crowded Milky Way region; but once the star has been picked up on a few different occasions, the field will be engraved in the memory and can be located in a matter of seconds. Astronomical societies such as the A.A.V.S.O. or B.A.A. usually provide at least two charts, one showing the general vicinity of the variable, the other giving a "close-up" view, showing perhaps a square degree of sky. All the stars lying within the variable's probable range of fluctuation are marked, and the ones with standard magnitudes, to be used for comparison in making estimates, are identified by letters or numbers.

Just occasionally, emergencies occur. A variable not on the observer's list may start behaving oddly and require estimates, or a nova may appear. In such a case, the only thing to do is to construct one's own guide chart, plotting all the stars in the field as accurately as possible and identifying them by symbols. It does not matter much if the actual magnitudes are unknown, since they can be looked up later on; the vital thing is to get estimates made while they are of value. If the nova is very bright, it can of course be estimated against other naked-eye stars in the same region; and this goes for the conspicuous variables like γ Cassiopeiae and Betelgeuse.

Fractional and step methods

To estimate a variable, examine its brightness in relation to the comparison stars, and select the ones that lie slightly on either side of the variable's magnitude. The next step is to decide just where it lies in the sequence, and it is here that the two main methods differ slightly.

The *fractional* method, which is somewhat the simpler of the two, is more suitable for the beginner to attempt. Here, the variable's magnitude is determined as a ratio between the brightness of two comparison stars A and B. If it lies midway between them, it is recorded thus in the standard shorthand: A 1 V 1 B, which means that the differences on both sides are equal. (A, the brighter star, is always written first.) It may, however, be judged $\frac{2}{3}$ fainter than A and $\frac{1}{3}$ brighter than B; in this case, the record will be A 2 V 1 B. If it is still closer to B in magnitude, the ratio might be judged in quarters instead of thirds; the entry will then be A 3 V 1 B.

The number of divisions that can be determined accurately depends on the difference between the magnitudes of the comparison stars. Even an experienced observer cannot judge values reliably to less than $\frac{1}{10}$; thus if A and B differ by only $\frac{1}{5}$ of a magnitude it is impossible to work in terms of more than two steps. Usually, however, the difference between adjacent comparison stars will be greater than this. If the variable is judged to be exactly equal to a comparison star, the magnitude can be derived directly.

At least two sets of estimates must be made, using different comparison stars, since these give a cross-check on accuracy. If the two reductions agree to $\frac{1}{5}$ of a magnitude, this is satisfactorily enough for most purposes; if there is a larger discrepancy, the observation must be checked with further estimates. So far as is humanly possible, all previous estimates must be ignored; so must one's natural expectancies of how the star will behave.

The *step* method is more sophisticated, and requires considerable training of the eye. In this case, the observer chooses several comparison stars and estimates the variable against each one separately. The magnitude difference is estimated in terms of "steps," which may be defined as the smallest difference of magnitude to which the observer's eye is sensitive. In most cases, it is about $\frac{1}{10}$ of a magnitude, and the precise value can be found by trial and error. If the variable is estimated to be 2 steps fainter than A and 1 step brighter than B, the entry would be A − 2, B + 1. Further stars should also be used if they are available.

There may at first seem to be little practical difference between the two procedural methods, but each has its advantage. If the difference between the variable and the comparison stars is about half a magnitude or more, the step method becomes unreliable. On the other hand, if the variable is brighter or fainter than any visible comparison star, then there is no "sequence" in which the fractional method can operate. The advantage of the step method is that reasonable estimates can be made against just one comparison star. Once proficiency has been gained, however, the observer will decide almost subconsciously which method to use under different circumstances.

When beginning variable-star work, a very useful exercise is to "estimate" stars whose magnitudes are known, both with the naked eye and with the telescope. In this way, one gets a good idea of what a magnitude "looks like," and by choosing stars that are closer in brightness, the detectable difference can be reduced to the smallest possible value. Experience and efficiency go hand in hand; the first results may be quite wild, but they will soon improve. When recording observations, it is necessary to give an estimate of the probable accuracy, and it is far better to be pessimistic than over-optimistic.

Some sources of error

The observation of variable stars is a field in which the amateur is attempting quantitative rather than qualitative standards. This means that exceptional care must be taken to allow for all the factors that can cause error and affect the result. Some of the more important ones are discussed below.

ATMOSPHERIC ABSORPTION. If naked-eye stars are being estimated, allowance must be made for dimming at low altitudes. A star near the horizon appears fainter than it would if near the zenith, so that the variable and its comparison stars should all lie at roughly the same altitude. If they are higher than 45°, dimming can be neglected; Table X provides the allowances necessary for lower altitudes:

TABLE X.   Atmospheric Dimming
(relative to zenith brightness)

ALTITUDE	43°	32°	26°	21°	19°	17°	15°	13°	11°	10°	6°	4°	2°	1°
DIMMING (MAG.)	0·1	0·2	0·3	0·4	0·5	0·6	0·7	0·8	0·9	1·0	1·5	2·0	2·5	3·0

This table ignores the presence of haze, which may dim very low stars by more than a magnitude and is bound to add extra uncertainty to the results. Telescopic variables are unaffected by this relative absorption, since all the stars in the field will lie at more or less the same altitude. Even so, it is wise to make estimates when the star is as high in the sky as possible.

MOONLIGHT, TWILIGHT, AND HAZE. Both moonlight and twilight spread a bluish cast in the sky. Naturally, this effaces the faintest stars, but of more consequence is the effect of the cast on stars of different colors. If a red and a blue star appear equal under dark conditions, the red one will appear superior in a bright sky, since the blue loses contrast with the background. A similar effect will be noticed if the observer happens to be surrounded by nocturnal illumination of any pronounced color. Haze and thin cloud, on the other hand, absorb red light more efficiently than other colors; so, under these conditions, a red star will be dimmed. These differences can amount to half a magnitude, or even more.

EYEPIECE AND RETINAL ERRORS. Neither eyepieces nor eyes are perfect over the whole of their surfaces. In an eyepiece, marginal stars appear somewhat blurred compared with central ones, and this blurring reduces their apparent magnitude. To minimize this defect, it is standard procedure with most observers to bring the variable and comparison stars alternately to the center of the field, holding them there just long enough to memorize their brightness. If the stars are close together, this precaution may not be necessary.

There is considerable controversy over the relative virtues of direct and averted vision. Some observers consider direct vision more reliable because

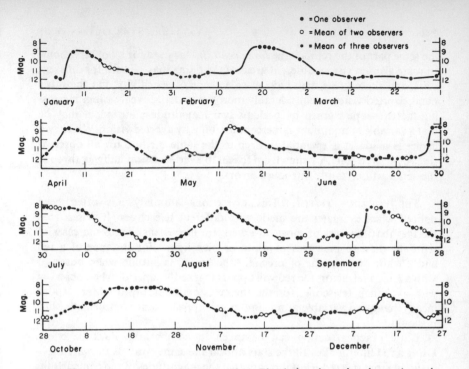

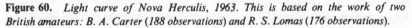

Figure 59. *Light curve of SS Cygni in 1964. This is based on the observations of three British amateurs: B. A. Carter (95 observations), J. S. Glasby (185 observations), and R. S. Lomas (103 observations).*

Figure 60. *Light curve of Nova Herculis, 1963. This is based on the work of two British amateurs: B. A. Carter (188 observations) and R. S. Lomas (176 observations).*

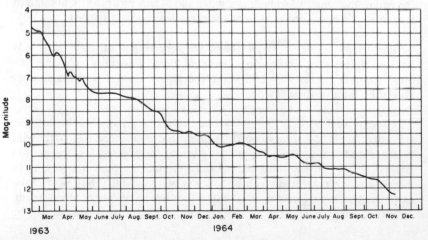

the same part of the retina—the *fovea centralis*—is used for all observations. In using averted vision, different areas of the retina may be used for different stars, and these areas may have a wide range of sensitivity. On the other hand, averted vision is considerably more acute than foveal vision, increasing the telescope's grasp by perhaps two magnitudes; indeed, it may be that a variable at minimum is perceptible only by averted vision. Whichever choice is made, it is essential to keep to the same method for all observations of a given star. A switch from foveal to averted vision halfway through the series will certainly introduce errors.

THE PURKINJE EFFECT. This mysterious anomaly has often been noticed when estimates are made with different telescopes. It arises from the fact that the eyes of most observers are more sensitive to increases of red light than to increases of white or blue light. For instance, if a red and a white star appear of precisely the same magnitude when observed with a 3-inch refractor, the red will appear markedly superior when observed with a 12-inch telescope, for the image is considerably brighter. It also means that a red variable at maximum will appear somewhat brighter than it should.

Little practical action can be taken to combat the Purkinje effect. Using a red filter makes all the stars appear the same color as the variable— assuming that it is red—but it means that the magnitudes of the comparisons will have to be redetermined. The most obvious precaution is to use the same telescope for all comparisons.

POSITION ANGLE EFFECT. If two equally bright stars of the same color are situated horizontally in the field of view, they appear equally bright; if they are situated vertically, the lower star usually appears the brighter of the two, sometimes by as much as a magnitude. This error can be minimized if each star is examined in turn at the center of the field of view.

BRIGHTNESS EFFECT. If two stars are very bright or very faint, it is hard to detect small differences between them. Generally speaking, the most accurate results are obtained in the region from 2 to 4 magnitudes above the limit of the telescope.

Clearly, variable-star observing has become a greatly refined art. On the other hand, it contains no difficulties that cannot be overcome with patience and enthusiasm. The actual stars to be selected for observation depend greatly on the available aperture. Some, the ones not sinking to below the 11th magnitude, can be followed with a 3-inch, but most of the interesting semiregulars need a 12-inch if they are to be caught at minimum. However, there is no reason why the unpredictable rises should not be caught with a much smaller instrument, so there is always plenty to do. An example of amateur work is shown in figure 59.

Nova-hunting

Estimates of a nova's decline and fall can be made by following the standard variable-star procedure. Figure 60 shows the combined work of two British observers, B. A. Carter and R. S. Lomas, on the fall of Nova Herculis, 1963. It is clear that the fading has not been regular; but other novae, such as Nova Persei, 1901, have been much more erratic. Some, such as Nova Herculis, 1934, have gone through perceptible color changes during their descent from glory. As with all the other objects in the night sky, we can never take anything for granted, and in many cases it falls to the patient amateur to investigate these deviations from the usual, and to hand over his findings to the more intensive scrutiny of the great observatories.

Nova-hunting, like comet-hunting, is a task that, though it might appear hopeless, offers tremendous compensations. It is fair to say that most novae have been discovered by chance, but just how often has chance been implemented by one's own initiative? If he is an earnest observer, the most "casual" glance at the sky or at a planet should be a moment at which the amateur is acutely attentive. Do any of the constellations look odd? Is there a delicate new belt on Jupiter? Did those meteors come from a known source? It is well worth spending five minutes checking up on something that may at first sight seem to lead nowhere, for it is just this that may discourage other observers from investigating further—and even if the trail is a false one, something new will have been learned. Often, too, a sixth sense comes into play, as is revealed by W. F. Denning's account, in *Telescopic Work*, of his discovery of one of his five comets:

> On July 11, 1881, just before daylight, I stood contemplating Auriga, and the idea occurred to me to sweep the region with my comet eyepiece, but I hesitated, thinking the prospect not sufficiently inviting. Three nights later Schaeberle at Ann Arbor, U.S.A., discovered a bright telescopic comet in Auriga! Before sunrise on October 4 of the same year I had been observing Jupiter, and again hesitated as to the utility of comet-seeking, but, remembering the little episode in my past experience, I instantly set to work, and at almost the first sweep alighted upon a suspicious object which afterwards proved itself a comet of short period.

He goes on:

> These facts teach one to value his opportunities. They cannot be lightly neglected, coming as they do all too rarely. The observer should never hesitate. He must endeavour to at least effect a little whenever an

occasion offers; for it is just that little which may yield a marked success—greater, perhaps, than months of arduous labour may achieve at another time.

So when the amateur, having concluded his observing program for the night, spends a few minutes scanning the Milky Way and its environment, he should consider just what opportunities are presented for the detection of a new star. The moon is the principal scene-changer, circling the ecliptic once a month and spending between two and three days in each zodiacal constellation. If a region of the Milky Way (e.g., in Orion or Sagittarius) has recently emerged from the moon's glare, it could conceivably harbor a nova which has not yet been detected; and around the time of full moon, the sky is less well scanned than when conditions are favorable, so that it is well worth putting up with the difficulties of observation in bright moonlight. Very faint naked-eye stars will be missed, of course, but of the 23 novae seen with the naked eye in the present century, 13 were brighter than magnitude 4·5 at discovery, so the chances of picking one up will still be fairly good.

Norton's Star Atlas is indispensable for this work; it shows stars down to about magnitude 6·5 and some fainter ones, so that any bright intruder can be clearly distinguished from any of the regular stars. Of course, there is always the possibility of its being a variable star near maximum (several variables can brighten up to the 4th or 5th magnitude), but the principal ones are represented in the atlas by small circles. One rule of thumb, though never to be used as more than a general guide, is that most novae are white or blue-white, while most LPV's are red. (Remember, though, that several "novae" have turned out to be *planets*; it is hoped that no reader of this book falls into that trap!)

To summarize: If a bright object is seen where neither star nor variable is marked in the star atlas, and if the positions of all likely planets can be accounted for, then the intruder is probably a nova. It should be reported at once giving its magnitude, position, exact time of observation, as described for a comet discovery (Chap. 16).

One sight that has given many amateurs, including myself, several shocks is a brilliant white object which on closer examination proves to be moving slowly across the sky. This trap can be set by any one of several bright satellites which are more remote than most of the artificial objects that now enliven the night sky. They move so slowly that at first glance they really do appear to be a new fixed star.

Yet not every "new" star is necessarily a nova. A member of the United Kingdom nova patrol, Dave Branchett, was sweeping in Scutum in the dawn of January 18, 1981, and described what happened then as follows:

I got up as usual to conduct some early-morning observations; these included Comet Panther and a few selected variable stars. Just before 6.15 U.T. cloud cover became evident from the southwest, so I decided, prior to dawn and total cloud cover, to conduct a nova search sweep with binoculars through Scutum. . . . On reaching the field of β and R Scuti, I noticed clearly an intruder star, placed to the south-west of β. My mind went blank, and I was dumb-struck. I checked my star chart and patrol print which goes down to about mag. 10½, but there was no star in this position.

I then estimated the position of the star and compared its magnitude with a field star and estimated the suspect to be about mag. 8.0. By this time cloud cover had reached the area and dawn was taking over. During the day I contacted various individuals who I knew would be interested in this star. The following morning was one of total cloud cover, not only for myself but for other fellow observers who had forsaken an hour or so of sleep to try and secure a second observation of the star. The morning of the 20th was fine, cold and clear. The Moon, nearly full, was over in the west in Gemini, its light drowning the background sky. Just after 6.00 U.T. I saw β Scuti rise. Soon it and the surrounding star field, including R Scuti, [were] high enough to be examined, but to my horror there was no sign of the intruding star. . . .

As one of the observers who joined in the search for this star, I too can record my disappointment as I found the star field in the lightening dawn, but with no sign of the intruder. Obviously, with no confirmation, it could not be announced. Yet, unknown to us, a photograph was obtained at the Royal Greenwich Observatory that morning, showing a 9th magnitude star in the correct position. Disappointment turned to elation—then to puzzlement. For the star simply vanished, lost among the myriads of faint ordinary stars that populate this particular region of the Milky Way. The photograph obtained at Greenwich, though definite enough to prove that the star had been in the sky on the morning of the twentieth, was not sufficiently good to allow an accurate position to be measured. The mystery will probably remain until, or unless, the star rises again. Was it an extraordinarily fast nova, or a strange new type of variable?

It has been estimated that upwards of 50 novae occur annually in the galaxy; this is based on the number that are detected in the Andromeda galaxy, which is believed to be similar to our own. Of these, only a tiny number are being found. One reason must be that many are distant and faint, but undoubtedly others are being missed. Statistically we should expect a marked increase in the number of faint novae compared with bright ones, in accordance with the increasing number of stars of fainter magnitude. For example, up to 1979 some 162 novae had been discovered since the middle of last century, of which five reached about magnitude 3–4 at maximum, and 12 reached magnitude 4–5. Assuming that no nova of magnitude 4½ or brighter had been missed (which is unlikely, in any case), then statistically about 30 novae of magnitude 5–6, 80 novae of magnitude 6–7, and about 300 of magnitude 7–8 should have been found in the same interval. In fact, only a third, a fifth, and a tenth, respectively, of these numbers have been found. G. E. D. Alcock, who looks not only for comets but also for faint novae with 11 × 80 binoculars, has commented: "I am most puzzled by the fact that so few novae have turned up in recent years. I am sure that many are still missed, especially those well away from the Milky Way; the type like Nova Herculis (1934 and 1963); T Coronae Borealis (1866); Nova Serpentis (1948); and Nova Pictoris (1925)—which is quite a list."

While few observers are likely to have the dedication to spend hundreds of hours every year in scanning the stars for a telescopic visitor, it does seem likely that a little more naked-eye effort by amateurs would have a beneficial effect on the rate of discovery of bright novae. It is futile to expect quick results—indeed, one may never make a discovery—but those rewarding acquaintances, the constellations, will more than compensate for that.

20

The Constellations

A separate book would be required to carry a full description of the spectacular objects to be found scattered across the night sky. The list given here is intended as merely a preliminary guide. Each item is marked in *Norton's Star Atlas*, and with a little care (and the assistance of a good finder) they can be swept up quite easily.

The stargazer's standard handbook, Webb's *Celestial Objects for Common Telescopes*, is once more easily available, having been reprinted in 1962 by Dover Books; unfortunately, it has not been brought up to date. Another useful compilation is Olcott and Putnam's *Field Book of the Skies*. *Norton's* itself includes useful lists of the brighter objects, and marks many others without giving details.

As a guide to the time of year at which any particular constellation is most suitably placed for observation, the period at which it is due south at midnight is given. An asterisk indicates that the group is so far south as to be wholly or mostly invisible from a latitude of 45°N; the standard three-letter I.A.U.† abbreviation is also included. An unlettered number for a nebula or cluster refers to the N.G.C.‡ The various objects are listed in order of increasing right ascension.

Andromeda (And; mid-October)

An important northern constellation, distinguished to the naked eye by the three bright stars running from the N.E. corner of the Great Square

†International Astronomical Union.
‡Dreyer's New General Catalogue, 1888.

of Pegasus toward Perseus. The Milky Way travels through its northern border.

Σ 3042   (7·0, 7·0; 5″·5; 85°); both white
Σ 3050   (6·0, 6·0; 1″·7; 262°); yellowish; binary system, stars closing
Σ 24   (7·2, 8·0; 5″·2; 284°)
π   (4·1, 8·0; 36″; 173°); white and blue
Σ 79   (6·0, 7·0; 7″·6; 193°); white and bluish
M.31   Easily found *np* Nu (*ν*), and visible with the naked eye
         In a small telescope it reveals a small bright nucleus, surrounded by an elliptical haze. This is the nearest spiral galaxy to our own, being a mere 2,200 million light-years away. Two small "satellite" galaxies (M.32 to the south and H.V.18 *np* at a greater distance) can also be made out.
H.VII.32 Open cluster of faint stars
R (var.)   5·0–15, 409ᵈ; remarkably deep red tint
W (var.)   6·7–14·5, 397ᵈ

Andromedid meteors are active during the period Nov. 17 to 30. Maximum about Nov. 27; slow-moving; radiant near Gamma (*γ*).

*Antlia, the Air Pump (Ant; late February)

An inconspicuous group near the southern rift in the Milky Way. It contains no star brighter than the 4th magnitude.

ζ¹ (5·9, 6·7; 8″·2; 211°)

*Apus, the Bird of Paradise (Aps; mid-May)

Distinguished by a small triangle of 4th-mag. stars near the southern celestial pole.

I 236   (5·7, 8·5; 1″·9; 110°)
θ (var.) 6·4–8·6, 119ᵈ

Aquarius, the Water Bearer (Aqr; late August)

A zodiacal constellation. An extensive but inconspicuous group lying between Pegasus and the bright southern star Fomalhaut. Contains prominent groups of small stars in the S.E. corner.

12   (6·0, 8·1; 2″·8; 192°); yellow and blue
41   (5·6, 7·6; 4″·9; 116°); deep yellow and blue; fine object

Aquarius (continued)

53	(6·0, 6·5; 5"·0; 320°); both white
ζ	(4·4, 4·6; 1"·8; 249°); binary, closing; at the limit of a 3-inch
107	(5·3, 6·5; 6"·5; 135°); white and bluish
M.72	A small globular cluster. The individual stars are so faint that a small instrument shows it simply as a circular nebulosity.
H.IV.1	Bright, elliptical planetary nebula. Found most easily with a moderate power, to distinguish it from a star; bluish; rather ill-defined
M.2	Large and bright globular cluster, visible in a good finder. A 4-inch resolves the outer regions into stars.
R(var.)	6·2–11, 387ᵈ

η Aquarids are visible during the first week in May; maximum about May 6. Meteors swift; not an intense shower.

δ Aquarids occur from mid-July to mid-August; maximum about July 29. Medium speed; rich shower in low latitudes.

Aquila, the Eagle (Aql; mid-July)

Distinguished by its 1st-magnitude star Altair, and lying in a rich region of the Milky Way, Aquila contains fine star fields and is well worth sweeping with a low-power. Altair forms the southern corner of the distinctive "summer triangle" (the other stars being Vega and Deneb) that characterizes the late summer sky.

5	(5·6, 7·4; 13"; 121°); white and bluish
Σ 2404	(5·8, 7·0; 3"·4; 182°); fine contrast of yellow and blue
23	(5·5, 9·5; 3"·4; 8°); difficult with 3-inch
π	(6·0, 6·8; 1"·4; 113°); stars yellowish
57	(5·2, 6·2; 36"; 171°); wide, colors curious
6709	Scattered group of 9th- to 11th-magnitude stars
R (var.)	6·2–12, 300ᵈ, period shortening

*Ara, the Altar (Ara, mid-June)

A small constellation lying south of Scorpio. It is in the Milky Way, and contains a number of fine clusters.

h4949	(6·5, 7·5; 3"·2; 267°)
6204	Small cluster of faint stars
6208	Scattered cluster of bright and faint stars
6250	Cluster containing 8th-magnitude and fainter stars

Aries, the Ram (Ari; late October)

A zodiacal constellation. The stars Alpha (a), Beta (β), and Gamma (γ) form a conspicuous triplet to the south of Andromeda's line. There is poor sweeping in this region, but it contains some fine double stars.

1	(6·2, 7·4; 2″·8; 166°); fine contrast of yellow and blue
γ	(4·2, 4·4; 8″·4; 360°); one of the finest pairs in the sky; both yellow; easy with low power
λ	(4·7, 6·7; 38″; 46°); white and yellowish
30	(6·1, 7·1; 39″; 274°); whitish and yellowish; wide
ε	(6·0, 6·4; 1″·5; 205°); both white; at the limit of resolution of a 3-inch
U (var.)	7–13, 370^d

ε Arietids are visible from October 12 to 23. Meteors are very slow; maximum about October 15.

Auriga, the Charioteer (Aur; mid-December)

A splendid constellation, distinguished by its yellow leader Capella (mag. 0·2), a star of interest in being very similar to the sun. This is one of the distinctive winter groups. The Milky Way passes through Auriga, in which lie three magnificent open clusters. There are many doubles; fine sweeping.

ω	(5·0, 8·0; 5″·8; 360°); white and bluish; pretty
14	(5·0, 7·2; 14″; 225°); cream and dull-blue
Σ 698	(6·2, 7·7; 31″; 346°); yellow and bluish
Σ 718	(7·2, 7·2; 7″·8; 74°); both white
θ	(2·7, 7·2; 2″·8; 332°); difficult with less than 4-inch
41	(5·2, 6·4; 7″·8; 355°); a pretty pair, both white
Σ 872	(6·0, 7·0; 11″; 217°); white and bluish
H.VII.33	Wide grouping of stars in a splendid region
M.38	Extensive cluster of bright stars ranged against star dust, with a cruciform outline. Closely south is a smaller cluster, H.VII.39.
M.36	Superb group of bright and faint stars
M.37	Magnificent cluster of faint stars, the individual members being glimpsed as powdery points against the sky
ε (var.)	Eclipsing type; range from magnitude 3·3 to 4·2; period 27 years—longest known for this kind of star
R (var.)	6·7–13·7, 458^d
UU (var.)	5·1–6·8, ±300^d

a Aurigids. There are two faint showers from this radiant: February 5–10 (slow); August 12–October 2 (very swift).

Boötes, the Herdsman (Boo; April-May)

The major star in this group, Arcturus (mag. −0·1), is the third bright-est in the sky, and is easily found by continuing the curve marked out by four of the bright stars in Ursa Major. The constellation lies well away from the Milky Way and contains no distinctive clusters; but there are many doubles.

Σ 1785	(7·2, 7·5; 3″·1; 148°); yellowish and bluish; neat binary pair
Σ 1816	(7·0, 7·1; 1″·9; 80°); yellowish pair
κ	(5·1, 7·2; 13″; 237°); white and lilac; attractive contrast
ι	(4·9, 7·5; 38″; 33°); white and gray pair in attractive field
Σ 1835	(5·5, 6·8; 6″·4; 195°); neat white pair
π	(4·9, 6·0; 5″·8; 110°); white pair
ζ	(4·4, 4·8; 1″·2; 308°); white, too close for 3-inch
ε	(3·0, 6·3; 2″·8; 340°); fine contrast of yellow and green; needs high power
39	(5·8, 6·5; 3″·3; 45°); white and bluish
ξ	(4·8, 6·9; 7″·0; 344°); yellow, bronze; binary system
R (var.)	6·0–13·0, 222ᵈ
34 (var.)	5·2–6·1, irregular

*Caelum, the Chisel (Cae; late November)

A small, inconspicuous asterism, west of Columba, containing little of interest. It used to be considered a part of the constellation Sculptor.

γ   (4·7, 8·5; 2″·9; 310°)

Camelopardus (or Camelopardalis), the Giraffe (Cam; late December)

A large but very obscure constellation near the north celestial pole, between Ursa Major and Cassiopeia. It contains no star brighter than the 4th magnitude.

Σ 485	(6·1, 6·2; 18″; 304°); white and bluish
1	(5·1, 6·2; 10″; 307°); white and bluish
2	(5·1, 7·4; 1″·6; 280°); yellow and bluish
Σ 1127	(6·2, 8·0, 9·2; 5″·5, 11″; 340°, 174°); triple; brighter stars white and gray
Σ 1625	(6·5, 7·0; 14″; 219°); both white
Σ 1694	(4·9, 5·4; 21″; 326°); yellowish and bluish
Z (var.)	10·2–14·5, irregular; subject to long standstills

Cancer, the Crab (Cnc; early February)

A small zodiacal constellation between Leo and Gemini; marked out by five 4th-magnitude stars, but containing its share of fine objects.

Σ 1177	(6·5, 7·4; 3″·5; 355°); white and bluish
ζ	(5·6, 5·9, 6·1; 1″·1, 5″·6; 348°, 82°); three stars, all yellow. A splendid sight, the close stars forming a binary system
φ²	(6·3, 6·3; 5″·0; 216°); white and grayish
ι	(4·4, 6·5; 31″; 307°); beautiful contrast of yellow and blue
σ²	(5·9, 6·4; 320°; 1″·5); both yellow; elongated with 3-inch
M.44	Known as Praesepe (the Beehive), and visible with the naked eye. A coarse cluster of bright stars, best seen in the finder since it is too extensive for ordinary telescopic fields, lying between Gamma (γ) and Delta (δ).
M.67	A splendid scattered cluster of bright and faint stars
R (var.)	6·0–11·3, 362ᵈ
X (var.)	5·9–7·3, 165ᵈ(?)

Canes Venatici, the Hunting Dogs (CVn; early April)

Lying below the curve of the "handle" of the Big Dipper (Ursa Major), this group is marked by only three conspicuous naked-eye stars. It lies in an extensive region of galaxies that extends from Ursa Major southward to Virgo, but few are bright enough to be seen well in small instruments.

2	(5·7, 8·0; 11″; 260°); deep yellow and blue; attractive pair
α	(3·2, 5·7; 20″; 228°); known as Cor Caroli; yellowish; a bright pair
25	(5·1, 7·1; 1″·7; 105°); white and blue; binary pair
M.94	A small cometlike nebulosity
M.63	An elliptical nebulosity with a brighter center. This is a distant spiral galaxy seen at an acute angle.
M.51	The famous Whirlpool Galaxy. In a small telescope, it is seen as two nebulae, one much larger than the other, almost in contact. Long-exposure photographs show it to be a spiral galaxy, with the smaller nucleus connected to the main system by an outflung arm.
M.3	A fine, bright globular cluster. The stars are too closely packed to be seen individually with a 3-inch, but they reveal themselves in larger instruments.
R (var.)	6·1–12·7, 333ᵈ
Y (var.)	5·2–6·6, 158ᵈ; a fine red star

Canis Major, the Greater Dog (CMa; early January)

A small but brilliant constellation lying southeast of Orion, near a rich part of the Milky Way. It cannot possibly be overlooked, for its leader, Sirius, is the brightest star in the sky, and it contains four 2nd-magnitude stars. It is rather low for observation in north temperate latitudes.

v^1 (6·0, 8·0; 17″; 263°); yellow, bluish

α (−1·4, 7·0; 10″; 85°). At its widest (11″·5 in 1975), it should be visible with an 8-inch reflector or a 6-inch refractor. This is a binary system, with a period of 50 years. The comes is a white dwarf star; it is about the size of Uranus, but almost as massive as the sun. It is only 1/2500 as luminous as Sirius A.

μ (4·7, 8·0; 3″·0; 339°); yellow and blue

M.41   An open cluster, lying 4° south of Sirius and visible with the naked eye. The brighter stars, some of which are orange, are arranged in distinctive curves. A splendid low-power sight.

H.VII.12   A beautiful powdery cluster of faint stars

Canis Minor, the Lesser Dog (CMi; mid-January)

Marked to the naked eye by its yellow leader, Procyon, east of Orion. Contains few objects of interest.

Σ 1103   (7·0, 8·5; 4″·3; 245°); white and grayish

Capricornus, the Goat (Cap; early August)

A zodiacal group lying south of Aquarius. A dull-looking constellation, but interesting for its leader, which is a naked-eye double.

a (3·2, 4·2; 376″; 291°); both yellow

π (5·1, 8·7; 3″·4; 145°); yellowish and bluish

o^2 (6·3, 6·8; 22″; 238°); white and bluish

M.30   A globular cluster appearing in a small instrument as a bright spherical nebulosity, centrally condensed

*Carina, the Keel (Car; late January)

The ancient, unwieldy constellation Argo Navis has been divided into Carina, Puppis (the Poop), and Vela (the Sails). Of these, Carina is the southernmost; it contains the second brightest star in the sky, Canopus (mag, $-0·9$). The original Greek-letter designations are retained; hence, we find Alpha (a) and Beta (β) in Carina, Gamma (γ) and Delta (δ) in Vela, and so on. The Milky Way passes through the region, which lies southeast of Orion and Canis Major.

C	(5·3, 8·0; 3"·8; 64°)
h4213	(6·0, 9·4; 8"·8; 327°)
v	(3·2, 6·0; 5"·0; 128°)
2808	A rich globular cluster, appearing as a nebulous blur in a small telescope
3114	A loose cluster of bright and faint stars
R (var.)	5·6–11·0, 309^d
S (var.)	4·5–10·0, 149^d

Cassiopeia (Cas; mid-October)

One of the most distinctive northern constellations; its characteristic M or W—depending on whether it appears above or below the polestar —can always be recognized. It lies in a rich part of the Milky Way, and the sweeping is superb.

σ	(5·4, 7·5; 3"·1; 327°); white and bluish; in a superb region
Σ 3053	(6·0, 7·3; 15"; 71°); yellow and blue
η	(3·7, 7·4; 11"; 298°); yellowish and purple; an attractive binary pair
ψ	(4·4, 8·9; 20"; 120°); yellow and blue; companion a close double
Σ 163	(6·2, 8·2; 35"; 36°); rich gold and blue; fine contrast
Σ 191	(6·2, 8·5; 5"·6; 191°); white and blue
ι	(4·2, 7·1, 8·1; 2"·4, 7"·4; 251°, 113°); a splendid triple star; yellow, blue, blue
M.52	A most beautiful cluster, somewhat triangular. A 3-inch shows it granular with faint stars.
H.VIII.78	A curious group of 9th-magnitude stars, shaped rather like a mushroom
H.VI.30	A large, faint, compressed cluster
	There are many other clusters in this constellation.
R (var.)	4·8–13·6, 431^d
S (var.)	6·2–15·3, 610^d
ρ (var.)	4·1–6·2, irregular
γ (var.)	2·0–3·3, irregular

*Centaurus, the Centaur (Cen; late March)

An extensive southern constellation lying partly in the Milky Way and containing some fine objects. The leader is a splendid binary star; a third, fainter member of the same system is the nearest star to the sun.

I 178	(6·3, 6·3; 1″·0; 94°)
D	(5·3, 6·5; 2″·9; 245°); both yellow; a fine object
γ	(3·1, 3·2; 1″·6; 2°); a fine binary pair
Q	(5·4, 6·8; 5″·2; 164°)
k	(4·5, 5·9; 7″·6; 110°)
y	(5·6, 5·8; 1″·2; 102°)
α	(0·3, 1·7; 14″; 200°); both yellow; perhaps the finest binary in the sky. A nearby 11th-magnitude companion, belonging to the same system and known as Proxima Centauri, is the sun's nearest neighbor.
T (var.)	5·2–10·0; 91ᵈ
3766	Rich, condensed cluster of bright and faint stars
ω	Appears as a hazy "star" to the naked eye; a magnificent globular cluster ablaze with faint stars—probably the largest and finest in the sky
5460	A loose cluster of bright stars

Cepheus (Cep; late September)

This group extends almost to the north celestial pole. It contains few bright stars, but there is some good sweeping in the southern region. An arm of the Milky Way intrudes from Cygnus.

κ	(4·0, 8·0; 7″·4; 122°); white and blue
Σ 2751	(6·0, 7·0; 1″·9; 344°); both white
β	(3·3, 8·0; 14″; 250°); greenish white and blue
Σ 2816	(6·3, 7·9, 8·0; 12″, 20″; 120°, 340°); yellowish, with bluish companions at either side
Σ 2840	(6·0, 7·0; 20″; 194°); white and blue; an attractive pair
ζ	(4·7, 6·5; 8″·0; 280°); yellowish and bluish
Σ 2893	(5·5, 7·6; 29″; 348°); yellowish and bluish
δ	(var., 5·3; 41″; 192°); yellow and blue; fine contrast The bright star is the prototype Cepheid variable, ranging from magnitude 3·8 to 4·6 in a period of 5½ days.
Σ 2950	(6·0, 7·2; 2″·3; 290°); yellow and grayish
o	(5·2, 7·8; 3″·0; 210°); yellow and greenish
T (var.)	5·1–10·5, 387ᵈ

μ (var.)   3·7–4·7, irregular. Herschel's "Garnet Star", shining like a drop
of blood. There is a rough period of between 5 and 6 years.

Cetus, the Whale (Cet; mid-October)

A dull, extensive constellation, extending from the west of Aquarius
northward toward Taurus. Since the ecliptic passes very close to its northern
boundary, the planets can sometimes lie in this constellation.

42   (6·2, 7·2; 1″·4; 30°); both white
Σ 147   (6·0, 7·3; 2″·9; 89°); white and yellowish
66   (6·0, 7·8; 16″; 232°); yellow and blue
γ   (3·7, 6·2; 3″·0; 295°); yellowish and grayish; curious colors
M.77   Spiral galaxy, appearing as a rather dim nebulous patch
T (var.)   6·6–7·7, 160^d
o (var.)   Mira; extreme range 1·7–9·6, period 330^d

*Chamaeleon, the Chameleon (Cha; late February)

A small group lying near the south celestial pole, containing five 4th-
magnitude stars.

ε   (5·4, 6·2; 1″·1; 310°)
3195   A small planetary nebula

*Circinus, the Compasses (Cir; April-May)

This small group is situated in the Milky Way, near Centaurus, and
so contains some good sweeping.

γ   (5·5, 6·0; 1″·3; 108°)
a   (3·4, 8·8; 16″; 232°); yellow and reddish
5715   A loose cluster of faint stars

*Columba, the Dove (Col; mid-December)

An inconspicuous constellation south of Lepus, marked to the naked
eye by the small triangle of Alpha, Beta, and Epsilon (a, β, ε). It contains
no objects of interest for a small telescope.

Coma Berenices, Berenice's Hair (Com; March-April)

Virtually an extensive naked-eye cluster of faint stars, north of Virgo.
Like its neighbors (Virgo, Leo, and Canes Venatici), it contains a great

Coma Berenices (continued)

number of dim galaxies. Only a few, however, are bright enough to be at all noticeable in a small telescope.

2	(6·0, 7·5; 3″·9; 235°); white and blue
24	(4·7, 6·2; 20″; 271°); yellow and blue
M.98	Faint galaxy, elongated in an E–W direction; very close to star 6
M.99	A large, pale galaxy on the other side of 6 from M.98
M.100	Circular nebulosity with little central condensation
M.85	Small nebulosity with a bright, almost stellar, central condensation. There is a 9th-magnitude star f.
M.88.	Elongated nebulosity, centrally condensed
M.64.	Rather faint elliptical nebulosity
M.53.	Small cluster of faint stars, much compressed
40 (var.)	5·5–5·9, 37^d

*Corona Australis, the Southern Crown (CrA; late June)

Marked by a curve of faint stars south of Sagittarius.

h5014	(5·8, 5·8; 1″·6; 221°); binary pair
κ	(6·0, 6·6; 22″; 359°)
γ	(5·0, 5·0; 2″·7; 33°); a fine pair

Corona Borealis, the Northern Crown (CrB; mid-May)

A distinctive semicircular group of stars east of ε Boötis.

ζ	(4·0, 4·9; 6″·3; 306°); white and turquoise; beautiful
σ	(5·7, 6·7; 6″·3; 231°); yellowish and grayish; a beautiful pair
R (var.)	5·8–14·8, irregular; long maxima with sudden falls
S (var.)	5·8–13·9, 361^d
T (var.)	2–9·5, irregular; the "Blaze Star"

Corvus, the Crow (Crv; late March)

A small constellation south of Virgo, easily identified by its distinctive trapezoid appearance. In this, it resembles Crater, its western neighbor.

δ	(3·0, 8·5; 24″; 212°); yellowish and bluish
Σ 1669	(6·1, 6·2; 5″·4; 308°); both yellowish
R (var.)	5·9–14·4, 317^d

Crater, the Cup (Crt; mid-March)

Marked by a trapezium of 4th-magnitude stars. There are few objects of interest for a small telescope.

Σ 1509   (7·2, 9·0; 33″; 15°); brighter star yellow

*Crux, the Cross (Cru; late March)

A splendid, compact constellation lying in that part of the Milky Way nearest the south celestial pole, and containing the famous black aperture known as "the Coalsack," which is simply a nearby dark nebula blotting out the stars beyond it. Crux is the smallest group in the sky, but with its clearly cruciform shape it is unmistakable.

α   (1·4, 1·9; 4″·7; 119°); a noble pair
ι   (4·7, 7·8; 26″; 25°)
μ   (4·5, 5·5; 35″; 17°)
4755 A magnificent open cluster of more than a hundred bright stars surrounding Kappa (κ). They range from the 7th magnitude downwards, and in a telescope of sufficient power a number of red and blue shades are visible. Sir John Herschel described it as resembling "a superb piece of fancy jewelry."

Cygnus, the Swan; also, the Northern Cross (Cyg; July-August)

A fine constellation which might aptly be called the Northern Cross, Cygnus lies in a superb region of the Milky Way and its low-power fields are encrusted with stars. There seems to be no end to its pairs, triplets, and clusters. Fresh combinations continually delight the eye, and the background to these patterns is powdery with half-glimpsed points of light. There are distinctive dark nebulae near Alpha (α) and Gamma (γ); and south of Gamma the Milky Way divides into separate streams.

Σ 2486   (6·0, 6·5; 8″·9; 210°); yellow pair in a rich field
β   (3·0, 5·3; 35″; 55°); mid-yellow, intense blue. One of the showpieces of the sky, being an easy pair that can be divided in any telescope
16   (5·1, 5·3; 38″; 134°); yellow pair, superb field
δ   (3·0, 6·5; 2″·1; 240°); difficult because of the brilliant primary; easiest in a twilight sky
Σ 2578   (6·6, 7·4; 15″; 127°); white and pale blue; a pretty pair
OΣΣ 191   (6·0, 8·0; 38″; 28°); gold and blue
ψ   (5·0, 7·5; 2″·9; 170°); white and pinkish

Cygnus (continued)

Σ 2671	(6·0, 7·4; 3″·4; 336°); white and ashen
61	(5·3, 5·9; 28″; 140°); both yellow; first stars to have their distance measured (in 1838)
Σ 2762	(6·0, 8·0; 3″·5; 316°); white and bluish
H.VII.59	A small aggregate of faint stars

Two views of Cygnus. *Taken with a Polaroid camera and ASA 3000 film. Left: Cygnus star trails, photographed on July 21, 1963. Below: Star clouds in Cygnus. This photograph was taken with a hand-driven equatorial mounting. (Thane P. Bopp, Kirkwood, Missouri.)*

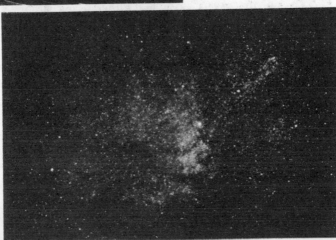

M.29   A small group of 8th-magnitude stars with fainter associates

H.V.14   Part of the Filamentary Nebula, well seen only in long-exposure
            photographs. A small telescope shows about 120° of a large
            nebulous circle that fits inside the margin of a low-power eye-
            piece. There are other faint nebulosities in the region.

H.I.192   Small nebulosity surrounding a 9th-magnitude star

ζ   There is very extensive, extremely faint nebulosity surrounding
            this star. It is best seen in the finder, or with binoculars.

7039   A small cluster of 9th-magnitude and fainter stars, lying
            between two stars of the 7th-magnitude

M.39   A coarse, triangular cluster of bright stars, with a double star
            at the center. Best seen in the finder

χ (var.)   2·3–14·3, 406ᵈ

R (var.)   5·9–14·6, 426ᵈ

RT (var.)   6·2–13·0, 190ᵈ

T (var.)   5·5–6·0, irregular

U (var.)   6·1–12·2, 462ᵈ

W (var.)   5·0–7·6, 131ᵈ + 125ᵈ (double period)

κ Cygnids. A short shower of slow meteors occurs on January 17; a
second shower, medium speed, radiates from August 10 to 20.

a Cygnids. A prolonged shower of swift, trained meteors occurs during
July and August.

Delphinus, the Dolphin (Del; July–August)

A compact and unmistakable constellation lying on the southern
border of the Milky Way, near Aquila; the arrangement of its five main
stars is somewhat fishlike. There are some rich fields here.

γ   (4·0, 5·0; 10″; 268°); yellow and turquoise; a charming pair

U (var.)   5·6–7·5, irregular

*Dorado, the Swordfish (Dor; mid-December)

An otherwise obscure far-southern group, made noteworthy by containing the Nubecula Major, or Greater Magellanic Cloud.

NUBECULA MAJOR. Appearing to the naked eye as a detached fragment of the Milky Way, this is the more prominent of two "satellite" star
systems revolving around the Galaxy. It is about 200,000 light-years away
(one-tenth of the distance of the Andromeda Galaxy), and since it is the

Dorado (continued)

nearest external system it is of great importance to professional astronomers. It contains many telescopic objects, the most prominent being the Great Looped Nebula, bright enough to be seen with the naked eye.

R (var.)   5·7–6·8,   360^d

Draco, the Dragon (Dra; mid-May)

An extensive constellation winding for almost 180° around the north celestial pole. The p region, that part north of Ursa Major, contains some dim nebulae. The easiest way of beginning identification is to find the two pairs of bright stars—Beta (β) and Gamma (γ); Eta (η) and Zeta (ζ)—which lie between Vega and β Ursae Minoris. After that, the rest of the Dragon's body can be traced.

Σ 1984	(6·2, 8·5; 6″·6; 274°); white and bluish
17	(5·0, 6·0; 3″·7; 116°); cream and bluish
v	(4·6, 4·6; 62″; 313°); both yellowish; well seen in the finder
ψ	(4·0, 5·2; 31″; 15°); yellow and lilac
40	(5·4, 6·1; 20″; 234°); yellow and pale yellow; the companion is known as 41
39	(4·7, 7·7; 3″·1; 6°); white and reddish; 7·1-magnitude star nearby
Σ 2348	(5·9, 8·1; 26″; 273°); fine contrast of yellow and blue
ε	(4·0, 7·6; 3″·5; 7°); cream and blue
R (var.)	6·3–13·9, 245^d
RY (var.)	5·6–8·0, irregular
H.IV.37	A bright planetary nebula looking exactly like a 5th-magnitude star slightly out of focus. Best picked up with a medium power, which exaggerates the disk. There is an 8th-magnitude star np.

Quadrantids. Known after the forgotten constellation Quadrans Muralis. An intense shower of bright, swift meteors occurring between December 30 and January 4, maximum on January 3.

ι Draconids. Short-lived shower of slow meteors, June 27–30, maximum June 28.

γ Draconids. Extensive shower lasting from June until August. Meteors slow; maximum June 25.

ζ Draconids. These occur between August 21 and 30; medium speed. Fainter meteors radiate from near Omicron (o) around August 22.

Equuleus, the Little Horse (Equ; mid-August)

A tiny asterism on the eastern border of Delphinus.

ε   (5·7, 7·1; 11"; 72°); yellowish and dull white. The brighter is a very close binary.

*Eridanus, the River (Eri; mid-November)

From the region west of Orion, this figure straggles southward toward its leader, Achernar (mag. 0·6), whose declination is −57°.

θ   (3·4, 4·4; 8"·2; 88°); a fine pair
f   (4·9, 5·4; 7"·8; 211°)
32   (4·0, 6·0; 7"·0; 347°); brilliant yellow and blue-green; a superb object
39   (6·0, 8·8; 6"·5; 146°); yellow and blue
55   (6·2, 6·7; 9"·3; 317°); both yellowish
H.IV.26   A small planetary nebula with a nucleus that is not perfectly central, but nearer the *sp* border

*Fornax, the Furnace (For; late October)

A large but unremarkable constellation lying in the dull region south of Cetus and the delta of Eridanus.

ω   (5·5, 8·0; 10"; 244°)
h3532   (6·5, 8·0; 5"·5; 145°)
H.V.48   Conspicuous nebulosity extended in the *sf* direction

Gemini, the Twins (Gem; early January)

The two principal stars, Castor and Pollux (mags. 1·6 and 1·2, respectively), can be found without difficulty some 40° northeast of Orion; the other bright stars take the form of an elongated parallelogram. There are many interesting objects in this group.

20   (6·0, 6·9; 20"; 211°); yellow and blue; fine field
38   (5·4, 7·7; 7"·0; 150°); yellowish, bluish
δ   (3·2, 8·2; 6"·2; 220°); pale yellow, reddish; delicate object with 3-inch
α   (2·0, 2·9; 1"·9; 151°); Castor; both stars yellowish. This is the brightest binary pair in the northern hemisphere; the stars have just passed periastron and are beginning to open up again. The widest separation, 6"·5, will be reached in about 80 years' time.

Gemini (continued)

κ   (4·0, 8·5; 6"·8; 236°); deep yellow, pale blue; beautiful

M.35   A striking, extensive cluster of bright stars, somewhat too large
           for normal fields, seen against a rich background. Visible with
           the naked eye

H.IV.45   A bright planetary nebula surrounding an 8th-magnitude star

η (var.)   3·2–4·2, 231ᵈ

R (var.)   5·9–13·8, 370ᵈ

Geminids. A prominent shower of swift meteors lasting throughout the
first half of December, reaching maximum on the 10th.

*Grus, the Crane (Gru; early September)

This group is not difficult to identify, for it lies immediately south of
Fomalhaut, the leader of Piscis Austrinus. There are other celestial "birds"
nearby: the Phoenix, the Toucan, and the Peacock.

θ   (4·5, 7·0; 1"·4; 50°)

D246   (6·1, 6·8; 8"·4; 257°)

S (var.)   6·0–15·0, 400ᵈ

Hercules (Her; mid-June)

This extensive and important constellation is not too easy to make
out, since it contains no star brighter than the 3rd magnitude; the best
signpost is the trapezium formed by Pi (π), Eta (β), Zeta (ζ), and Epsilon
(ε), which lies between Vega and Corona Borealis. Once identified, Hercules
provides an almost inexhaustible store of fine doubles, of which only a
few can be listed here.

κ   (5·0, 6·0; 29"; 14°); yellow and bronze; fine field

Σ 2063   (5·7, 8·2; 16"; 194°); white and bluish

Σ 2104   (6·2, 8·0; 5"·9; 20°); yellowish, clear blue; a very pretty pair

a   (var., 6·1; 4"·4; 110°); golden-yellow and greenish; a fine pair.
           The primary varies from 3rd to 4th magnitude.

ρ   (4·0, 5·1; 3"·8; 317°); white and grayish

Σ 2194   (6·2, 8·5; 16"; 9°); gold and blue

Σ 2245   (6·8, 7·0; 2"·6; 295°); a neat pair, both white

95   (4·9, 4·9; 6"·2; 259°); yellowish and white; beautiful

100   (5·9, 5·9; 14"; 183°); both white, a superb equal pair

M.13   The largest and brightest globular cluster in the northern sky, just visible with the naked eye. A small telescope shows it as an extensive spherical nebulosity, brighter at the center; a high magnification allows some of the marginal stars to be detected by averted vision. A superb sight in a large telescope.

Σ 5N   A small planetary nebula, like a bright star out of focus; one of the few distinctive objects overlooked by Sir William Herschel, and later discovered by F. G. W. Struve. A high magnification is needed to make the bluish disk obvious.

M.92   A smaller version of M.13. The stars are more compressed, and there is only a suspicion of marginal resolution if a small telescope is used.

30 (var.)   4·7–6·0, irregular
S (var.)   5·9–12·5, 300^d
U (var.)   6·2–13·3, 406^d
a (var.)   3·1–3·9, irregular

*Horologium, the Clock (Hor; mid-November)

A barren far-southern group nearly devoid of interesting telescopic objects.

R (var.)   6·3–15·0, 401^d

Hydra, the Water Monster (Hya; mid-March)

The longest and largest constellation in the sky, with an area of 1,303 square degrees; it is also one of the most obscure. Named after Hydra, the nine-headed serpent, or monster, of Lake Lerna, slain by Hercules. Its "head" lies north of the equator, between Leo and Canis Minor; but it extends eastward almost to Scorpio. Considering its size, it is poorly stocked with objects.

Σ 1245   (6·0, 7·0; 10″; 26°); pale yellow and reddish
$ε$ (3·8, 7·8; 3″·6; 270°); yellow and blue
Hh376   (5·8, 5·9; 9″·1; 212°); white and bluish; the companion may be variable
$β$ (4·4, 4·8; 1″·2; 360°); a fine close pair
54   (6·0, 7·5; 9″·0; 129°); yellow and bluish
H.IV.27   A bluish planetary nebula of the same apparent size as Jupiter, with a bright nucleus. Lies 2° south of Mu ($μ$).

R (var.)   4·0–10·0, 386^d
U (var.)   4·5–6·0, irregular
V (var.)   6·0–12·5, 532^d

*Hydrus, the Water Snake (Hyi; late October)

A much smaller serpent than Hydra, the monster, Hydrus lies near the south celestial pole; its name is distinguished in the genitive form by the masculine ending *i* (Hydri), as against Hydra's feminine form, Hydrae.

h3475   (6·5, 6·5; 3"·3; 38°); an attractive pair
h3568   (5·7, 7·7; 15"; 224°)

*Indus, the Indian (Ind; mid-August)

A straggling southern constellation in a dull region.

θ (4·7, 7·1; 6"·0; 275°)

Lacerta, the Lizard (Lac; late August)

An obscure group lying between Cygnus and Andromeda, containing some fine Milky Way fields.

Σ 2894   (6·0, 8·2; 16"; 194°); white and bluish
8   (6·0, 6·5; 22"; 186°); yellowish; two faint stars *sf*
H.VIII.75   A bright cluster set in a rich region

Lacertids. These rather faint meteors radiate during August and September; no definite maximum.

Leo, the Lion (Leo; February-March)

A splendid zodiacal constellation, in form very reminiscent of a crouching lion. The ecliptic passes just south of its leader, Regulus (mag. 1·3), which suffers periodical occultation by the moon. Leo contains a number of faint galaxies.

γ (2·4, 3·8; 4"·3; 122°); both golden-yellow; a magnificent binary
          pair
49   (6·0, 8·7; 2"·4; 158°); white and bluish
54   (5·0, 7·0; 6"·3; 110°); white and blue
83   (6·3, 7·3; 29"; 150°); yellow and reddish
88   (6·4, 8·2; 15"; 330°); pale yellow and bluish
90   (6·0, 7·3; 3"·4; 209°); white and bluish
R (var.)   5·4–10·5, 312^d
M.95   Circular nebulosity
M.96   Circular nebulosity, less well defined than M.95; two faint
           nebulae *nf*

M.65 &   Two elongated nebulosities lying in the same low-power field
M.66

Leonids. An unpredictable display, usually faint, but prominent in 1961 and 1965 and magnificent in 1966. Meteors, swift-moving, occur between November 9 and 17 with a maximum about November 16.

Leo Minor, the Lesser Lion (LMi; late February)

A small group to the north of Leo. It contains a few galaxies, but none bright enough to be of general interest.

Lepus, the Hare (Lep; mid-December)

Easily found, since it lies directly south of Orion and contains a conspicuous trapezium of bright stars. Here is found the famous "Crimson Star" R.

ι	(4·2, 10·5; 13″; 335°); white primary; companion difficult with 3-inch
κ	(5·0, 7·5; 2″·6; 360°); yellowish and bluish
a	(4·0, 9·5; 35″; 156°); a group of four faint stars f
M.79	A globular cluster, appearing in a small telescope as a bright, central condensed patch
R (var.)	6·0–10·4, 430^d. The celebrated star which its discoverer, John Russell Hind, described in 1845 as "resembling a blood-drop on the background of the sky; as regards depth of color, no other star visible in these latitudes could be compared with it." In a small telescope, its striking tint can be well seen only near maximum.

Libra, the Balance (Lib; mid-May)

A dull group but easily found, since it lies between Virgo and Scorpio. Its brightest star, Beta (β), has a somewhat greenish tinge, which is not, however, obvious to the casual glance.

μ	(5·4, 6·3; 2″·0; 350°)
Σ 1962	(6·3, 6·4; 12″; 187°); both white, a fine pair
H.VI.19	A globular cluster, large and dim
ι (var.)	4·3–5·0, irregular

*Lupus, the Wolf (Lup; mid-May)

Marked by a conspicuous group of 3rd- and 4th-magnitude stars along the western border of the Milky Way, south of Scorpio.

h4715 (6·1, 6·6; 3"·0; 278°)
π (4·7, 4·8; 1"·5; 75°); fine sight in a 4-inch
κ (5·2, 6·5; 27"; 144°)
μ (4·8, 5·2; 1"·6; 145°)
ε (4·0, 9·0; 26"; 175°)
ξ (5·5, 6·0; 11"; 49°); a fine pair
η (4·0, 8·0; 15"; 22°); the primary may be variable
5822   A small cluster of 8th-magnitude and fainter stars
5986   Bright globular cluster, but unresolved in a small telescope; visible in the finder

Lynx (Lyn; mid-January)

One of the dullest northern constellations, filling the gap between Auriga and Ursa Major. There are many fine doubles, but most of them are faint.

12   (5·2, 6·1; 1"·7; 100°); both white; a third star at 8"·6, 310° makes it triple
Σ 958 (6·0, 6·0; 5"·1; 257°); both yellowish; beautiful
19   (5·3, 6·6; 15"; 315°); white and bluish
38   (4·0, 6·7; 2"·9; 230°); both yellowish

Lyra, the Lyre (Lyr; early July)

Small in size but rich in objects, Lyra lies in the Milky Way and contains a great many pairs, triplets, and beautiful fields. Its leader, Vega (mag. 0·0), is the brightest star in the northern hemisphere, and during late summer in north temperate latitudes passes almost directly overhead.

α (0·0, 9·0; 60"; 175°); the companion is hard to see against the blaze of the primary. A 3-inch shows many other stars in a low-power field.
Σ 2380 (6·7, 8·2; 26"; 10°); cream and blue; a pretty pair
ε (4·0, 4·5; 208"; 173°); a naked-eye pair, though very difficult to resolve. This is the famous double-double, i.e., each star is itself a binary pair: ε^1 (4·6, 6·3; 2"·9; 355°); and ε^2 (4·9, 5·2; 2"·3; 100°). All four stars are white except the 6·3, which is bluish.
(4·2, 5·5; 44"; 150°); white and cream; fine low-power object

Lyra. *The constellation photographed on July 30, 1962. (Stephen A. Walther, Stevens Point, Wisconsin.)*

Σ 2470   (6·7, 8·2; 14″; 267°); white and bluish
Σ 2474   (6·7, 8·0; 17″; 259°); cream and pale blue; in the same field as
          Σ 2470 forming a second double-double
M.57   A remarkable object: an annular planetary nebula, bright, bluish,
          and slightly elliptical, looking like a smoke ring—hence the name
          Ring Nebula
M.56   A globular cluster, unresolved in a small telescope but bright, with
          a curious V of 7th-magnitude stars *nf*
R (var.) 4·0–4·7, 46^d

Lyrids. One of the few spring showers, swift meteors occurring between April 16 and 22 (maximum, April 20); occasional Lyrids are also seen in May. Records of "April meteors" go back for some 2,500 years, so they are clearly distinctive.

*Mensa, the Table (Men; mid-December)

An obscure group very near the south celestial pole, and containing part of the Nubecula Major. It was originally christened Mons Mensae

Mensa (continued)

(for Table Mountain, south of Cape Town), presumably to immortalize Sir John Herschel's expedition to the Cape of Good Hope; but his arduous researches in the southern skies deserve greater recognition than this barren offering.

2058   A conspicuous nebula, with a central condensation, lying in the Nubecula Major

*Microscopium, the Microscope (Mic; early August)

An almost imperceptible group south of Capricornus; contains only one star brighter than the 5th magnitude.

α   (5·0, 8·5; 22″; 164°)

Monoceros, the Unicorn (Mon; early January)

Contains few conspicuous stars, but lies in a splendid region of the Milky Way to the east of Orion and is replete with memorable telescopic objects.

8(ε)	(4·0, 6·7; 13″; 27°); rich yellow and bluish; magnificent field
11(δ)	(5·0, 5·5, 6·0; 7″·2, 2″·5; 132°, 105°); all white; a glorious triple; one of the showpieces of the skies
Σ 921	(6·0, 8·2; 16″; 4°); yellowish and bluish
15	(6·0, 8·8; 4″·0; 100°); dull-white and blue; fainter pairs nearby
Σ 1183	(5·5, 7·8; 31″; 326°); yellowish and white
H.VII.2	A splendid open cluster of bright and faint stars, including 12 Mon (yellow)
H.VI.27	A fine, bright open cluster
M.50	Attractive cluster of faint stars, in a glorious region
H.VI.37	A great heap of powdery stars
U (var.)	5·9–8·0, 92^{d}
V (var.)	6·0–14·0, 334^{d}

*Musca, the Fly (Mus; March-April)

A small group in the vicinity of Crux, containing some attractive Milky Way fields.

h4432	(5·7, 6·5; 2″·5; 302°)
h4498	(6·2, 7·9; 8″·7; 61°)
β	(3·9, 4·2; 1″·3; 6°)

θ (5·8, 8·0; 5″·7; 186°)
4833   A globular cluster, too condensed for resolution

*Norma, the Square (and Rule) (Nor; mid-May)

Lies in a rich region of the Milky Way, to the south of Scorpio, and offers much telescopic work. A low-power will reveal many clusters and attractive combinations of stars.

ι^1 (5·5, 8·5; 10″; 252°)
ε (4·8, 6·5; 24″; 335°)
6067   A fine open cluster
6087   A scattered cluster of bright stars
6115   A concentration of stars in a region of unbounded splendor
T (var.) 6·2–13·4, 243^d

*Octans, the Octant (Oct; August)

The south-polar constellation, but containing no conspicuous marker as does the north; can be almost equally well observed at any season of the year, though lacking in interesting objects. The 5th-magnitude star Sigma (σ) is only 1° away from the celestial pole.

λ (5·5, 7·7; 3″·1; 67°)

Ophiuchus, the Serpent Bearer (Oph; mid-June)

An extensive constellation north of Scorpio. The boundary is well marked out with stars, the interior obscure; but it contains many fine objects, especially the southern part, which intrudes into a magnificent region of the Milky Way.

ρ (5·7, 6·4; 3″·4; 350°); yellowish and reddish
36   (5·6, 5·7; 4″·3; 180°); both rich yellow; a splendid pair
39   (5·5, 6·0; 11″; 355°); deep yellow and blue
Σ 2166   (5·6, 7·4; 28″; 283°); white and deep blue
61   (5·5, 5·8; 21″; 94°); both white, very neat
τ (5·3, 6·0; 1″·9; 273°); both cream, just divided with 3-inch
70   (4·3, 6·0; 3″·4; 78°); both golden-yellow; a superb pair
Σ 2276   6·0, 6·3; 6″·8; 258°); both white
M.12   A fine, bright, condensed cluster, partly resolved in a small
           telescope
M.10   Globular cluster; appears as a bright nebulosity with a blazing
           center
M.19   Globular cluster, resolved with large instruments

Ophiuchus (continued)

H.I.45 & H.I.147	Two faint nebulae seen in almost the same low-power field
M.9	A globular cluster seen as a bright nebulosity; a faint star to the south
M.14	An extensive globular cluster, resolvable with large instruments. There is a wide double *p* and a coarse triple *f*.
H.VIII.72	A fine open cluster, just visible with the naked eye, consisting of 8th-magnitude and fainter stars; it has two distinct nuclei. The very ground of the Milky Way is seen glittering with minute points.
R (var.)	6·2–14·4, 302ᵈ
RS (var.)	4·3–12·3, irregular; subject to occasional outbursts
X (var.)	5·9–9·2, 335ᵈ

Orion, the Hunter (Ori; mid-December)

The grandest constellation in the sky, most happily placed on the celestial equator so that it can be appreciated by observers all over the world. Famous for the Great Nebula, visible with the naked eye, it also contains many fine doubles, and lies on the western border of a magnificent part of the Milky Way. Oddly enough, it contains no distinctive clusters. The star Delta (δ) lies only 20′ south of the celestial equator.

Orion. *Photographed on February 22, 1963, with a hand-driven Polaroid 110A camera, exposure time 20 minutes, using an f/4.7 lens, 5-inch focal length. (Thane P. Bopp, Kirkwood, Missouri.)*

Σ 627	(6·3, 6·5; 20″; 256°); a neat pair, curious colors
ρ	(4·7, 9·0; 7″·1; 63°); yellow and bluish; companion not easy with 3-inch
β	(0·1, 6·7; 9″·4; 203°); Rigel; a good atmospheric test; primary blue-white
33	(6·0, 7·3; 2″·0; 26°); both white
λ	(4·0, 6·0; 4″·2; 43°); yellowish and grayish
Σ 747	(5·6, 6·5; 36″; 223°); white and bluish; split in the finder; in a superb region
Σ 750	(6·0, 8·0; 4″·3; 59°); primary white; beautiful field
σ	In a fine group of two pairs: (4·0, 10·0; 11″; 236°); (7·0, 7·5; 13″; 56°); the faint star rather difficult with 3-inch
ζ	(2·0, 5·0; 2″·8; 159°); white and grayish; pretty
52	(6·2, 6·2; 1″·3; 209°); reddish
W (var.)	5·9–7·7, 200ᵈ
M.42	The Great Nebula in Orion, visible with the naked eye as a greenish haze around Theta (θ). A small telescope converts this into a convoluted veil, with bright condensations, streamers, and an especially distinct black intrusion into the center from the north. At the tip of this intrusion is the multiple star Theta (mags. 6·0, 6·5, 7·0, and 8·0), which is easily seen. This nebula is about a thousand light-years away from the sun. Ramifications extend over many degrees, but these are invisible except by photography, although local concentrations can be made out, especially around the star Iota (ι). The star Theta, also known as the Trapezium, contains some fainter objects, as shown in figure 61; these form interesting test-objects for users of moderate telescopes.

Figure 61. *The Trapezium. Stars A, B, C, and D can be seen in a very small telescope. Stars E and F have been glimpsed with a 3-inch, but rarely. G and H are very elusive.*

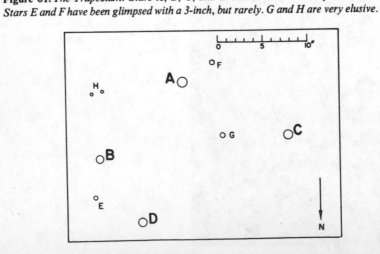

Orion (continued)

H.V.28   A faint, extensive nebula closely $f\zeta$; also a part of the Great
           Nebula
M.78   Another faint concentration with a definite "combed" structure
a (var.)   0·0–1·4, irregular; a very rough period of about 5 years
U (var.)   5·3–12·6, 374^d

Orionids. A prominent shower of swift meteors active between October
9 and 29, with maximum about October 20.

*Pavo, the Peacock (Pav; mid-July)

A far-southern asterism containing some distinctive naked-eye stars.

ξ (4·3, 8·1; 3"·5; 154°)
L8550   (5·8, 5·8; 2"·7; 91°)
L8625   (5·8, 6·1; 8"·1; 136°)
6752   A bright globular cluster, partly resolved with small instruments.
Y (var.)   5·7–8·5, 233^d

Pegasus (Peg; August-September)

A constellation marked to the naked eye by the Great Square of
Pegasus formed by α, β, and γ Peg and α And. The Square lies west of
Andromeda, but the constellation actually extends much farther westward.
Considering its size, it is poorly stocked with objects of general interest,
most of the doubles being wide or faint.

1   (4·5, 8·6; 36"; 311°); deep yellow and bluish
3   (6·0, 7·4; 39"; 349°); white and pale blue; fine field
Σ2978   (6·8, 8·0; 8"·4; 146°); white and bluish
M.15   A noble globular cluster, visible in the finder; seen nebulous with
           a low-power, blazing in the center. With a high-power, a small
           instrument achieves traces of resolution around the margin.
ε (var.)   Average magnitude 2·5, but suspected of variation
β (var.)   2·4–2·9; irregular, with a rough period of 35 days

Pegasids. A brief shower of swift meteors, radiating from near η about
May 30.

Perseus (Per; early November)

So far as north temperate observers are concerned, this constellation,
occupying a magnificent region of the Milky Way before it sweeps southward through Auriga toward Orion, offers some of the richest fields to be
found anywhere in the sky. Its leader Mirfak can be found by extending

eastward the line of Andromeda; here we find the dark-eclipsing variable Algol, and clusters and doubles are ranged against the sky in dazzling profusion.

Σ 268	(6·9, 8·2; 2″·7; 129°); white and tawny
η	(4·0, 8·5; 28″; 301°); intense yellow and deep blue; fine contrast
Σ 331	(5·3, 6·7; 12″; 85°); bluish and yellowish
Σ 336	(6·5, 8·0; 8″·2; 9°); yellow and blue
40	(4·2, 9·5; 20″; 237°); primary white
ζ	(2·7, 9·3; 12″; 208°); primary white; two faint companions
ε	(3·1, 8·3; 9″·0; 10°); white and bluish
Σ 533	(6·0, 7·5; 20″; 60°); reddish and bluish; grand field
Σ 552	(6·3, 6·5; 9″·0; 114°); splendid pair, both white
M.76	A small nebulosity with two distinct nuclei
H.VI.33 & H.VI.34	The famous Double Cluster, visible with the naked eye as a concentration in the Milky Way. A low-power reveals two brilliant open clusters, each larger than the moon, set side by side against a background granular with faint stars. This is a superb object, one to which the observer returns again and again; one never tires of surveying this dazzling array of suns, and the only pity is that few telescopes can include both clusters fully in the same field of view.
M.34	A loose cluster of bright stars, visible with the naked eye
H.VI.25	A beautiful cloud of faint stars; requires careful observation
H.VII.61	A coarse cluster of bright stars
ρ (var.)	3·3–4·1, irregular

Perseids. The most famous and reliable of all meteor showers. Swift meteors are noticed by the casual observer during the second half of July and much of August; a well-defined maximum occurs on August 10–11, when the radiant lies near Eta (η). A shower radiates from the region of Beta (β) between July 25 and August 4, but since these too are swift they can be confused with the main Perseids unless observed carefully.

ε Perseids. A shower of swift meteors occurring between September 7 and 15.

*Phoenix (Phe; early October)

There is little of note in this constellation, which lies between Achernar and Fomalhaut.

β	(4·1, 4·2; 1″·3; 357°); a binary pair
ζ	(4·1, 8·4; 6″·8; 245°)

Phoenix (continued)

θ (6·3, 6·9; 4"·2; 272°)
625   A faint nebula with central condensation

*Pictor, the Painter (Pic; mid-December)

An undistinguished constellation marked by the brilliant star Canopus, which lies nearby.

ι (5·0, 6·0; 12"; 58°)
I5   (6·4, 8·5; 2"·9; 269°)

Pisces, the Fishes (Psc; late September)

A zodiacal constellation. The head is marked by a small group of stars south of the Great Square of Pegasus, but the northeastern reaches of this extensive constellation are obscure. The sun lies in Pisces at the vernal equinox, when it moves into the northern hemisphere.

35   (6·2, 7·8; 12"; 149°); white and bluish
51   (5·0, 9·0; 27"; 82°); primary white
55   (5·5, 8·2; 6"·6; 193°); deep yellow and clear blue
65   (6·0, 6·0; 4"·5; 297°); both yellowish; a fine pair
ψ^1 (4·9, 5·0; 30"; 160°); both white; splendid low-power field
ζ (4·2, 5·3; 24"; 63°); white and grayish
a (4·3, 5·2; 1"·9; 291°); curious tints, somewhat greenish; a binary pair, closing beyond the range of a 3-inch
M.74 A spiral galaxy, visible in a small telescope as a dim nebulosity

*Piscis Austrinus, the Southern Fish (PsA; late August)

Very low in temperate latitudes, but easily distinguished by its bright leader, Fomalhaut (mag. 1·1).

β (4·4, 7·8; 30"; 172°)
γ (4·5, 8·5; 4"·3; 264°)

*Puppis, the Poop (Pup; early January)

One of the fragments of Argo Navis, a splendid constellation, lying beyond Monoceros in the Milky Way and forming the south and east borders of Canis Major. It has been ill-treated by celestial cartographers, and objects quoted in old catalogues as belonging to Puppis may now be found in any of the neighboring groups. Considerable revision is necessary, and it is worth taking trouble in identification, for the views will more than reward the effort involved.

D31	(5·0, 6·5; 13″; 317°); yellow and blue
D32	(5·7, 7·0; 8″·4; 276°)
h3966	(6·5, 6·5; 7″·0; 141°)
D49	(5·7, 6·5; 9″·4; 52°)
n (Hh269)	(6·0, 6·0; 9″·0; 108°); both yellowish
k	(4·5, 4·6; 9″·9; 318°); fine pair; both yellowish
2	(6·2, 7·0; 17″; 340°); white and pale blue
5	(5·3, 7·4; 3″·4; 6°); yellowish and tawny
r	(5·4, 6·2; 4″·0; 190°)
H.VIII.38	A brilliant, extensive, naked-eye group, somewhat diamond-shaped, containing the small pair Σ 1121
M.46	A large mass of faint stars, sprinkling the field of view with tiny points of light
H.IV.64	A small planetary nebula resembling a slightly hazy 8th-magnitude star; found most easily with a moderate-power
M.93	A small cluster of bright and faint stars
H.VII.11	An extensive group of faint stars easily found np 19 Pup
L$_2$ (var.)	3·4–6·2, 140^d

(Many other clusters and groups can be found by sweeping this wonderful region.)

*Pyxis, the Compass (Pyx; early February)

Some authorities consider this small constellation to have been a part of Argo; it is certainly in the right position, to the east of the Poop. It was introduced by the French astronomer Nicolas de Lacaille during his pioneer work on the southern skies in the mid-nineteenth century, after which the German observer and mathematician Johann Bode added a Log and Line to the Ship's outfit. This constellation (Lochium Funis), together with some others of Bode's invention, was forgotten almost as soon as it was introduced. Later, he actually installed a "typewriter" a few degrees east of Sirius!

H.VII.63	A small cluster of faint stars
2818	A small planetary nebula lying in a coarse group of stars

*Reticulum, the Net (Ret; mid-November)

A small far-southern group, lying between Achernar and Canopus.

θ (6·2, 8·0; 4"·5; 4°)

h3670   (5·9, 8·4; 32"; 99°); yellow and blue

Sagitta, the Arrow (Sge; mid-July)

A tiny asterism north of Aquila, obscure to the naked eye, yet lying in a superb region of the Milky Way. It is worth more attention than it is usually accorded.

Hh630   (6·5, 8·6; 28"; 301°); reddish and blue

ζ (5·7, 8·8; 8"·5; 313°); white and blue; a fine pair

θ (6·0, 8·3; 11"; 327°); yellowish and grayish; a nearby 7th-magnitude
         star completes a fine triple

M.71   A large cluster of very faint stars, appearing nebulous in a small
        telescope

*Sagittarius, the Archer (Sgr; early July)

This splendid zodiacal constellation, with its neighbors Ophiuchus and Scorpius, marks out the richest region in the sky for the observer of clusters. Sagittarius contains no fewer than fifteen of the objects catalogued by Messier; and the star clouds of the Milky Way are brilliant, even finer than those in Cygnus. It is therefore unfortunate that this part of the sky lies too far south for satisfactory observation from north temperate latitudes. Observation is handicapped, too, by the region's appearance during the short nights of midsummer.

h5003   (6·0, 7·0; 4"·8; 180°); bronze and blue; lies in a starry region

κ^2 (6·0, 7·3; 1"·1; 230°); elongated with 3-inch

M.23   A fine open cluster of 9th-magnitude stars arranged in festoons;
        partly resolved in the finder

M.8   A coarse, rather spherical cluster of bright stars wrapped in
        luminous haze, with an open cluster f; known as the Lagoon
        Nebula

M.24   An extensive cluster, visible as a nebulous patch, lying on the
        edge of a region of indescribable richness. There is a brilliant
        naked-eye cloud of stars between this object and Mu (μ).

M.18   A small cluster of faint stars

M.17	A wide, bright, nebulous streak with the *p* end overlaid by dark matter; contains two 9th-magnitude stars; known as the Omega Nebula
M.28	A small, bright, globular cluster with a blazing center
M.22	A superb globular cluster, partly resolved with a small telescope; visible in the finder and lies between two 9th-magnitude stars
M.54	An unusual globular cluster; center so condensed that it looks like an 8th-magnitude star with a hazy surround
6723	An extensive and bright nebulosity
M.55	A large, bright globular cluster, with little central condensation
M.75	A rather faint globular cluster resembling M.54
RR (var.)	6·0–14·0, irregular
RU (var.)	5·5–14·0, 354^d
RY (var.)	6·3–14·0, 241^d

Scorpius, the Scorpion (Sco; early June)

A magnificent zodiacal constellation that cannot possibly be mistaken, for it not only lies in the Milky Way, but also contains the brilliant orange star Antares (mag. 0·9) whose name means "Mars-like." The tail of the "scorpion" hangs low in the sky, and many of its objects cannot be well seen from latitudes higher than 40°N due to inevitable horizon haze.

ζ	(4·4, 7·2; 7″·4; 62°); cream and bluish; the primary is a close binary
β	(2·9, 5·2; 14″; 23°); white and grayish
L6706	(6·5, 8·0; 15″; 86°); white and bluish; a pretty pair
ν	(4·2, 6·5; 41″; 336°); yellowish and grayish; the companion is a close double
σ	(3·1, 7·8; 20″; 272°); yellowish primary
h4850	(6·5, 7·0; 6″·9; 352°); both golden; a pretty pair
α	(0·9, 6·8; 3″·0; 275°); reddish and greenish. Notoriously difficult when Antares is low in the sky, but easily seen with a 3-inch when high. It is easier in strong twilight or moonlight than against a dark sky, when the primary's glare is accentuated.
M.80	A brilliant globular cluster with a blazing center
M.4	An extensive globular cluster immediately *p* Antares; visible in the finder; partly resolved
6231	A coarse cluster of bright stars, with fainter members, lying in a glorious region (especially to the north)

Scorpius (continued)

6242	An irregular cluster of bright and faint stars
M.62	A globular cluster with a blazing center
6281	A curious trapezoidal cluster in a fine field
6388	A globular cluster with a blazing center
M.6	An open cluster, visible with the naked eye. In a small telescope, the stars are seen to radiate from the center, suggesting the appearance of an open flower; brighter stars are ranged around the perimeter. This is certainly one of the finest objects in the sky.
H.VI.13	A small, condensed cluster of faint stars, lying in a superb region
M.7	A brilliant, scattered mass of bright stars, visible with the naked eye. They fill a low-power field of view and include one of yellow tint.
RR (var.)	5·6–11·3, 279ᵈ
RS (var.)	6·0–12·7, 320ᵈ

Scorpiids. Very slow-moving meteors which radiate from near Antares between June 2 and 17.

*Sculptor (Scl; late September)

A drab constellation south of Cetus and Aquarius, containing only four 4th-magnitude stars.

D253	(6·0, 7·0; 6″·8; 267°)
ζ	(5·5, 6·5; 6″·1; 165°)
ε	(5·0, 8·5; 4″·5; 47°); white primary
H.V.1	Conspicuous extended nebula
H.VI.20	Compressed cluster of faint stars
H.I.281	Bright nebula
R (var.)	6·2–8·8, 376ᵈ
S (var.)	6·3–13·4, 365ᵈ

Scutum, the Shield (Sct; June-July)

A small but rich constellation occupying the part of the Milky Way between Aquila and Sagittarius. It used to be more extensive than it is today—the constellation realignment of 1930 robbed it of Messier's objects 16, 17, 18, and 24—but enough remains to afford magnificent sweeping. Many of the "clusters" here are mere concentrations in the endless strata of stars.

6682   This marks the position of two sprinklings of faint stars. A rich region lies to the southwest.

M.26   A small cluster, consisting of a few bright stars seen against a haze of fainter ones

M.11   The Wild Duck cluster. A superb fan-shaped cluster, granular with stars, with a 9th-magnitude star at the apex and a small double to the north. The surrounding region glitters with faint stars.

H.I.47   A small but distinct nebulosity

R (var.)   6·3–8·6, 144^d

Serpens, the Serpent (Ser; May-June)

This constellation consists of two separate groups: the head (Caput) to the west of Ophiuchus, the body (Cauda) to the east. The stars are listed as for a single constellation.

Σ 1919   (6·1, 7·0; 24″; 10°); yellowish and bluish
Σ 1931   (6·2, 7·6; 13″; 167°); yellowish and bluish
δ   (3·0, 4·0; 4″·0; 176°); white and grayish; fine pair
59   (5·5, 7·8; 3″·9; 317°); yellow and bluish
Σ 2375   (6·2, 6·6; 2″·4; 116°)
θ   (4·0, 4·2; 22″; 103°); both white; a superb pair in a fine field
M.5   A splendid globular cluster, partly resolved
M.16   Hexagonal cluster of bright stars and star dust
R (var.)   5·5–13·4, 357^d

Sextans, the Sextant (Sex; late February)

A barren group south of Leo.

35   (6·1, 7·2; 6″·4; 235°); yellow and grayish
H.I.163   Bright, centrally condensed nebula

Taurus, the Bull (Tau; November-December)

A fine winter zodiacal constellation. Its leader, Aldebaran (mag. 0·8), glows red to the northwest of Orion, and the Pleiades and Hyades clusters are readily distinguishable with the naked eye.

Σ 401   (6·5, 7·0; 11″; 270°); both white; a fine pair

Taurus (continued)

χ   (5·7, 7·8; 20″; 25°); white and lilac
62   (6·2, 8·0; 29″; 290°); white primary
Σ 548   (6·0, 8·0; 14″; 36°); yellowish and bluish
a   (0·8, 11·2; 121″; 34°). The comes has been seen easily with a 3-inch, but with such an aperture it is certainly a good test of vision and steadiness.
118   (5·8, 6·6; 4″·8; 205°); white and bluish
Σ 730   (5·8, 6·7; 9″·8; 142°); yellowish and bluish

PLEIADES. A well-known naked-eye group. Six stars are easily visible to normal sight, and some observers have recorded a dozen or more, but the cluster is too widespread to be a satisfactory telescopic object. It is best seen in the finder, or with binoculars. A chart of the brighter members is shown in figure 62. There are faint nebulae associated with the group, very conspicuous photographically but obscure to visual observers. Like the nebula around Xi (ξ) Cygni, they tend to be more easily visible with small apertures and very low magnifications.

HYADES. A very loose sprinkle of naked-eye stars to the west of Aldebaran. There is no association, however, for Aldebaran is much closer to the sun than is the cluster.

M.1 The Crab Nebula. Visible with a small telescope as a misty, elongated patch, this is the remains of the supernova observed in A.D. 1054 by Chinese astronomers.

Figure 62. *The Pleiades.*

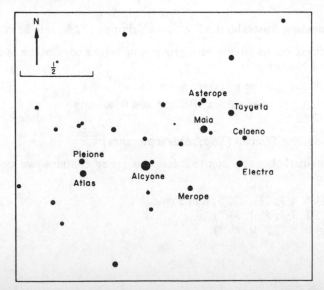

Taurids. A diffuse, long-lasting shower of slow meteors with a complex system of radiants. Noticeable activity lasts between October 26 and November 16.

*Telescopium, the Telescope (Tel; mid-July)

An uninteresting group south of Sagittarius.

6868   A small, centrally condensed nebula. There is a nearby nebula, dimmer and larger; this is N.G.C. 6909.

Triangulum, the Triangle (Tri; late October)

A small but well-defined group south of Gamma (γ) Andromedae. The three principal stars form an obvious elongated triangle; but it must not be confused with Aries, further to the south.

ι (5·0, 6·4; 3"·9; 74°); yellow and blue, fine contrast; also known as 6
Σ 239   (7·0, 8·0; 14"; 212°); yellowish and pale lilac; faint but neat
M.33   A nearby external galaxy, associated with the one in Andromeda. This is a notoriously difficult object, for it is both large and faint, and too high a magnification renders it invisible from lack of contrast. A good finder will show it best, and under good conditions it can be glimpsed with the naked eye. Telescopically, it appears as a barely perceived, extended haze of about the same diameter as the moon.
R (var.) 5·8–12·0, 270^d

*Triangulum Australe, the Southern Triangle (TrA; late May)

A conspicuous far-southern group lying on the edge of the Milky Way, south of Norma.

L6477   (6·4, 6·6; 2"·1; 151°)
6025   A fine open cluster of bright and faint stars

*Tucana, the Toucan (Tuc; mid-September)

Unremarkable as a constellation, this group contains two noteworthy objects.

β (4·5, 4·5; 27"; 170°); a fine pair
κ (5·1, 7·3; 5"·4; 346°)
δ (4·8, 8·1; 7"·0; 282°)

Tucana (continued)

104 Popularly known as 47 Tucanae, this is a magnificent globular cluster lying on the *f* border of the Nubecula Minor. To the naked eye it appears as a hazy star; a low-power eyepiece reveals it as a blazing mass. In terms of magnificence, it is second only to Omega (ω) Centauri.

330 A very compressed globular cluster lying within the Nubecula Minor

362 A fine, bright globular cluster, visible to the naked eye

NUBECULA MINOR. This is the smaller Magellanic Cloud, a satellite galaxy of our own star system. It contains a great many interesting objects, especially star clusters and variables, but because it is somewhat more distant than the Nubecula Major it offers a less attractive hunting ground for a small telescope.

Ursa Major, the Great Bear (UMa; mid-March)

Also known as the Big Dipper, or the Plough, this is the best-known northern constellation; its sequence of seven bright stars is known to everyone. To places in latitudes higher than 40°N, the group is circumpolar, never setting below the northern horizon; at 30°N, however, only the star Alpha (*a*) remains visible at the constellation's passage below Polaris, the polestar. Its boundaries extend much farther in the southern and western directions than is usually supposed, and they enclose a number of galaxies, mostly faint.

Σ 1349 (6·8, 8·0; 19″; 165°); yellowish and bluish

23   (3·8, 9·0; 23″; 271°); white and lilac; fine contrast

Σ 1415 (6·1, 7·0; 17″; 167°); both white

Σ 1520 (6·5, 7·8; 13″; 345°); white and bluish

ζ (4·4, 4·9; 2″·6; 135°); the first binary star to have its orbit calculated; now near its greatest separation; period 60 years

57   (5·2, 8·2; 5″·5; 357°); white and ashen

Σ 1561 (5·9, 8·0; 10″; 252°); primary yellowish

65   (6·0, 8·3; 3″·7; 36°); white and blue

ζ (2·1, 4·2; 14″; 150°); both greenish white. This is Mizar, forming a naked-eye pair with Alcor, the nearby 5th-magnitude star. An 8th-magnitude star lies between the two.

H.I.205 A fairly conspicuous nebulosity with a central condensation

M.81 & Two bright galaxies which can be included in the same field of

M.82   view; visible in the finder. M.81 appears as a circular nebulosity; M.82 has the effect of a narrow, curved streak, somewhat like a scimitar blade. This is because we are seeing the galaxy at an acute angle.

M.97   A large, dim planetary nebula, best seen with a low power
R (var.)   6·5–13·0, 302^d

Ursa Minor, the Little Bear (UMi; May)

The north polar constellation, also known as the Little Dipper, invaluable for containing the polestar, Polaris, Alpha (*a*), the most useful star in the northern sky. At the present time Polaris is less than a degree from the true pole, which will approach closest, due to precession, in A.D. 2095, at a distance of only 26′. Thereafter, the slow circling of the earth's axis will carry the celestial pole through Cepheus and Cygnus and, by A.D. 15,000, into Hercules. No star as bright as Polaris lies near the track, although Alpha (*a*) Cephei will make a fairly conspicuous marker in A.D. 8000. In 2000 B.C., Alpha (*α*) Draconis lay fairly near the pole.

a   (2·1, 9·0; 18″; 218°); white and blue; very easy with a 3-inch under good conditions
$π^1$ (6·1, 7·0; 31″; 80°); yellowish and bluish

*Vela, the Sails (Vel; mid-February)

A Milky Way constellation north of Carina and east of Puppis.

δ   (2·0, 6·6; 3″·0; 160°); a delicate object
H   (4·9, 7·7; 2″·7; 339°)
h4188   (5·5, 6·5; 3″·0; 287°)
h4220   (5·5, 6·0; 2″·1; 210°)
S   (6·2, 6·5; 13″; 219°)
2792   A small but conspicuous planetary nebula
2932   A compressed cluster of bright and faint stars

Virgo, the Virgin (Vir; mid-April)

This extensive constellation accommodates the ecliptic after it has passed through Leo, and its blue-white leader, Spica (mag. 1·0), lies in a region of sky otherwise devoid of bright stars. A mass ·of faint galaxies lies between Epsilon (*ε*) Virginis and Beta (*β*) Leonis, extending northward into Coma Berenices.

Σ 1627   (5·9, 6·4; 20″; 196°); both yellowish, barren field
17   (6·2, 9·0; 19″; 337°); white and bluish

Virgo (continued)

γ   (3·6, 3·7; 4″·8; 306°); both white; a superb binary pair with a period of 180 years. They are now closing.

Σ 1689   (6·7, 9·0; 29″; 198°); yellowish and bluish

θ   (4·0, 9·0; 7″·2; 343°); white primary

Σ 1904   (6·5, 6·5; 9″·6; 346°); a neat pair

M.84 &   Two nebulosities in the same field, M.84 being slightly the
M.86   brighter

M.49.   Faint nebulosity positioned between two telescopic stars

M.59 &   Seen in the same field; moderate apertures reveal many faint
M.60   nebulae in this region

R (var.)   6·2–12·6, 145^d

S (var.)   6·3–13·2, 380^d

SS (var.)   5·9–10·0, 359^d

*Volans, the Flying Fish (Vol; mid-January)

A far-southern constellation, the name being abbreviated from Piscis Volantis. It lies between Beta (β) Carinae and the Nubecula Major.

γ   (3·9, 5·8; 14″; 299°)

ζ   (3·9, 9·0; 17″; 116°); yellow and blue

ε   (4·5, 8·0; 6″·1; 22°)

Vulpecula, the Fox (Vul; late July)

Like its neighbor Sagitta, this is a small group which is often overlooked because it contains no bright stars. But, lying in the Milky Way to the south of Cygnus, it contains some superb star fields.

Σ 2445   (6·3, 8·0; 12″; 263°); white and grayish

Σ 2769   (6·5, 7·5; 18″; 301°); both white

H.VIII.21   A neat sprinkle of 9th-magnitude and fainter stars inside a curious curve of brighter stars

M.27   The Dumb-bell Nebula. With a small telescope it appears as a disk of faint luminous haze, with the interior intensified in the rough form of a cotton-reel, or spool. Visible in the finder; and there are several doubles in the low-power field. It is probably the most conspicuous of all planetary nebulae.

H.VIII.20   Surrounding star 20; a coarse group of bright and faint stars lying in an extensive starry region

H.VII.8   A fine, delicate open cluster

PART IV

ASTROPHOTOGRAPHY

21

An Introduction to Astrophotography

The first successful astronomical photograph, a picture of the moon, was taken by John Draper of New York, only a few months after the first workable photographic process was announced by Daguerre in Paris. But it was not until around 1880 that emulsions became sufficiently fast and reliable to be a serious competitor to the eye. From that time on, the visual observation that had formed the basis of modern astronomy was, in most fields, gradually supplanted by the more sensitive and impersonal photographic film or plate.

The last great visual telescope to be built, the 40-inch refractor at Yerkes, was installed in 1897. The first of the new generation of telescopes, the 60-inch reflector at Mount Wilson, was essentially a photographic instrument, as have been practically all the new twentieth-century telescopes. Essentially, they are nothing but giant cameras. The photographic plate and other image-processing devices can record stars, nebulae, and galaxies thousands of times fainter than any objects that could be seen visually.

At the beginning of photographic history, improvements in emulsion sensitivity were slow. To record a worthwhile number of stars, the amateur using a small telescope or camera might have needed an exposure of several hours, guiding on the sky as the earth turned. Not surprisingly, amateur astronomical photography was confined to the sun, moon, and perhaps the brighter planets. Since the war there has been an astonishing improvement. Modern emulsions are so rapid that all the naked-eye stars can be recorded by a 35mm camera in a second or so. An amateur with a 12-inch reflector can record stars as faint as the 16th magnitude with an exposure of just a few minutes. Superb color photographs of clusters, nebulae, and galaxies are now commonplace.

This is not to say that effective astrophotography is easy. Although there are many good photographs around, there are far more failures. Astrophotography is largely about technique. Like visual observing, it improves with practice. The danger (although devotees would probably not call it danger!) is that the technique can eclipse the subject. By this I mean that the pursuit is more about photography than about astronomy. There is no doubt that some astrophotographers do view their work in this light. There is nothing to say that they should not. On the other hand, there is huge scope for photography as an observing tool if simple, reliable, and relatively inexpensive methods are used.

The photographic process

All photographic processes depend upon the fact that some silver compounds (notably those called the *halides*: Silver bromide and silver iodide particularly) respond to certain wavelengths of radiation. When this energy impinges on them, their molecular structure changes. This is a chemical change, not a visible one: The silver which has been exposed to radiation responds in a different way than does the unexposed silver when the film is developed. If a suitable developing agent is used, this exposed silver is turned dark, and the depth of the blackness is a function of the amount of exposure it has received, as well as of the degree of development. Therefore the bright parts of the subject are rendered dark on the photographic emulsion, and the result is a *negative*.

In normal black-and-white photography, the unexposed white silver halide is dissolved out of the film by using a so-called fixing agent. The result is a transparent negative, which can be put into an enlarger and printed on sensitive paper by the same process, resulting in a positive print. It is possible to use a reversal developer and produce a positive on the original film, which is convenient if a slide is wanted. However, the negative/positive process allows a limitless number of prints to be obtained from the original negative.

A black-and-white film has only a single layer of sensitive emulsion. A color film has three, each one treated to respond to one of the primary colors of the visible spectrum. It is as if one were taking three negatives simultaneously. Once the film has been processed and reversed, a color slide results. Alternatively, the film can be left as a color negative and printed in an enlarger to produce a color print. Color development can be carried out at home, using special kits, but the newcomer to astronomical photography is strongly recommended to keep to the relatively straightforward black-and-white system. Processing is quick and inexpensive, so that the negatives can be inspected within an hour or less of the photographs having been taken, and failures (of which there are bound to be many) do not hurt the pocket so seriously. Basic darkroom techniques are covered in many books on the subject.

Films and developers

Silver halides in their normal state are sensitive only to shortwave radiation—blue, violet, and shorter wavelengths beyond the range of the human eye. An emulsion containing sensitive salts of this kind is said to be *ordinary* or blue-sensitive: It can be handled in yellow or red light in the darkroom with complete safety. The emulsions used for most enlarging paper are of this type, but of course they are much less sensitive than are camera films. Some emulsions have been treated so as to extend their response into the green and yellow region of the spectrum, and these are called *orthochromatic*. They can be handled in red light. Orthochromatic films are sometimes used by professional photographers for special purposes, but practically all the popular films sold in photographic shops are *panchromatic* (i.e., sensitive to all colors). A panchromatic film should be handled in total darkness, although very faint blue-green illumination is sometimes permissible.

The sensitivity of an emulsion is carefully measured and controlled, and is indicated either by an ASA or a DIN number—or both, for they are complementary. In the ASA system, a doubling of the number means a doubling of the film speed. Doubling of the film speed in DIN rating increases the number by 3. Equivalent ratings of some popular films appear below:

ASA	DIN	Film
32	16	Panatomic-X (Kodak)
50	18	Pan-F (Ilford)
125	22	Plus-X (Kodak), FP4 (Ilford)
400	27	Tri-X (Kodak), HP5, (Ilford)

The first two films are considered slow, the second two medium speed, and the last two fast. In other words, in normal use, Tri-X film would achieve the same negative density with less than half the exposure required for Plus-X and with about 1/12 the exposure required for Panatomic-X. It may be asked, therefore, what is the point of making slow films at all? The principal reason for offering a choice is that when an emulsion is tuned up to a high degree of sensitivity, it develops in a coarse way and is what photographers call "grainy." It will stand much less enlargement before the individual clumps of silver in the negative begin to show in the print. In some branches of photography this matters, in others it does not and even can be desirable if certain artistic effects are being sought. In astronomical photography, emulsion speed is almost always at a premium, but grain cannot always be tolerated, particularly when the moon and planets are being photographed.

However, most astronomical photography makes rather special demands on the process. In ordinary photography, exposures of more than a fraction of a second are rare. In astronomical work on the other hand—the sun alone excepted—exposures of *less* than a second are given only when the moon or Venus is the subject, and even then not always. In the case of stellar photography exposures may be as long as ten minutes or even more. The difficulty is that commercial emulsions are not designed to work efficiently at such long exposures. In theory, a film will record an object that is ten times as faint as another if the exposure is ten times as long. This works if the exposure times are only a fraction of a second, but the relationship breaks down if the light-level is very low. This effect is known as *reciprocity failure.* If it takes one second to record a 6th magnitude star, it may take 30 seconds rather than 10 to record another star 10 times fainter. There eventually comes a point where further exposure is futile.

Professional work mostly is done using special emulsions that are pre-sensitized to record only certain regions of the spectrum and that have a very low reciprocity failure. The Kodak series of so-called spectroscopic emulsions, rated I–V in order of decreasing sensitivity, is widely used, and some are available on 35-millimeter film, to special order. Many successful amateurs, however, have relied entirely on ordinary commercial films, boosting their power of recording very faint objects by various methods. There are two lines of attack: to hypersensitize the film or to force the development.

Hypersensitization. This can be done in two general ways, but they are not particularly straightforward, and general mention is all that will be made here, since the amateur who wishes to hypersensitize his films will have progressed far beyond the scope of this book. One way is literally to make the emulsion more sensitive. Manufacturers do not push their films to the limit, since in this state their storage characteristics are affected. Methods include: soaking the emulsion in liquid ammonia to remove some of the inhibiting chemicals applied in manufacture; heating or "baking" the film for some hours beforehand, at a temperature of 50–70°C; exposing the emulsion to a hydrogen atmosphere (in practice, using a mixture of hydrogen and nitrogen known as "forming gas," to reduce the risk of explosion). Another way is to tackle the reciprocity cause at source. A film emulsion becomes increasingly insensitive to very faint light because it contains molecules whose electrons are in rapid motion. This internal energy tends to repel

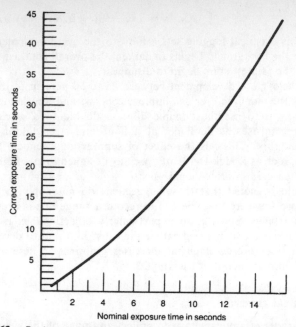

Figure 63. *Reciprocity failure curve for a typical fast photographic emulsion.*

light photons, and there comes a point where there are so few photons impinging on the emulsion that none can reach the light-sensitive molecules; hence, there is effectively no exposure. Freezing the emulsion lowers the energy of the electrons and improves the chances of the photons being able to do their work. This fact has encouraged the technique of "cold camera" photography, wherein the film is frozen, during the exposure, by solid carbon dioxide. Gains of several magnitudes during a given exposure have been reported by the use of some of these methods.

Forced development. The choice of developer influences the quality of the final negative, but the way in which it is used has perhaps even more effect. Theoretically at least, a star photograph offers the ultimate in contrast: a white point of light against a black background. The stellar photographer therefore develops for the best contrast; he is not interested in halftones. Increasing the development time from that recommended is one way of increasing the contrast, since the times given on the leaflet accompanying the developer refer to the best development for a whole range of tones. If the development is prolonged, the "half-dark" parts of the negative tend to go on blackening until they are as black as the true highlights, and this is good as far as the astronomer is concerned, since it means that the fainter stars in his negative are increasing in prominence. However, there comes a point where the light regions of the negative (i.e., the background sky) also begin to darken. This chemical fogging sets a limit to the overdevelopment method since, once the background begins to darken, the overall contrast is reduced again and the faintest stars begin to disappear.

Nevertheless, overdevelopment between 25 to 50 percent of that recommended by the manufacturer can improve contrast and limiting magnitude. Another way is to use the "double dilution/double time" technique—in other words, to use the developer at half normal strength for twice the development time. This has the effect of suppressing chemical fog. Some developers, such as Kodak D-19, are specially designed to give high contrast, and are often used in professional work.

It should be noted that these suggestions do not apply to lunar and planetary work, where the object is to record a range of tints representing the disk markings. Since grain is particularly objectionable, it is usually advisable to use medium-speed rather than fast film and to develop in the usual way, since overdevelopment increases graininess. These aspects are considered in more detail in Chapter 23.

Astronomical cameras

The purpose of any camera is to project an image of the view on to the sensitive emulsion and to allow the correct exposure to be given. It therefore consists of a film transport mechanism, a lens (or mirror system), and a shutter. To these essentials must be added the means of focusing the image sharply and a viewfinder with which the camera is pointed at the object. There is no operational difference between a hand camera and a photographic telescope, even though they may appear to have little in common. For telescopic photography, a camera body alone can be purchased, either new or secondhand, and adapted to fit the telescope.

If a wide area of sky is to be recorded, then an ordinary camera lens is used. The size of a given angle of sky on the negative is directly proportional to the focal length. If one unit (whatever it happens to be) is to correspond to $1°$ in the sky, then the focal length of the imaging system must be 57 units. Therefore a short-focus system is needed if a large area of sky is to be photographed, while a long-focus system will be preferred for photographing a small area. To record lunar and planetary detail would require an impractically long focal length, and the primary image must be magnified by an eyepiece or other suitable lens.

In addition to the question of image scale, aperture and focal ratio raise special considerations that must be examined in their proper place.

As regards the camera shutter, short and precisely determined exposures are not often needed, with the exception of solar photography, which was mentioned in Chapter 7. For most other purposes the shutter is manually opened and closed. The focal-plane shutters of standard single-lens reflex cameras are not entirely suitable, since their own momentum, combined with the mirror-lifting mechanism slam, can cause the telescope to shake. One solution is to arrange for the mirror to lift separately in advance of the exposure, or to remove it altogether. Another is to have a supplementary shutter of the leaf or Compur type with which to make the actual exposure, after the main shutter has been opened and the vibration has died away. Some observers have simply used a hat or a piece of cardboard held over the mouth of the telescope tube.

Focusing is straightforward if an ordinary camera is used, since the lens is set at "infinity." If the camera is fitted to a telescope, or if a nonstandard lens is being used, the razor-edge method described on page 138 is satisfactory provided there is no likelihood that refocusing will be necessary once the camera has been loaded with film. A still better procedure is to use some form of reflex focusing. The focusing screens on some miniature cameras can be adapted to use a visual magnifying system. Alternatively, a home-made retractable periscopic-type viewer can be used, as described on page 364. It is very difficult to focus a faint image accurately on the usual etched screen of a miniature camera.

The image can be centered in the camera frame either by using the viewfinder, in the case of a single-lens reflex camera, or else by using a well-adjusted finder on the main telescope.

Closely allied to the problem of image location is the problem of guiding. Apart from small-camera work, which can be trusted to the motor drive of most equatorial mountings, and for other exposures of only a few seconds' duration, it is necessary to follow the image and correct the telescope for any drift. One way is to use a powerful guiding telescope rigidly attached to the main tube. Another is to use an eyepiece focused on a star just outside the field being photographed. This is the more rigorous method, particularly if a reflecting telescope is being used, since it is not unknown for a lightly constructed tube to flex, or a mirror to shift in its cell, resulting in an image shift that will not be duplicated in a separate guide telescope. Since the method is rather inconvenient, however, most amateurs have preferred to use a separate guide telescope and to pay particular attention to their mutual stability.

Telescopes

Either a reflector or a refractor may be used, but in practice reflecting telescopes are more suitable, for not only are they available in larger apertures, but they are perfectly achromatic. In color work, the nonachromatic nature of an object glass is a serious drawback, since the purple secondary spectrum gives an exaggerated flare. This is because the layer in the emulsion that is sensitive to blue light peaks at about 4200Å, where visual acuity becomes poor but where the residual color in an ordinary achromatic lens is important. If black-and-white film is being used, this blue can be suppressed by using a yellow filter such as the Kodak Wratten 8.

A lens scores over a mirror in its ability to cover a field of several degrees. The image scale is now small, and secondary spectrum is unimportant compared with the other benefits. Ordinary camera lenses are available in focal lengths of from about 20 millimeters to 1000 millimeters or more, and excellent astronomical photographs can be taken with them. Some ex-government aerial survey lenses have also been found to work well.

Mountings

Virtually all astronomical photography requires equatorial motion, and the characteristics of stability and accurate guidance are essential no matter what work is being undertaken. Unlike the eye, which can follow the slow wanderings of a star or planet in the field of view, the photograph records everything that happens during the exposure, and a shaking stand or poor alignment will result in blurring or trailing of the image.

The quality of guidance required is dictated by the work being undertaken. In lunar and planetary photography, where the aim is to record the finest detail that the aperture and atmosphere will permit, the drive must be accurate enough effectively to hold the image stationary on the emulsion during the second or two of the exposure. A very slow overall drift during the session does not matter, since the image can be relocated at will between exposures. In the case of long-exposure work with a short-focus camera, where the resolving power of the system is a least several seconds of arc, a continuous slow drift is serious but a short-period oscillation (such as might be caused by a faulty worm) need not matter if its extent is smaller than the size of the star images.

Long-exposure photography at the telescopic focus (usually of deep-sky objects) is the most demanding of all. Not only must the polar axis and drive be accurately adjusted, but there must be no short-period errors. The image must be monitored with a guiding eyepiece and any drifting corrected immediately.

22

Wide-Field Photography

A photographic film consists of a plastic base that is coated with a cream-colored emulsion. As we have seen, this emulsion consists of a number of delicate silver compounds that are unstable in the presence of light. The brighter the light, or the longer it is allowed to impinge on the film, the more numerous the developable grains and the denser and blacker the resulting layer. When taking an ordinary photography it is most important to give the film exactly the right exposure. If it is too short, the darker regions of the subject will not have had time to make even a slight impression on the emulsion, and the corresponding areas on the negative will be completely blank. Conversely, with overexposure the brighter parts of the subject will appear too dense, with much of their detail obscured. This is because the conversion of the emulsion grains by light is not an instantaneous process; as long as light keeps shining on the film, the emulsion will continue to blacken until it is saturated.

This gradual conversion means that, in theory at least, the longer the exposure the fainter the stars that are recorded. For example, suppose that the camera in question has a 1-inch aperture lens (it should be noted that, just as with visual observation, the clear aperture determines the brightness of a star image). Using a fast emulsion, an image of the bright star Vega might be recorded with an exposure of 1/250 of a second. Since Vega is about 250 times as bright as a star of the 6th magnitude, the latter should be recorded with a 1-second exposure. Assuming that the camera is guided to follow the stars and bearing in mind the brightness ratio of about 2½ between adjacent magnitudes, we could expect to photograph the 7th magnitude in 2½ seconds, the 8th in about 6 seconds, the 9th in about 16, and the 10th in 40 seconds. The curious fact is that this lens, used as the objective of a telescope, would have a visual limit no fainter than this; so prolonging the exposure beyond about 40 seconds should begin to bring into view stars that could not be detected visually at all.

This argument ignores the reciprocity failure of all common emulsions: Several minutes rather than 40 seconds may be required to record a 10th magnitude star. But the reciprocity failure does not affect the essential difference between the eye and the emulsion. The eye has no cumulative power; assuming that it is dark-adapted, its response does not improve with prolonged gazing. Compared with the eye, the emulsion is a slow starter; but since it can maintain a reasonable pace for a long time, it will eventually overtake the human retina.

But prolonging the exposure in order to record faint stars also has an effect on the images of the brighter stars. For example, the image of Vega on a photograph recording stars to the 10th magnitude has been overexposed by some 10,000 times. As a result, the light has diffused into the grains for some considerable distance around the central image, giving the appearance of a disk rather than of a point of light. The stars of intermediate brightness show disks of different size, according to their magnitude. This means that brightness as well as position can be measured on a photograph.

Aperture and focal ratio

When taking wide-field photographs of the sky using a lens of relatively short focus, one obtains images on the film as close to being points of light as they can be. "Bad seeing" has no effect on the image, since the scale is so small. This means that results are fairly repeatable and consistent, given comparable atmospheric transparency, and it is easier than in other branches of astrophotography to make the general statements that follow.

Effect of aperture. Camera lenses are normally rated by focal ratio and focal length rather than by aperture. This is because most photography is of extended objects, and the intensity of the image of an extended object on the emulsion is a function of focal ratio alone. Any f/5 lens will give as bright an image of, say, a human face or a landscape as any other f/5 lens, regardless of the linear aperture or the focal length. This is because as the linear aperture increases, so the focal length, and hence the image scale, must increase too. Doubling the area of the lens also doubles the area of the image, so that the intensity remains the same. Hence the exposure with an f/5 telephoto lens of 1000mm focal length (linear aperture 200mm) is the same as with a standard 50mm focal length lens working at f/5 (linear aperture 10mm).

The situation is entirely different in the case of stellar photography. Assuming that the star image formed by any lens, regardless of its focal length, is a "point" (the diameter of the image of a faint star on the negative should be about 0.002 of an inch or less), then the brightness of the point is proportional to the area of the lens alone. A lens of double the aperture will produce an image four times as bright, and will photograph the same star in a quarter of the time, or photograph a star four times as faint in the same time. Taking as a basis the assumption that a 1-inch aperture lens will record a magnitude 9 star in 30 seconds, the following theoretical table can be constructed:

TABLE XI. Exposure Times

Aperture		Exposure to reach mag 9 (seconds)	Faintest star reached in 10 mins (magnitude)
in	*mm*		
½	12·5	120	10·75
1	25	30	12·25
2	50	8	13·75
3	75	3·5	14·6
4	100	2	15·25
6	150	0·8	16·2
8	200	0·5	16·75
10	250	0·3	17·25
12	300	0·2	17·6
16	400	0·1	18·25

This table was based on some negatives taken on Tri-X film, using a Pentax camera with a 105mm lens at f/5, with exposures ranging from a few seconds to 10 minutes. Curiously enough, in an issue of *The Astronomer* (No. 208, August 1981), an article by leading English astrophotographer Ron Arbour discussed a photograph taken on Tri-X using his 16-inch f/5 reflector, with an exposure of 10 minutes. The limiting magnitude on his negative was 18.3, in excellent agreement with the table above. It should be noted that the effect of reciprocity failure was ignored, and that the very short exposures given in the second column may be overestimated by a factor of several times. It is clear from this that impressive numbers of stars can be recorded even when using a very small lens, and with quite short exposure times.

Effect of focal ratio. It has been pointed out that focal ratio has no effect upon the brightness of a star image. But it does influence the result if a diffuse object such as a large faint nebula or a comet's tail is being photographed. For a given linear aperture, shortening the focal length (and hence reducing the focal ratio) decreases the size of an extended object and therefore increases its intensity. An f/3 lens will record a greater extent of a

Stellar Photographs. *These were taken by H. N. D. Wright with the astrocamera shown on page 343. (a) The Hyades (very scattered) and Pleiades (top right). (b) Clusters in Cygnus. (c) The Milky Way in the Aquila-Sagitta region, with the effect of the star clouds well brought out.*

comet's tail, in a given time, than will an f/6 lens, regardless of linear aperture, although the f/6 lens will show fainter stars if it is the larger of the two. To achieve the same intensity with the f/6 lens, the exposure time would have to be increased in accordance with the ratio familiar to photographers: i.e., exposure time is proportional to the square of the focal ratio. Therefore the exposure with the f/6 lens would need to be four times as long as with the f/3.

This is the theoretical effect of focal ratio, and it would apply exactly if the sky were perfectly dark at night. But the sky is never absolutely black. Many amateurs are pestered by urban and suburban glare, which can raise the background brightness to such a level that 6th and 5th magnitude stars may be lost from naked-eye view. Even in the most favored sites, however, a faint energy discharge from molecules high in the atmosphere diffuses a very dim glow into the sky: This background must be considered as an extended light source, and it is subject to the focal-ratio law: The faster the lens, the more intensely the background will be imprinted on the photographic emulsion.

The stellar photographer is after contrast between his subject and the sky background. The object of forced development is to blacken the stars proportionally more than the clear areas of the negative. A bright sky background obviously operates against this aim. There comes a point, either sooner or later, when the sky begins to darken faster than faint star images accumulate. If the background is faint, it will take a longer time to overcome the inertia of the emulsion and begin to darken than if it is bright. The brightness of the background is partly a function of the sky quality at the observer's site, and partly an inverse function of the square of the focal ratio. At f/3 it is four times as intrusive as at f/6. Therefore, if the sky is bright enough to start fogging the negative before star images have stopped accu-

Astrocamera. *This astrocamera, belonging to H. N. D. Wright, has an f/4 lens with a 5-inch focus, and is mounted on a simple equatorial stand. It is guided during the exposure by visual tracking through the attached elbow telescope.*

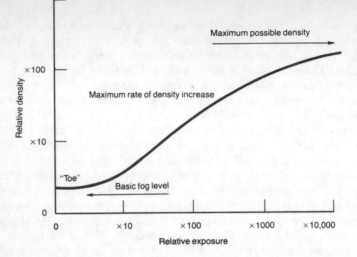

Figure 64. *Characteristic curve for a typical fast photographic emulsion.*

mulating (e.g., before reciprocity failure makes it pointless to continue the exposure, or when further guiding becomes impractical), the limiting magnitude of the negative can be improved if a lens of the same diameter but larger focal ratio is substituted.

Indeed, in the very bright sky of a town, fogging can be so serious that actual reduction of aperture (in other words, stopping the lens down) can improve the limiting magnitude. This may sound strange, but it follows from the way in which a photographic emulsion responds to light. Figure 64 shows what is known as the *characteristic curve* for a typical fast emulsion. The line indicates the way in which the negative darkens as the exposure to a light source is lengthened. It will be noticed that the upper part of the curve begins to flatten out; this is simply because darkening cannot carry on indefinitely, but reaches a limit beyond which further exposure to light can have no further effect on the density. But the lower end of the line is more significant. It will be noticed first of all that it does not pass through "zero" density. In other words, every emulsion has a background fog level, even if it is not exposed to light at all. Secondly, the emulsion does not begin to darken to light at the same rate as it does later on in the exposure; if it did, the line would be straight all the way up, and there would be no "toe," as it is called. The toe represents the reciprocity failure: At very faint light levels, doubling the exposure does not increase the negative density by the same ratio as it does if the light level is higher, further up the curve. Therefore, provided the sky fog level can be held down at the toe, increasing the exposure will keep on bringing fainter stars into view. Once the sky fog has sufficient intensity to come off the toe and on to the straight part of the graph, the overall contrast of the negative will begin to drop. It is this curious property of the photographic emulsion that justifies reducing the aperture of a lens to improve its limiting magnitude. If the characteristic curve had no toe, then stopping down would have no effect upon the relative build-up of star images and sky fog.

Using fast film, even in country skies that are completely free from artificial light, exposures of more than about 6 minutes at f/2 begin to show the effects of fogging. The photographer must determine the optimum exposure for his own location, but in fairly favorable suburban conditions a 10-minute exposure at f/5 on Tri-X begins to show noticeable fogging, the limiting magnitude being about 11½ with a 105mm focus lens. Arbour, as mentioned above, has also found 10 minutes a satisfactory exposure with his large f/5 Newtonian in rather poor town conditions. Under the best possible conditions, a maximum useful exposure in minutes of $1.6 \times (\text{focal ratio})^2$ has been suggested.

Lens characteristics, and field of view

The scale of the photograph at the focal plane, in, for example, millimeters per degree, varies with the focal length of the lens being used. Since most work will be done using standard 35mm cameras and lenses, it may be useful to indicate the field of view given by lenses of different focal length, when used with the standard frame size of 24mm × 36mm. For convenience, metric units are used:

LENS FOCAL LENGTH (mm)	SCALE, mm°	35 mm FRAME COVERAGE (DEGREES)
35	0·61	39 × 59
45	0·79	31 × 46
55	0·96	25 × 37·5
90	1·6	15 × 23
105	1·8	13 × 20
135	2·4	10 × 15
200	3·5	6·9 × 10
300	5·2	4·6 × 6·9
500	8·7	2·8 × 4·1
1000	17·5	1·4 × 2·1

It will be seen from this that the popular 135mm lens covers a convenient format: 10° of declination and 1ʰ of right ascension at the celestial equator.

A feature of many, if not all, of the common 35mm format lenses is their failure to illuminate the entire frame evenly when used at full aperture. This is partly because modern lenses are complex and very thick. With the lens opened to its widest aperture, open the back of the camera and point it toward the daylight sky or a light surface. If the eye is at or near the center of the frame, the outline of the aperture stop is circular, as it should be. Now move the eye to the extreme corner of the frame. The stop will probably have a lenticular shape, caused by vignetting at the edges of the front and rear elements. To discover how severe this loss of illumination is, turn the aperture diaphragm control until the opening, viewed from the frame corner, has a regular outline, either circular or slightly elliptical. This is the aperture down to which the lens must be stopped in order to obtain even illumination to the extreme corners of the negative, and it will probably be about $f/5$ instead of the $f/2$ or so at full aperture.

It is always easy to distinguish a sky photograph that has been taken at full aperture with a fairly wide-angle lens, since the margins and corners of the frame will be almost devoid of stars! But since, except for recording faint nebulae or the barely resolved star clouds of the Milky Way, $f/5$ is probably the optimum aperture for this work anyway, this defect is of no account. It is satisfying to take a photograph that is crowded with stars to the very corners, even though it may involve a longer exposure. There is another argument, too, in favor of using a lens stopped down from its widest setting. Although most modern camera lenses work at $f/2$ or $f/2.8$ in normal focal lengths, this is not their most satisfactory setting. It is extremely difficult to design a lens of this speed that combines perfect definition with the flat-field and off-axis characteristics necessary to give images of equal quality all over the frame. Designers are well aware of the fact that most pictures are taken at apertures of between $f/4.5$ and $f/11$. A camera lens usually works best at about two stops below its maximum opening. With most lenses, even the star images at the center of the field are noticeably crisper at $f/5$ than at $f/2$.

The smaller ex-government lenses can give good results with 35-millimeter film. The Ross Xpres 5-inch $f/4$, used at $f/5$, gives excellent definition. Another useful lens is the 8-inch $f/2.9$ Pentac, again stopped down, but this is much heavier and bulkier. The larger lenses, such as the Kodak Aero-Ektar type, are very heavy, and are not really suitable for small-format work. Many of these lenses were designed for use with a red filter so as to penetrate atmospheric haze, and in unfiltered light show some excess color. If a nonstandard lens is being adapted for use, it should be fitted into one end of a piece of brass or aluminium tube, the other end of which has a thread to screw into the camera body. A lathe is almost essential for this work, but the thread on a close-up focusing ring can be incorporated if no screw-cutting facility is available. The lens can be focused by the razor-edge method described on page 138, with final trials on a piece of film. It is advisable to make the attachment a permanent part of the lens, rather than having to check and refocus each time the lens is used.

Star magnitude and image size

The image of a star at the focal point of a lens or mirror should measure only a few microns (thousandths of a millimeter) across. Diffusion within the emulsion will cause the smallest developed star image to be about 50 microns (0.002 inch) across. As the exposure continues, diffusion causes the image to increase in size. By measuring the diameter of the star images, therefore, we have a way of relating their magnitudes, and star fields photographed using both camera lenses and telescopes have been measured for this purpose.

Professional astronomers use sophisticated automatic equipment that measures the size of each image using an adjustable diaphragm. But this does not mean that simpler methods are useless. Although they may be of lower accuracy, they can still give results that approach the accuracy of the visual observer. The late W. E. Pennell, of Lincoln, England, described a method that is simple, straightforward, and usually gives results to within about 0.2 of a magnitude. It was described in the *Journal* of the British Astronomical Association (Vol. 80, No. 5, 1970).

The negative is first enlarged. Either a positive print is made in the usual way, with a magnification of about × 10, or else it is projected onto a screen. Assuming that it is of a variable-star field or of some other field in which a few stars have known magnitudes, the diameters of these standard stars are measured using a magnifying glass and engineers' dividers or vernier cali-

Figure 65. *Image diameter/magnitude curves, derived from two negatives taken on fast panchromatic film with a light yellow filter, using a 280 mm lens at f/5.6, exposure 10 minutes. (After W. E. Pennell.)*

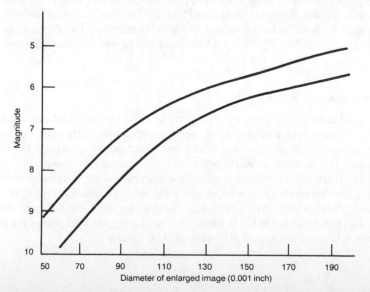

pers. Only experience can enable the true diffusion disk to be distinguished from the faint outer halo that appears around bright stars. Very faint and very bright stars are difficult to measure; using a 280mm focus lens at f/5.6 with Tri-X film and an exposure of 10 minutes, stars between magnitudes of 5.5 and 9.5 were the easiest to measure.

The diameters and magnitudes of the standard stars are then plotted on a graph as shown in figure 65, and a smooth line is drawn through the points. The curve shown here is taken from Pennell's paper. Once the graph has been drawn, the magnitudes of other stars can be derived by measuring their images and placing them on the curve. Of course, photographs taken on different nights (or even on the same night, if the fields were at different altitudes) are not likely to be comparable, although the shape of the curves should be similar. Nevertheless, the method was used successfully by Pennell in establishing comparison-star sequences for new variables.

Photographic estimates of this type would be useful for observing variable stars that suffer particularly seriously from observer bias, and also for variables of small range, such as many semiregular and irregular stars. A series of short-exposure negatives of an eclipsing binary as it passed through a minimum could be measured in random order, so eliminating the tendency to expect the star to be rising or falling—a difficulty which afflicts the visual observer when he is making a series of estimates.

Star color can have some effect on the magnitude derived in this way. Panchromatic film, despite its name, does not respond to light in exactly the same way as the eye does. In figure 66, it can be seen that all black-and-white emulsions respond best to blue and violet light, and even panchromatic emulsions refuse to respond to extreme red light unless they are specially sensitized. Therefore, red stars will appear somewhat too faint if their magnitude is measured against yellow and white stars. The shortwave difference in sensitivity matters less, since no star is intensely blue; but Pennell found that the light yellow Wratten 8 filter brought the overall color response of Tri-X closer to that of the eye.

Mountings and drives

Small cameras are usually mounted on an existing equatorial instrument, either on the tube or somewhere on the declination axis. Care must be taken that no part of the instrument or stand—or observatory—obstructs the field of view. If the drive is fairly good and the instrument reasonably well aligned, there will be no need to guide the camera visually if the exposure is only a few minutes long. It may be worthwhile, at the outset, to indicate the precision with which the camera must follow the stars, for different focal lengths. The following table is based on the assumption that image trailing on the film must not exceed 0.05mm, or 0.002 inch.

TABLE XII. Star Trailing

FOCAL LENGTH (mm)	MAXIMUM PERMISSIBLE TRAILING	MAXIMUM PERMISSIBLE STATIONARY EXPOSURE (SECS)
35	5′	19·5
55	3′	12·5
90	2′	7·6
135	1′ 15″	5·1
200	51″	3·4
500	21″	1·4
1000	10″	0·7

Note that the maximum exposure value is for a star on the celestial equator. This can be multiplied by the cosecant of the declination to find its value elsewhere in the sky.

It is clear from this table that as the focal length increases, the precision of the mounting's drive and alignment becomes more and more important. In a sense, alignment is the more serious of the two, since nothing can be done to improve it once the exposure has begun. Faulty alignment of the polar axis causes a star to drift in both right ascension and declination, but the declination drift is the more important. For example, if the polar axis is displaced by 1° from the true pole, the declination drift during a 10-minute exposure can be anything from effectively zero (if the field being photographed lies in the same plane as that containing the polar axis and the true pole) to a maximum of over 2′ if it forms a right angle with this plane. Therefore, even using a 200mm lens, the alignment of the polar axis needs to be reasonably good. Longer focal length cameras will need to be guided

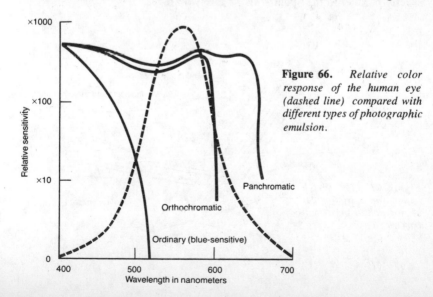

Figure 66. *Relative color response of the human eye (dashed line) compared with different types of photographic emulsion.*

using the main telescope unless the axis is set with considerable accuracy and the drive is very smooth and precisely rated.

This requirement, which implies a well-equipped facility, conflicts somewhat with the spirit of simple astrophotography. Taking good sky pictures with modern cameras and film is fun, instructive, and may even be useful (though that need not be the prime consideration). Many people have adequate photographic equipment, but relatively few have anything suitable on which to mount it if exposures of up to 10 minutes or so are to be attempted. Yet there is at hand a very simple homemade mounting that costs almost nothing and can produce really excellent results. It was first described by G. Y. Haig in 1975, and has come to be known as the "Scotch" mount. Consisting essentially of two boards hinged together, one being driven by a screw while the other remains fixed, with the line of the hinges pointing toward the celestial pole, it is capable of tracking as accurately as an expensive equatorial mounting.

The mounting is shown in figure 67. The two boards are linked together with door hinges; a pair of spreader arms, as shown, increases their rigidity. The lower board is secured to a base so that the altitude of the line passing through the hinges is equal to the latitude of the site. This base can be fixed to a stand or tripod, or placed on some solid support, as required. A length of screwed rod passes through a threaded hole in the lower board, and its rounded end bears on the underside of the upper board. By turning the rod, we have a tangent drive. The upper (moving) board carries a mounting for the camera so that it can be directed to the appropriate part of the sky. It also carries a low-power telescope with a crosshair in the eyepiece. The direction of this telescope must be adjustable so that it can be aligned exactly parallel to the "polar axis"—the two hinges.

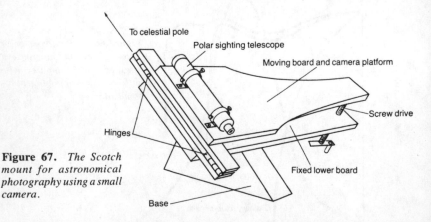

To celestial pole

Polar sighting telescope

Moving board and camera platform

Screw drive

Hinges

Fixed lower board

Figure 67. *The Scotch mount for astronomical photography using a small camera.*

Base

The principle of the tangent drive was discussed on page 50. It is sufficient here to give a table showing the distance that the screwed rod needs to be from the axis of the hinges, for different threads and rotation speeds:

TABLE XIII. Tangent drive for a Scotch mount

PITCH OF DRIVE ROD IN TURNS/IN	SPEED OF TURNING IN RPM	HINGE-ROD DISTANCE (IN)
16 (⅜-in Whit or NC)	1	14·28
	2	28·57
18 (⁵⁄₁₆-in Whit or NC)	1	12·70
	2	25·40
20 (¼-in Whit or NC, ⁷⁄₁₆-in & ½-in NF)	1	11·43
	2	22·86
24 (⁵⁄₁₆-in & ⅜-in BSF or NF)	1	9·52
	2	19·05

The rod can be turned by a synchronous motor, but the original mountings, which were designed for ultimate simplicity, were turned by hand at 1 rpm by advancing a notched wheel containing n teeth by one notch every $60/n$ seconds, the interval being sufficiently short to prevent any trailing from appearing on the film. I was amazed at the quality of the photographs obtained in this way by young amateurs, using old-fashioned roll-film cameras with apertures as small as f/11.

The alignment of the polar axis is, as we have seen, increasingly critical as the focal length of the camera is increased. Even setting the line of the hinges within 1° of the pole (the safe limit of accuracy using a 90mm lens with a 10-minute exposure) is difficult without some optical help. My first experiments with this mounting were disappointing for this very reason, and the polar sighting telescope attached to the upper board transformed its performance. This must first be aligned with the axis of the hinges. To do this, set it approximately parallel by eye, and then arrange the mounting so that either the polestar itself or some very distant well-defined object is on the crosshair. Now rotate the upper board backward and forward through the largest possible arc, which should be 180° or so, and adjust the direction of the telescope until the object does not drift from the crosshair during the rotation. It must then be parallel to the polar axis.

To point the telescope at the celestial pole, it is necessary to know the pole's position in the star field. Figure 68 shows the brighter stars near both celestial poles. The north polestar, magnitude 2, is about 50′ from the true pole at the present time, while the southern polestar, σ Octantis (magnitude 5.5) is 57' away. There should be no difficulty in bringing the mounting's polar axis to within about 10' of the celestial pole in just a few seconds.

If a long-focus lens is being used in conjunction with a driven equatorially mounted telescope, it will probably be essential to guide on a star during the exposure. A high-power eyepiece is used so that a shift of only a few seconds of arc is immediately obvious. The first requirement is a reliable adjustment on each axis. Most modern instruments incorporate a variable-frequency drive with which to control the R.A. motion, but the facilities on the control box vary. Some simply have a single adjustable knob with which the motor's rate can be varied, but this is unsatisfactory as a slow motion, since the original setting is destroyed if the motor is made to run fast or slow in order to correct drift. The most satisfactory (or least unsatisfactory) arrangement is to make the motor run slightly fast, and to switch it off for an instant when the image has drifted far enough, letting the Earth's rotation bring it back again. A box with separate adjustments for the rate and the corrections is much more satisfactory. The typical small worm-and-wheel declination adjustment will give poor service, and a tangent-arm facility is far superior. If the mounting is properly aligned, however, there will be little need to touch the declination control during the exposure.

Ordinary human hair can be used to form the eyepiece crosshairs. They are much too thick for work such as micrometer measures, but if a bright star is used for guiding purposes, the image can be thrown out of focus so that the intersection of the hairs is seen outlined against the defocused star. Any drift of the star is seen as a destruction of the symmetry of the disk and the cross, an effect to which the eye is very sensitive.

Experience with ordinary refracting telescopes will have introduced the observer to the nuisance of dewing, and this can be a plague to the photographer who enjoys, or endures, a maritime climate such as that in the British Isles. Dewing of a refractor's lens can be inhibited by fitting a dewcap three or four times as long as the aperture. This solution is useless with a wide-angle camera, since the margins of the field will be lost. One answer is to make the dewcap in the form of a funnel, but this does not supply such efficient protection, especially when the camera is pointing in the region of the zenith. The most direct solution is to warm the air in front of the lens, and hence the surface of the lens itself. Dew cannot be deposited on any surface that is even a fraction of a degree above the air temperature. This protection can be achieved by wiring three or four flashlamp bulbs in series and connecting them to a power transformer (or, more expensively, a dry cell) so that they are under-run by about 50 percent. With the connections well insulated, the bulbs are wrapped in cooking foil and mounted around the front rim of the lens. At least a shallow dewcap also is required in order to hold in the warm air; dewcaps and lenshoods should be made of plastic, cardboard, or other poorly conducting material in order to minimize the loss of heat. If current is to be obtained from the public power supply, ensure that the transformer is well insulated and grounded.

A final note about shutters may be useful. Since even the use of a cable release can jolt the camera, it is always best to open the shutter with a very loose, very light cap over the dewcap. This is then lightly whisked away at the beginning of the exposure, and gently replaced at the end, before the shutter is closed.

Principles and methods having been described in some detail, it is time to consider the useful or interesting work that can be done with small or moderate lenses and 35mm film.

Star charting

Many amateurs, having taken a few photographs that show a great many stars, entertain the project of charting the entire visible sky by these means. Very few, however, have completed such a task. It depends partly, of course, on local conditions. Moonlight, combined with prolonged cloudy or hazy spells, can mean that difficult regions below the celestial equator are completely missed during one cycle of the sky. But with persistence it should be possible to photograph in two years the stars that come above, say, 20° of the horizon. W. E. Pennell, from his garden near Lincoln, England, made a photographic chart of the sky using a 90mm lens and Tri-X film in a single year; even though conditions on the east coast of England can be very poor.

A chart made using panchromatic film would have a great advantage over most of the photographic atlases available today, many of which were made with blue- or red-sensitive emulsions, and so do not show the normal visual appearance of colored stars, particularly red ones, which can appear over a magnitude too faint if photographed on blue-sensitive film. This does not mean that such atlases are useless for the visual observer, but it does make things difficult when the magnitudes of the stars happen to be more important than their positions. Panchromatic film, as we have seen, has slightly different sensitivity from that of the eye, but with a few highly colored exceptions the photographic magnitudes can be relied upon to within about 0.2 of a magnitude if a light yellow filter is used to reduce the excessive blue sensitivity.

In some ways, a 105mm lens would be ideal for the purpose, since it covers 1^h in right ascension at the equator, and 10° in declination, with considerable overlap (overlap is important; charts with a common edge are most annoying—since the object being sought is always to be found there!). If a hemisphere were being covered using this format, the plan might look like this:

Dec. 0°–10°: 0ʰ–1ʰ, 1ʰ–2ʰ, etc. (24 photographs)
Dec. 10°–20°: 0ʰ–1ʰ, 1ʰ–2ʰ, etc. (24 photographs)
Dec. 20°–30°: 0ʰ–1ʰ, 1ʰ–2ʰ, etc. (24 photographs)
Dec. 30°–40°: 0ʰ–1ʰ, 1ʰ–2ʰ, etc. (24 photographs)
Dec. 40°–50°: 0ʰ–1ʰ20ᵐ, 1ʰ20ᵐ–2ʰ40ᵐ, etc. (18 photographs)
Dec. 50°–60°: 0ʰ–1ʰ30ᵐ, 1ʰ30ᵐ–3ʰ, etc. (16 photographs)
Dec. 60°–70°: 0ʰ–2ʰ, 2ʰ–4ʰ, etc. (12 photographs)
Dec. 70°–80°: 0ʰ–3ʰ, 3ʰ–6ʰ, etc. (8 photographs)
Dec. 80°–90°: 0ʰ–12ʰ, 12ʰ–24ʰ (2 photographs)

The atlas would therefore consist of 152 photographs. Assuming that each negative has to be taken twice (to allow for bright satellites or aircraft trails, and other faults), a total of about 300 35mm frames will be needed. The film and developing costs will be about $20 or £10; not much to pay for recording perhaps two million stars! If printed on an 8 × 6-inch format, the scale would be 2½°/inch, somewhat larger than that of *Atlas of the Heavens*.

A project such as this might be attractive to the determined amateur, who could feel that the end product would be useful as well as satisfying. Numerous variable stars would be caught, and one or more of the photographs might contain the pre-discovery image of a comet or nova. Star photographs should be printed on the most contrasty paper available; photographic shops do not always stock this material, and it may be necessary to order it from a wholesaler's.

Nova patrolling

The great sensitivity of modern films has made it possible to carry out useful patrol work with *unguided* cameras. A 30-second exposure at f/2 records stars down to about magnitude 9, and observers in the United Kingdom have been covering the Milky Way in this manner for some years. Honda, in Japan, has had considerable success in discovering novae photographically by comparing photographs taken on different nights. Even if no nova is found, the stationary camera technique is a rapid way of obtaining data on numerous variable stars, which can be evaluated at leisure. A 30-second exposure with a 50mm lens gives a slight trailing effect, which helps to distinguish specks in the emulsion from real stars.

Comet photography

Each year at least two or three comets attain a magnitude of 8 or 9 and are therefore within the range of small cameras. They are frequently sufficiently far from the sun to be high above the horizon and thus can be photographed from urban backyards. The much rarer bright comets deserve better attention. A nebulous tail, as we have seen, can only be photographed with a small focal ratio, which itself demands a darker sky than will be found in a town. On such occasions, the Scotch mount comes into its own. Not only is it completely portable, but it can be accurately aligned on the celestial pole in a very short time. A synchronous motor driven through a step-up oscillator from the automobile battery can be used if the hand-winding method is considered too primitive.

Although comets show appreciable orbital movement in front of the stars, this will not be noticeable during a 10-minute exposure with a short-focus lens. If the tail is long, the camera must be arranged so that the head is toward one edge or corner of the frame. Comets are particularly rich in blue light and can make pleasing color photographs. The head is usually overexposed, and it is advisable to take a series of exposures, the shortest being only a few seconds long, in order to reveal detail near the nucleus, which is burned out in an exposure intended to show the extensions of the tail.

Photography of very faint comets presents problems similar to those encountered in photographing nebulae, discussed in Chapter 24.

Meteor photography

This is one of the most attractive, but at the same time one of the most frustrating, departments of photographic work. It has the advantage, however, that the amateur can concentrate his efforts on the few nights in the year when a reasonably active shower is expected. There is little point in exposing film for sporadic meteors on nonshower nights. Fireballs (mentioned below) are a special case.

The photographic requirements for useful meteor work are rather different from those for stars. A meteor, although a starlike point, is moving across the emulsion at a linear speed that depends upon both the angular velocity of the meteor and the focal length of the lens. Let us suppose that a zero-magnitude meteor covers a path of 50° in half a second, probably a typical velocity for a swift meteor. Using a 55mm focus lens, the effective image size of a star (assuming that its diameter on the film is 50 microns, or 0.002 inch) is equivalent to 3′. The meteor will effectively be a spot of this size moving across the emulsion at a rate equal to its own diameter in 0.0005 of a second. This, therefore, is the exposure that emulsion receives. Since a lens of 1-inch aperture requires an exposure of perhaps 0.004 of a second to photograph a zero-magnitude star, the meteor will probably not be recorded on the negative.

To improve the chances of recording a meteor, assuming that the emulsion sensitivity is already at its maximum attainable value, we must (*a*) photograph slow rather than fast meteors, (b) increase the linear aperture of the lens, or (c) reduce its focal length. Obviously, nothing much can be done about (a) except to remember that meteors near the shower radiant have a lower angular velocity than those far from it, so the farther the camera is pointed away from the radiant, the faster the meteors will be moving and the fainter will be their impression on the film. Factors (b) and (c) argue for the largest and fastest possible lens; in theory, at least, the actual focal length has no effect on limiting meteor magnitude. In practice, reducing the focal length increases the area of sky covered, and so enhances the chance of a meteor being caught. Even so, using f/2 lenses of between 35mm and 55mm focal length, the chances of recording meteors fainter than about magnitude −1 are poor unless the angular velocity is unusually slow.

It is worth noting that reciprocity failure has no effect on meteor recording since the exposure is instantaneous. Forced development is therefore the most useful way of trying to improve emulsion sensitivity. Reciprocity failure works to the photographer's advantage, since it inhibits the formation of sky fog. Nevertheless, exposures with fast film at f/2 cannot usefully be maintained for longer than about five minutes, unless the location is unusually dark: One must balance the chances of losing a meteor through more frequent film changing or through background fogging.

As far as the mere recording of meteors is concerned, it is not necessary to drive the camera during the exposure. But an undriven photograph is worthless for position determination (the main point of meteor photography) unless the instant of the meteor's appearance and the times of the beginning and ending of the exposure are known. With this information, the meteor's path in relation to the appropriate point on the star trails can be determined. But this means that the photographer must spend all his time watching the sky area being photographed. If the camera is driven, he can pay attention to more general observation; it is surprising, too, how many visually unrecorded meteors appear on photographs! Serious meteor photographers have employed batteries of several cameras so as to cover a large area of sky simultaneously.

The observation of fireballs using all-sky cameras is a field of work that has been developed by amateurs in Europe following the professionally run networks that operated in Canada, the United States, and Czechoslovakia from the 1960s onward. The United States Prairie Network, which has now closed down, consisted of 16 camera stations spread over seven states, and the result of several years of nightly photographic monitoring was the recovery of just one fallen meteorite out of thousands of bright fireballs that were photographed. These networks have proved that the proportion of fireballs that reach the ground intact is very small indeed. Amateur fireball patrols are being carried out in several ways, the cheapest and most convenient

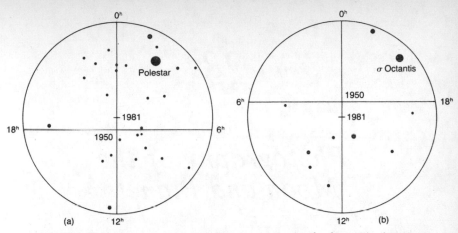

Figure 68. *Star fields around the celestial poles.* (a) *North pole; the position is not easy to locate with great precision because of the absence of nearby guide stars easy to identify with a small telescope.* (b) *South pole; although there is no very bright pole star, its present position can be fixed with considerable accuracy by the two bright binocular stars with which it forms a triangle. Both fields are about 2½° across; the faintest stars shown are about magnitude 10.*

being to use an all-sky "fisheye" lens. No monitoring need be done. The camera is pointed at the zenith, and the shutter is opened when darkness falls. After a convenient time has elapsed, the film is wound on and another exposure made. Since these lenses work at about f/11, sky fogging is negligible over an exposure of an hour, but the image scale is sufficiently large to give measurable trails of the brighter stars and any fireball that happens to occur. The method relies on the fireball being recorded visually by someone, somewhere—but this is likely enough, since they are bright enough to be noticed by members of the public, if not by other amateur astronomers.

Even using this fairly modest equipment, information about the true path of the fireball, its probable mass, its point of disintegration, and its likely terrestrial impact point can be determined if it is caught by two photographic stations.

Photography of meteors has the advantage over visual work that extremely accurate paths can be derived (it has, it must be admitted, the drawback that only the brightest meteors can be recorded with standard amateur equipment). The usefulness of a photograph is multiplied if a rotating shutter is arranged to occult the lens at a known speed—about 20 breaks per second is a suitable rate. If this is done, not only the meteor's path through the atmosphere but its true velocity and hence its orbit around the sun can be calculated. Two-station work is far from easy to organize, however, and many hours of fruitless exposures may be necessary before the sought-after bright meteor passes through exactly the right corridor in the upper atmosphere.

23

Photography of the Moon and Planets

The amateur astronomer who wishes to record the finest possible detail on the moon and planets must use his eye. Although some very fine photographs have been taken with backyard telescopes, a close study of them will reveal nothing that could not have been seen with a good telescope of one-third or one-half the aperture. In fact, very few photographs taken from anywhere on the earth's surface, no matter how large the telescope, show planetary detail beyond the range of a skilled eye and a 10- or 12-inch instrument. There are some exceptions: A few of the photographs of Jupiter taken at the Pic du Midi Observatory in the Pyrenees, with the 1-meter reflector there, even reveal markings on some of the Galilean satellites! However, such extraordinary freaks of performance only emphasize the relatively lowly status of most planetary photography. Only the ultraviolet markings of Venus, which have been recorded with a 6-inch reflector, can truly be said to represent an advance over visual observation.

Why is this? The explanation lies neither in the telescope nor in the photographic emulsion but in the atmosphere. To take a large-scale lunar or planetary photograph involves an exposure of appreciable duration—perhaps one or two seconds—and the chances of the air being perfectly still and optically homogeneous during this interval are very small. Almost inevitably, the image will suffer from a tremor or two. This is enough to blur the photograph and to cause the loss of the finest details. On the other hand, the swift reaction of the eye can grasp the best moments of vision and ignore the poor ones. The emulsion's wholesale, cumulative manner—so useful in stellar photography—places it at a disadvantage in high-resolution work.

It was stated that the atmosphere must carry the blame for this state of affairs, but this is not quite fair. If such a thing as a fast, grainless photographic emulsion existed, then tiny images on the negative could be enlarged to reveal all the detail present in the focal plane, and, being tiny, the exposures could be reduced to the point where the chances of obtaining perfect seeing were greatly improved. But a grainless emulsion does not exist. Some emulsions have a finer grain (or greater resolving power) than do others, but their speed is lower. Therefore one is faced with either a large image on a coarse film or a smaller image on a fine film, both of which require about the same exposure. Although perhaps the majority of successful amateur and professional workers have elected to use medium-speed film and moderate image size, equally good photographs have been taken using material such as Tri-X and the larger primary image that this allows.

The question of contrast, which on average tends to be lower with fast emulsions, is a very important one. It is of little use being able to resolve planetary detail if the film cannot then discern the different intensities of that detail. Such a photograph may show an agreeably sharp edge to the planet's disk—where the contrast is high, since it is projected against a black background—but the disk itself may be almost featureless. Ideally, an emulsion should be able to separate out the various tones on a planet's surface into a series of tints varying from black to white. If it could do this, it could begin to reveal details that are invisible to the eye; in other words, it would be exaggerating the contrast. In practice, this is not possible, except in a limited sense by the use of colored filters. Planetary markings, however much they may vary in tint and intensity from one part of the disk to another, do not show high contrast near the limit of resolution. Much of this low contrast is due to the blurring caused by diffraction within the telescope itself, as well as to light scattered by dirt in the atmosphere and (more seriously) by dusty and degraded optical components. Successful planetary photography probably represents the highest state of the art. Optics, atmosphere, and telescope drive must all be in prime condition.

These hard, if not insurmountable, facts have done nothing to discourage amateurs from tackling the problem. In fact it sometimes appears as if, in the United States at least, the brighter planets are photographed at least as frequently as they are drawn! From the purely observational point of view, this is a pity when one considers the amount of planetary detail that is being missed. On the other hand, astronomy's loss is photography's gain, and the remarkable improvement in the overall quality of astrophotography during the past two decades shows how much effort has gone into it. But the advance in amateur planetary photography has been less general and spectacular than in the photography of deep-sky objects, simply because of the inherent difficulties already mentioned. Compared with the acuity of the visual observer, the planetary photographer is trying to achieve the unattainable.

Photographic telescope. *Cmdr. Hatfield's tubeless 6-inch reflector. It is on a German mounting, with an 18-inch pulley wheel to take the synchronous motor drive. The telescope and mounting are homemade.*

The beginner would be well advised to tackle lunar photography first, since even a relatively poor lunar picture can be impressive. It is worth remembering that the disk of Jupiter, at mean opposition distance, is equivalent in size to a lunar crater 45 miles across, while Mars's diameter is usually less than half this. There is a large-scale contrast on the moon, and the shadows give the image some bite. Planetary photography, for which the most suitable candidates are Venus and Jupiter, is barely feasible with instruments of less than 10 inches aperture, although some good photographs of Venus in the crescent phase have been obtained with a 6-inch reflector. On the other hand, Venus appears virtually featureless in all but ultraviolet light.

Telescopic resolution and image size

Manufacturers determine a value of resolving power for all their emulsions; for a film of average speed it may be of the order of 1000 lines/inch (40 lines/mm), while for slow, fine-grain films it can be considerably higher than this. The difficulty in applying such values to practical photography lies in the difference between the high-contrast targets used for evaluating films in the laboratory, and the much lower contrast of planetary detail. It has already been emphasized that a film cannot resolve two features, no matter how far apart they are, if it cannot differentiate between their intensities. Therefore the quoted resolving power of an emulsion is an unreliable guide. The general approach of planetary photographers is to calculate image size on the assumption that the film's resolving power is between two and five times worse than that claimed by the manufacturer.

To ensure that all the information contained in the telescopic image is being transferred to the film, the telescope's resolving power must be represented, in the final image plane, by a distance not less than the film's resolving power. The linear distance between two just-resolved points is a function of the focal ratio of the object glass or mirror:

Focal ratio	Limit of resolution at prime focus	
	inch	*mm*
4	·00010	·0027
6	·00016	·0040
8	·00021	·0054
10	·00026	·0067
15	·00039	·010
20	·00053	·013
50	·0013	·034
75	·0020	·050
100	·0026	·067

If we take the working resolving power of medium-speed film as 500 lines/

inch (20 lines/mm), then it appears as if a focal ratio of 75 will match the telescopic and emulsion resolving power. At f/75, the moon and planets will have the following real image sizes, at mean opposition or elongation distance, in telescopes of various apertures:

	8-INCH		10-INCH		12-INCH		16-INCH	
	inch	*mm*	*inch*	*mm*	*inch*	*mm*	*inch*	*mm*
moon	3·9	100	6·5	166	7·9	200	10·5	266
Mercury	0·015	0·39	0·025	0·65	0·031	0·78	0·041	1·0
Venus	0·065	1·7	0·11	2·8	0·13	3·3	0·17	4·4
Mars	0·044	1·1	0·073	1·8	0·087	2·2	0·12	3·0
Jupiter	0·098	2·5	0·16	4·2	0·20	5·0	0·26	6·6
Saturn (globe)	0·037	0·94	0·062	1·6	0·074	1·9	0·1	2·5
(ring)	0·095	2·4	0·16	4·1	0·19	4·9	0·26	6·5

The image in an f/8 Newtonian will therefore have to be magnified about 10 times. Using the formula on page 378, the distance between the enlarging eyepiece and the film plane is found to be 11 times the focal length of the eyepiece. A 12.5mm (½-inch) Orthoscopic would be satisfactory, with an eyepiece-image distance of just under 6 inches. A Barlow lens, which is normally designed to give an amplification of only × 2 or × 3, would not be suitable. Some amateurs have enlarged the image even more than this, using effective focal ratios of 150 or even 200. Such amplification is usable only when the seeing conditions and telescope drive permit an unusually long exposure.

A camera for planetary work

At f/75, it is possible to take photographs of Venus in about 1/20 second on Plus-X, but the moon and the other planets require exposures of 1/5 second or more. In none of these cases is it practicable to use a standard single-lens reflex camera without modifying the shutter release mechanism to avoid mirror slam. Neither is it easy to focus a relatively dim astronomical image on the normal focusing screen. Special magnifying focusing attachments, which eliminate the etched focusing surface of the standard roof prism, can be attached to some cameras but are expensive and in any case do not solve the problem of vibration. Furthermore, it is wasteful to have to expose a whole 35mm frame on the small image of a planet. By dispensing with the interlocking shutter/winding mechanism of the standard camera, the film can be advanced by only two or three sprocket holes at a time, permitting a hundred or more exposures to be taken on a single length of film. It is therefore worth considering the use of a special camera attachment, which should embody the following features:

(*a*) Direct focusing, using a low-power eyepiece or magnifier to examine the image.
(*b*) Vibration-free shutter of the leaf type, found in old-fashioned cameras, usually of the roll-film bellows types. Alternatively, and less satisfactorily, a hand-operated cardboard shutter held in front of the telescope tube can be used, in conjunction with a standard focal-plane shutter that serves only to exclude light from the film when the camera is being loaded and between exposures.
(*c*) A facility for optional film advance.

A camera embodying these principles, based on the one used by the well-known planetary photographer H. E. Dall of Luton, England, is shown schematically in figure 69. The front of a 35mm camera body, with interlocking winding mechanism removed, is closed by a leaf shutter, operated either by a cable release or, as in Dall's case, by a pneumatic bulb and tube closing the shutter plunger. The latter way reduces the risk of vibration

Figure 69. *A camera for planetary work, based on a design by H. E. Dall.*

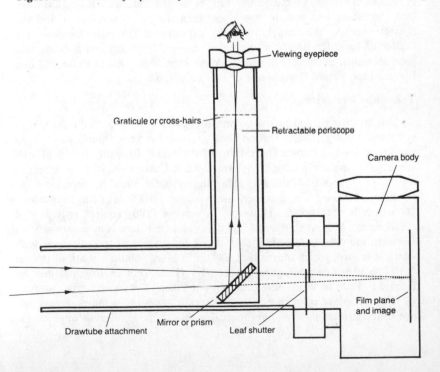

being imparted to the telescope. A tubular extension fits onto the telescope drawtube and carries a retractable periscopic unit, in which a small prism reflects the light up to an eyepiece of about 25mm focal length. This is either focused on crosshairs or on a very finely ground screen with a clear spot in the center to which the image is brought; the length of the periscope is adjusted so that the distance from the center of the prism to the emulsion is the same as the distance to the screen or crosshairs. Thus, when the image is sharp in the eyepiece, it must be in focus on the film. The enlarging eyepiece is located at the appropriate distance from the film further up the tube that fits into the telescope drawtube.

Once focusing is accomplished, the periscope is moved up out of the way of the light path before an exposure is made. Normally, two or three photographs would be taken before the focus and image location is rechecked. It is extremely helpful to have a fairly powerful finder on the telescope so that the planetary image can be brought within range of the prism; at f/75 the total coverage of a 35mm frame is only a few minutes of arc.

Some miniature cameras are specially designed for scientific work and have given good results without any modification. The Miranda Laborec is one, with a low-vibration mirror-lift and shutter. The advantage of using an automatic reflex system is that the photograph can be delayed until a period of reasonably steady seeing occurs. On the other hand, there is no guarantee that the seeing will remain steady between the eye perceiving it and the shutter exposing the film. Ultimately, exposing at the right moment is a matter of luck. Provided the seeing is reasonably steady, with occasional good moments, a sequence of several dozen exposures, which can be shot in a few minutes, should contain one or two acceptable images.

Telescope and drive

Due to their perfect achromatism, reflectors will be found preferable to refractors for high-magnification photography. Even good-quality eyepieces can introduce small color aberrations of their own. Probably the ideal telescope for planetary photography would be a Cassegrain with an effective focal ratio of about 80, since no enlarging eyepiece would be necessary. No commercial Cassegrain, however, has a larger final focal ratio than about 20, which is not enough. However, to achieve f/80 requires only a × 4 enlargement. Moreover, the residual color of an eyepiece is inversely proportional to the cube (approximately) of the focal ratio of the telescope with which it is used, which means that at f/20 it is only about a tenth of that at f/8. The very fine photographs taken by H. E. Dall in England and by Ronald S. Price in Texas have been obtained using Cassegrain-type systems.

The secondary color of a refractor seriously degrades the planetary image contrast when used photographically, due to the emulsion's appetite for blue light. A yellow or red filter removes the haze but necessitates perhaps a doubling of the exposure time. With most amateur-owned refractors, which will have apertures of 6 inches or less, the image is already on the small side even at f/80. It must regretfully be concluded that, in the apertures normally encountered, refractors are not capable of high-quality planetary photography.

Most photographic work has been done with Newtonians. Due to the rather high image magnification required, it may be found that a microscope objective gives better results than a normal eyepiece. A × 10 objective would be the correct choice for an f/8 telescope. An eyepiece, fairly obviously, is designed for visual use, to feed the magnified image into the eye, whereas a microscope objective is intended to examine and refocus the enlarged image of a small object whether that object is a planetary disk or something mounted on a slide. Good results also have been achieved with compact catadioptric systems, although their contrast is not improved by the relatively large central obstruction.

The drive requirements are critical in one sense, tolerant in another. If the film is to have every chance of achieving the optical resolution limit, then the guiding must be precise to within $0''\cdot4$ or better for a 6-inch telescope, or $0''\cdot2$ or better for a 12-inch. These amounts are half or less of their theoretical resolving limit, since a drift *equal* to the resolving limit would blur two adjacent image structures together, resulting in loss of detail. The values quoted here correspond to the image shift that occurs with a stationary telescope in the space of 1/40 and 1/80 of a second respectively. It follows from this that, although the drive may be slow or fast by an amount equivalent to a total error of several minutes per day, it must be extremely smooth. The typical, small coarse-toothed wheel encountered on low-cost commercial telescopes will not be good enough. A worm speed of at least 1 rpm, and preferably 2 rpm, is needed. A tangent drive, whose short arc is of no account in planetary photography, will be the cheapest way of achieving such quality.

An obvious counterargument to these demands is that acceptable photographs have been taken with much cruder equipment than this. This fact does not, however, mean that still better ones could not have been achieved with superior means! The planetary photographer, if he is to have consistent success, must leave to chance only those factors over which he can have no control. The nights and moments of first-class seeing are too rare to be vitiated by a lack of attention to detail elsewhere.

The moon

The earth's rotation causes most celestial objects to circle the sky once in

Lunar photographs by Lee S. Najman, New York, taken on Kodak high contrast copy film, using a home-built 12½-inch (32cm) f/6 Newtonian, either at prime focus or amplified to f/60. The film was developed in highly diluted Agfa Rodinol (1:150) for 18 minutes at 21°C, with very brief agitation at 6-minute intervals. Najman finds that this technique reduces the relative density of the highlights, permitting straightforward printing without the need to "dodge" the exposure.

The Moon, aged 4, 6, 8, 17, and 19 days.

Theophilus, Cyrillus, and Catharina under evening illumination.

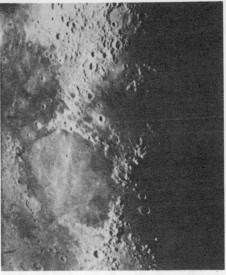

Mare Serenitatis under morning illumination.

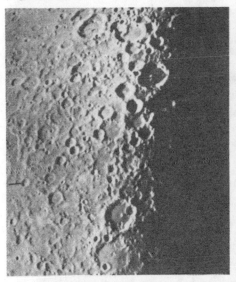

The southern terminator at First Quarter, showing Hipparchus and Albategnius (lower) and Walter, with its inner mountain catching the sunlight (upper).

The southeastern highlands under evening illumination. Note the Altai Mountains (lower left).

These photographs of Mars, Jupiter, and Saturn were taken by Lee S. Najman (12½-inch [32cm] Newtonian at f/60) in the winter of 1977–1978. Mars and Jupiter were photographed on Koda-chrome (exposure 4 seconds and 3 seconds respectively); Saturn on High Speed Ektachrome (exposure 8 seconds).

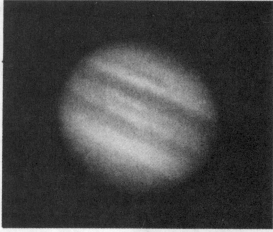

23^h 56^m—the sidereal day. The most obvious exceptions to the general rule are the sun and the moon. The sun takes about 24 hours—4 minutes longer than do the stars. The moon's rate is slower still. On average, it rises 50 minutes later each night due to its eastern motion around the sky. Therefore, when observing and photographing the moon, the telescope drive must be made to run about 3 percent slow, equivalent to a drift in front of the stars of about 1″ per second of time. This motion in R.A. is at a maximum when the moon is at the north and south points of the ecliptic, in Taurus/Gemini and Sagittarius. There is also a motion in declination, the maximum rate being about 0″·5 per second of time, which occurs when it is near the equinoxes, in Pisces and Virgo. It should be allowed for in critical work, if the exposure exceeds a second or so. For most of the time, however, motion in declination can be disregarded.

The required exposure for the moon varies not only with phase but with the distance of the region from the terminator. A shot of the full moon requires about ⅕ of a second at f/75 (medium-speed film being assumed throughout), whereas near the terminator in the crescent phase (when much sunlight is scattered away from the earth) the exposure may have to be a couple of seconds. In fact, the tone range from highlights to extreme terminator is too great to be handled properly by any emulsion or printing paper. If the exposure is correct for the terminator, the areas under high sun will be burned out. The best that can be done is to adopt an average exposure and to suppress the exposure of the terminator region in the enlarger by "dodging": masking the underexposed region using a constantly moving strip of card held above the enlarging paper. Such a technique can be mastered only by practice. A different method, used by Lee S. Najman, is described in the caption to the accompanying photographs.

Near the center of the moon, a second of arc is equivalent to a crater just over a mile across. Therefore even a 6-inch reflector should be able to record craters as small as about 1½ miles across, while linear features narrower than this will be caught. Photography using red, green, and blue filters (e.g., Kodak Wratten 25, 58, and 47 respectively) will enable suspect areas to be checked for transient and permanent colored spots. It should be noted that most ordinary panchromatic emulsions record red light beyond about 6,400Å rather poorly, and the exposure using the number 25 filter may have to be lengthened as a result. Kodak SO-410 film has a better red response, and may be more suitable for three-color work.

Lunar eclipses are interesting to photograph and can produce beautiful color slides. To record features in the deep shadow of a total eclipse using a fast film (e.g., Ektachrome 400), an exposure of about 15 seconds at f/5 may be needed. However, few photographs resemble the visual appearance, since the small exposure latitude of color film exaggerates the different

Central region of the moon. *A photograph of the region from the cleft of Hyginus (lower) to Arzachel (top right). On the original negative, which was taken with a 6-inch reflector, the lunar image was 1½ inches across. The photograph was taken by Cmdr. H. Hatfield, R.N., who made the telescope and mirror himself.*

brightness across the shadow; many total eclipses look like partial ones! Achieving a more realistic picture may require deliberate underexposure even though detail may be lost from the darkest part of the eclipsed moon.

Ptolemaeus. *Seen soon after first quarter. Ptolemaeus is at bottom right; to the south is Alphonsus, with a conspicuous dark patch inside its eastern wall. The Straight Wall is visible as a dark line in the upper right-hand part of the picture. Photographed with a 12-inch reflector at f/30, 1 sec., Ilford G-30 plate, on May 28, 1966. (H. R. Hatfield.)*

The inferior planets

Mercury is even more of a challenge to the photographer than to the visual observer, and the bare recording of the phase is something of an achievement since Mercury's disk is always very small and must be observed low in the sky, in twilight. A red filter helps to improve the contrast with the bright background. Mercury is surprisingly bright, needing an exposure of only about ½ second at f/75.

Venus, on the other hand, makes an attractive picture, and as far as sheer brilliance is concerned it is the easiest planet to photograph. Unfortunately the best photographic opportunities occur in the twilight sky, when it is not at a particularly high altitude; daylight photography, though possible, is hindered by scattered sunlight. An orange or red filter is necessary then, and this increases the necessary exposure. If one desires to take useful photographs showing cloud markings, an ultraviolet filter must be used in a very transparent and preferably fairly dark sky, since ultraviolet reduces the contrast between the planet and the background.

If an "ordinary" emulsion is used, then the planet in effect is being photographed in blue light without using a filter at all. Kodak III-0 is a special blue-sensitive scientific emulsion, one of several types used by professional astronomers. Another mainly blue-sensitive film is Kodak Contrast Press Ortho. Both of these types (or, indeed, any panchromatic film) must, however, be used in conjunction with an ultraviolet filter such as the Schott UG-2, if one is to have the best chance of recording the cloud forms. The biggest problem with this work—the absorption of this radiation by the glass lenses in the eyepiece—becomes serious below about 3,500Å, so the "window" through which ultraviolet photography can be carried out is, in practice, between about 3,500 and 4,000Å. A long-focus Cassegrain, which would give a reasonably large image of Venus without needing an enlarging eyepiece, would be ideal for the purpose since the wavelength range could then be extended to below 3,000Å, at which point atmospheric absorption becomes serious.

Relatively few amateurs seem to have experimented in this field, but back in 1956 some photographs showing terminator deformations and faint shadings, as well as bright cusps, were obtained in England by T. W. Rackham, using medium-speed film and a 6-inch Newtonian working at f/93. He used an ultraviolet filter and a 1-second exposure. With large professional equipment, these markings are fairly easy to record.

Mars

The fleeting visibility of this attractive planet makes it necessary to seize every chance that occurs. Pictures taken at 30-minute intervals will reveal its rotation, but a continuous series over a fortnight or so will be required if all longitudes are to be recorded. Using an orange filter to increase the contrast of the dark markings, exposures of from ½- to 1-second at f/75 will be about right. An orange filter emphasizes any surface obscurations due to dust storms, while a blue filter darkens the reddish surface, emphasizing the white polar caps and any atmospheric hazes. White haze is often seen at the limb, even in equatorial regions, and may be detected more easily photographically than visually.

Jupiter

The giant planet offers so much for the amateur that a photographic program could usefully be devoted to it alone. Not only is it observable for about nine months of the year, with an agreeably large disk, but its surface is always a tangled mass of detail that offers scope for useful measurements. Although the finer details visible to the visual observer will not be discernible, the more obvious features can have their latitudes and longitudes measured with a precision approaching that attainable at the eyepiece. If such measures are going to be made, however, it is essential to give sufficient exposure to record the limb of the planet. Unlike Mars, which due to atmospheric haze often appears brighter at the limb than on the disk, the edges of both Jupiter and Saturn fade away very markedly. Due to the slightly "warm" color of the belts, a light blue filter improves the contrast of the markings.

The surface of Jupiter is so highly reflective that, despite its great distance, the necessary exposure is only about 1 second at f/75. The four Galilean satellites will require a slightly longer exposure, and it will be a matter of some difficulty to obtain a satisfactory photograph of both together. Satellite shadows are, however, fairly easy to photograph, and with care the ingress or egress of a satellite, when it appears bright just inside the darker limb of the planet, may also be caught.

Saturn

The globe of Saturn is far less detailed than that of Jupiter, and most amateur photographs will show no more than a brightish equatorial zone. The rings, of course, claim most attention. The first aim must be to record the different brightness of the outer rings A and B; the next step is to record the Cassini division. Ring C is too dim to be photographed satisfactorily, since even the bright rings and the disk require an exposure of 2–3 seconds at f/75.

An interesting project would be to capture as many satellites as possible on one photograph. The time would have to be carefully chosen, so that the satellites were well clear of the planet. An exposure of perhaps 30 seconds, with a 6-inch, should record several of the brighter moons, but the image of the disk and rings would of course be greatly overexposed. To achieve a realistic result, a properly exposed photograph of the planet can be taken at the same time, and the two negatives combined in the darkroom by printing first one and then the other on the same sheet of paper. A similar technique may be helpful in obtaining firm images of the Jovian satellites with the planet.

The minor planets

Literally hundreds of minor planets are within the range of a 6-inch reflector, if fast film and a 10-minute exposure at the prime focus are used; the limiting magnitude of such a combination is about 16. Since the average westward motion of a minor planet at opposition is about 6″ in 10 minutes,

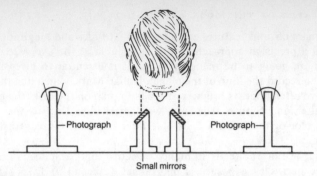

Figure 70. *A simple stereoscopic viewer with which to examine a pair of photographic prints.*

the telescope will not need to be guided specially, since the drift will not exceed the object's image diameter. Exposures made an hour or two apart, if almost but not quite superimposed in the enlarger, will show the minor planet's double image looking wider than those of the stars, thereby demonstrating its motion. A quite spectacular effect can be achieved if these two photographs are printed separately and placed in a simple stereoscopic viewer, since the minor planet then stands out in front of the stellar background. Care must be taken not to put the photographs in the wrong way round, for in this case the stars will appear to be nearer than the minor planet!

The greatest challenge to the photographer of minor planets is recording a fast-moving object that is too faint to be seen visually in a guide telescope. Some of the interesting recently discovered minor planets are of this type. For example, object 1978CA, of about magnitude 15, was moving across the sky in early March 1978 at a rate of about 8″ per minute. Because of the high inclination of its orbit, motion was mainly in declination, and the English amateur Brian Manning succeeded in photographing it by setting his 10-inch Newtonian on the field, opening the shutter, and adjusting the tangent-arm declination slow motion to move the tube in increments of 4″ every half minute during the 5-minute exposure on Tri-X film. The resulting photograph showed 1978CA as a starlike spot, while the stars had trailed by the amount of its passage. If the telescope had simply followed the stars, the minor planet would have moved across the emulsion too quickly for its image to have been recorded at all, since the effective exposure would have been only about 30 seconds. A rapid motion in right ascension, of course, can be followed by an appropriate adjustment of the drive rate. It is through the enterprise of amateurs who try to tackle new and difficult tasks, whether in the visual or photographic field, that the science has progressed in the past and will continue to develop in the future.

24

Deep-Sky
Photography

The photography of clusters, nebulae, and galaxies is almost entirely a technical exercise. It cannot really be incorporated into a research program. The two branches of astronomical photography that have been considered so far—star charting and high-magnification planetary photography—have some observational validity in that they can be used to obtain star magnitudes in the first case and to record ephemeral markings in the second. Deep-sky work does not seek to do more than obtain a picture of what is, in all practical respects, a still life. The end product is a photograph that is to be enjoyed aesthetically, or compared with others in order to notice the differences that result from equipment and technique rather than from real changes in the object.

This chapter will, therefore, briefly cover some of the technical aspects in which this branch of astrophotography differs from those that have been covered in the previous pages.

Optical systems and image scale

Practically all deep-sky photography is performed at the prime focus, and therefore the focal ratio of the system is of some importance. Newtonian reflectors of from f/5 to f/8 are probably the most suitable. Refractors suffer from the problem of uncompensated color, and at their typical focal ratio of 15 they are too slow for satisfactory portrayal of nebulae and the larger galaxies. Similarly, the equivalent focal ratio of most Cassegrains is too large, although the compact f/10 catadioptric systems have given very good results. It must be remembered that if a nebula (which is an extended object) is being photographed, the exposure required to produce a certain

The Ring Nebula (M57) in Lyra, photographed by Lee S. Najman, using a 12½-inch (32cm) Newtonian at f/6, on July 16, 1977. This 5-minute exposure was made on Kodak 103a-F film, using a Wratten 25 filter.

density on the negative is a function of focal ratio, not aperture. Increasing the aperture increases the image scale, and therefore the resolution, but does not alter the density. On the other hand a larger aperture, almost regardless of focal ratio, will record faint stars more readily. At f/2, any miniature camera can photograph the famous North America nebula in Cygnus (N.G.C. 7000), but at f/8 it is practically invisible, regardless of the aperture. This important principle was demonstrated long ago by one of the greatest astrophotographic pioneers, E. E. Barnard, who took some of the finest Milky Way photographs of all time, using a 6-inch aperture portrait lens of 36 inches focal length (f/6):

> *One is too apt to think that, for astronomical investigations of any value nowadays (when single telescopes have run up to a hundred thousand dollars or so in cost), we must have great expensive instruments—the results, in a sort of way, being expected to bear a certain proportion in point of value to the cost of the telescope. . . . What, therefore, must be thought of the performance of an ordinary "magic lantern" lens only 1½ inches in diameter and 3 or 4 inches focus?*
>
> *During the fall of 1894 the writer made a series of experiments with such a lens in photographing certain regions of the heavens. . . . Some of the cloud forms of the Milky Way were shown in ten minutes' time, which had required from two to three hours' exposure with the 6-inch lens! With it a large and previously unknown nebula was photographed about the star ν Scorpii.*
>
> *With this lens a photograph was made of the constellation of Orion with about an hour's exposure. To my astonishment this picture*

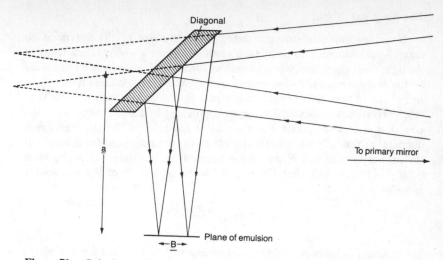

Figure 71. *Calculating the size of diagonal required for photography at the Newtonian focus.*

showed an enormous curved nebula covering almost the entire figure of Orion. This tremendous nebula is the largest—in point of extent—in the entire heavens. It cannot be seen with any telescope in existence.

Field of view

It is well known that a parabolic mirror can give perfect definition only if it is pointing directly toward the object; i.e., if the object is at the very center of the field of view. Away from the center, images are increasingly degraded by the twin aberrations of coma and astigmatism. Astigmatism is relatively unimportant, being more serious in wide-field systems, but coma can begin to intrude quite near the center of the field if the focal ratio is small. For example, if an f/4, 6-inch reflector were being used for photography, the useful field of view would be little more than a degree across; in a 12-inch f/4, it would be only half as large. At f/6, on the other hand, images would be reasonably good across fields of 3° and 1½° respectively. It is worth remembering that the linear field of good definition depends only upon the focal ratio of the mirror, not upon its aperture. The acceptable field width for photography at f/4 is about 0.6 inch; at f/6 it is 1.8 inches; and at f/8 it is larger than any reasonable diagonal or drawtube could accommodate. Bearing in mind that a 35mm frame measures about 1.7 inches across a diagonal, an f/6 or slower system is needed if the whole negative is to be covered in sharp focus.

The diagonal mirror

If a Newtonian telescope is being used, the size of the flat must be carefully calculated. It is pointless planning to cover a whole 35mm frame if the light from the primary is vignetted at the margins, either through having an undersize diagonal or too narrow a drawtube. The diagonal needs to be rather larger than that required for visual use, since the field of view is so much larger; but this extra size can be minimized by having the plane of the emulsion as close as possible to the side of the tube. If the required linear diameter of the fully illuminated field is B, the diameter and focal length of the primary are D and F, and the distance from the diagonal to the focal point is a (figure 71), then the size of the minor axis m of the diagonal is given by

$$m = \frac{Da + B(F - a)}{F}$$

For example, a 2⅛-inch minor axis diagonal will be required for use with a 6-inch, f/6 Newtonian, if the film is 5 inches from the center of the flat. The fact that this diameter is well over the desirable limit for minimum diffraction effects is of no account in prime-focus photography, since the mirror's theoretical resolving power will not be approached within a factor of 5–10 times.

The following table sets out the fields of view that will be encompassed by a frame of 35mm film, using various Newtonians:

TABLE XIV. Field of view (in degrees) at prime focus

Aperture (in)	f/5	f/6	f/8	f/10
6	1.8 × 2.7	1.5 × 2.3	1.1 × 1.7	0.9 × 1.4
8	1.3 × 2.0	1.1 × 1.7	0.84 × 1.3	0.67 × 1.0
10	1.1 × 1.6	0.90 × 1.3	0.67 × 1.0	0.54 × 0.81
12	0.90 × 1.3	0.75 × 1.1	0.56 × 0.84	0.45 × 0.67
16	0.67 × 1.0	0.56 × 0.84	0.42 × 0.63	0.34 × 0.50

Tube and fittings

In general, solid-walled tubes are preferable to framework ones for astronomical photography, since all stray light is excluded from the light path. The influence of tube currents is less important than with visual work, due to the much lower resolution threshold.

The rigidity of the optical components with respect to each other, and of the tube as a whole to the mounting, are both very important. During an exposure of, say, half an hour, the slowly changing orientation can introduce flexure in a weak tube, resulting in an image shift. If a separate guiding telescope, rather than an off-field guiding eyepiece, is being used, this shift will pass unnoticed until the film has been developed. Furthermore, lack of rigidity between the tube and the declination axis can mean that the drive is not being followed properly, demanding unnecessary use of the slow motions.

Points to which particular attention should be paid are the firm seating of the primary in its cell and rigidity of the diagonal mounting.

Guiding methods

Above all else, a deep-sky photograph is made or marred by the quality of the guiding. If a relatively short-focus camera is used, the acceptable guiding error can be up to several minutes of arc—sufficient to be left to the care of any reasonably accurately aligned mounting with a standard drive. With planetary photography, extremely accurate following is necessary, but only over a duration of a very few seconds. With long-exposure photography at the focus of a Newtonian, however, the image must remain within a few seconds of arc of the initial position on the film for a period of up to half an hour, or even more.

The first step in any successful work of this nature is to align the polar axis within a very few minutes of arc of the celestial pole (see below). The second is to equip the mounting with efficient slow-motions. The usual way of making fine adjustments in right ascension is by varying the frequency of the current operating the synchronous motor. As far as declination adjustments are concerned, the small worm and wheel often fitted to Newtonians is far too coarse for the delicate control required. A tangent-arm slow motion should therefore be fitted with a screw of sufficiently fine thread and an arm of sufficiently long radius to allow controlled adjustments of a very few seconds of arc. Some photographers, preferring the robustness of a mechanical adjustment, have incorporated a similar system on the polar axis, leaving the driving worm to run at its preset rate.

However the adjustments are to be made, it is essential that the controls are comfortably in the observer's hands throughout the exposure, and that no vibration is imparted to the telescope when they are operated.

If a guiding telescope is used, its focal length should be, if possible, about the same as that of the primary mirror. If this image is then viewed with an eyepiece of about 9mm focal length, it is effectively magnifying the resultant negative scale by about 30 times (the normal viewing distance of an object

These fine photographs were taken with an aperture of only 6.6 inches by Herman R. Dittmer of Seattle. The instrument, a home-constructed Maksutov, normally works at f/10, but its speed can be increased to f/7 by placing a special focal-reducer before the final focus. The film used was Kodak 103a-F; exposures ranged up to two hours.

The galaxies M.81 (below) and M.82 in Ursa Major.

Left below, *The famous Horsehead Nebula in Orion. The bright star near top right is ζ Orionis.*

Below, *The large but faint galaxy M.33 in Triangulum.*

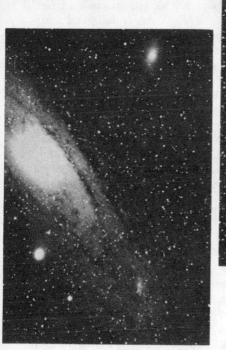

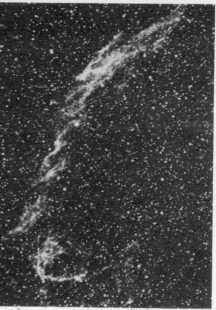

A part of the Filamentary
Nebula in Cygnus.

Above, *The Andromeda Galaxy, M.31, showing its companion galaxies M.32 (below)
and N.G.C. 205. The bright patch at lower right is a condensation of gas and dust in one
of the galaxy's outer spiral arms.* Below, *The Pleiades, showing the extensive dust
clouds illuminated by the brighter stars.*

held in the hands being about 10 inches or 25cm). With a 6-inch, f/8 system, the desirable magnification is therefore about × 150. Higher powers have been recommended by some workers, but if a bright star is defocused and outlined against the crosshair in the eyepiece, even a small image wander is readily seen. It is very convenient if the guide telescope can be displaced by a few degrees from the direction in which the telescope is pointing so that a bright star can be acquired should there not be one within the field being photographed. An offset guider will also be necessary if a comet is being photographed with its head toward one corner of the negative.

The most rigorous guiding method is to view a star just outside the field, using a small prism or mirror and a high-power eyepiece fixed at right angles into the drawtube in front of the camera. This eliminates the effect of any flexure in the linkage between the main telescope and the guide telescope. However, it does restrict the choice of guide star to those in the immediate vicinity of the field, and these may all be faint. Most practical astrophotographers have preferred the guiding-telescope method.

Polar alignment

The effect of any misalignment between the polar axis and the Earth's axis is to draw the star images out into trails if the telescope is simply left to track in right ascension, and to make them appear to rotate around the guide star if the following is monitored and corrected. This rotation is most serious if the field being photographed is near the pole, and least so if it is on the celestial equator. For most purposes an accuracy of 5' should be aimed at. This is not difficult if the instrument is permanently mounted, but can necessitate irritating, even fatal delays if a portable telescope is being transported to a special dark site. In this case, undoubtedly the most convenient way is to use a small sighting telescope fixed to the declination axis crossplate (German mounting) or to the fork, and accurately aligned with the polar axis by the method described on page 43; a diagonal may be necessary to bring the eyepiece to a suitable position. Instead of judging the position of the true pole in relation to the stars, the telescope can instead be deliberately offset from parallelism with the polar axis by an angle equal to the distance of the appropriate polestar from the pole (northern polestar 50', southern polestar [σ Octantis] 57'), and at the right ascension of the star (02^h and 20^h respectively). If this is done, centering the star on the crosshair when the right ascension circle shows the correct sidereal time will mean that the polar axis is automatically aligned. However, it may be found easier to achieve accurate alignment between the axis and the sighting telescope than to set them at a small angle of deviation with equal precision.

When aligning an equatorial from scratch, with no polar sighting facility, the best course of events is to set the altitude of the axis as accurately as possible, using a set-square or wedge and a spirit level. With care, an accuracy of at least ± 30′ can be achieved. The azimuth is then set by observing the meridian passage of a star near the celestial equator, using the highest available power. If the star drifts upward (in an inverting telescope), the upper end of the polar axis is pointing east of the pole, and vice versa; to fix ideas, a drift in declination of 15″ per minute means than the azimuth error is 1°. Errors of azimuth are most important when the object is on the meridian, and least important when it is 6^h of right ascension away from the meridian. Errors of altitude have the opposite emphasis: To check these, observe the southerly or northerly drift of a star in the eastern or western sky. If an eastern star drifts to the south, or a western star to the north, the polar axis is not pointing high enough.

Most deep-sky photography is performed on objects near the meridian, where they are at their highest altitude, and so precision of azimuth setting should be the first consideration.

Films and development

The first practical experiments should be made using fast commercial black and white film, such as Tri-X or HP5. The speed-increasing methods suggested on page 334 can come later, although it may be worth experimenting with the double-dilution/double-development technique. A very energetic developer such as Kodak D-19 may bring out the fainter stars, at the cost of clogging up the highlights if delicate nebulae are being photographed. The great advantage of this type of film is its relative cheapness, particularly if bought in bulk rather than in single lengths, and the speed

The Orion Nebula, M.42. *A photograph taken on December 7, 1970, using a 10-inch reflector at f/6. Exposure 30 minutes on Tri-X. (W. E. Pennell, Lincoln, England.)*

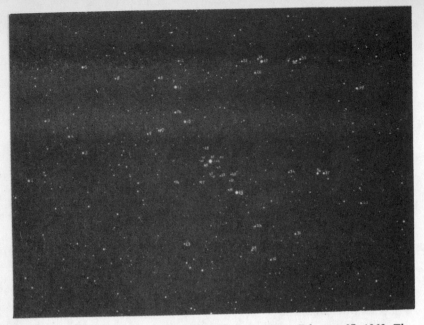

Nova Herculis, 1963. *Photograph taken by D. S. Brown on February 27, 1963. The nova, which is at the centre of the field, was still relatively bright. A 6½-inch Schmidt camera was used; the exposure time was 15 minutes. The faintest stars shown are of the 13th magnitude. Comparison stars are identified by numbers.*

with which the results can be seen. Although reciprocity failure begins to limit its performance, a 10-minute exposure, which will be quite long enough for the first attempts, will nevertheless reach stars and nebulosity of impressive faintness.

Users of color film will quickly find that emission nebulae (such as the Orion nebula), which appear white or greenish to the eye, acquire a vivid red tint on the photograph. This is because the red line of hydrogen acts more strongly on the film than do the other lines of the hydrogen spectrum, due to the lower reciprocity failure of the red-sensitive layer in the emulsion. Hypersensitising the film reduces this effect; alternatively, color filters can be used to restore the balance, but the exposure must be lengthened. If negative color film is used, the prints can be made using a filter that gives the most satisfactory color rendering.

Obvious choices for the first attempts include the Orion Nebula (which has such a large tonal range that a single exposure cannot reveal details in both the core and the outer regions), the bright emission nebulae in Sagittarius, the Dumbbell nebula in Vulpecula, and condensed clusters such as the

Hercules cluster in the northern hemisphere and the splendid globulars in Centaurus and Tucana in the southern. An obvious and appealing project would be to photograph all the accessible Messier objects, perhaps extending the work to the brighter galaxies, nebulae, or clusters in the N.G.C.; the enthusiast with a moderate aperture, able to reach about magnitude 16 in a few minutes, could search the outer regions of the Andromeda galaxy (M.31) for novae, as well as monitor innumerable faint variable stars when they have fallen beyond the limit of visual detection.

V

OPTICAL WORK FOR AMATEURS

25

Principles
and Equipment

Before World War II there was a considerable telescope-making cult, both in Britain and the United States, consisting of people who were determined to produce a complete astronomical telescope with their own hands. This was, at least partly, an indication of the difficulty of buying a good commercial article at a reasonable price—yet the result was that a number of skillful amateur opticians emerged, capable of making first-class mirrors of up to 12 inches aperture.

These days, the would-be observer can buy a good Newtonian mirror for such a reasonable price that he is unlikely to consider the alternative of making one for himself; the job may take many months, and he has no guarantee of producing a good result. Mirror-making, however, is a fascinating task (the great danger is that it may prove more absorbing than observing!), and the amateur who has made two or three optical systems will be able to reap the true rewards of his labor, for within his capabilities now lies a whole host of telescope designs, mostly unavailable commercially, save at great cost, and just begging to be tried out.

This section, then, is aimed at the beginner in optics who wants to make an astronomical mirror—largely as an experiment in glass-working—and we shall consider the manufacture of a 6-inch mirror of 48 inches focal length.

As a rough guide, the difficulty of making a mirror increases as the square of the aperture, and inversely as perhaps the cube of the focal ratio. By this rule of thumb, if we consider a 6-inch f/8 mirror as representing one unit of difficulty, increasing the aperture to 8 inches makes the task about twice as formidable, while the increased focal length presents a greater mounting problem. All this means a more lengthy project and a greater risk of abandonment, while the difference between the performance of 6-inch and 8-inch telescopes is very small compared with the advantage of either over the binoculars or terrestrial telescope that commonly form the observer's first instrument.

Let us consider the function of the Newtonian mirror. A source of light, be it an electric lamp or a star, can be considered as emitting its radiation in a series of wavelets traveling outward from the source, as ripples in a pond move in circular rings from the point at which a stone is dropped. If we consider a point on one of these rings, we find that it moves away from the source in a straight line, and that its distance from a near neighbor on the same ring will steadily increase as the ring expands. We can therefore consider light from a nearby source as being divergent. But if the distance from the source is very great, the section of each ring as it passes into the observer's eye or telescope becomes closer and closer to being a straight line. The effect of divergence increases, and adjacent points move apart more and more slowly. A source is said to be at "infinity" if no divergence in its rays is noticeable with the aperture being used to study it, so that the light rays can be considered as moving in parallel straight lines. Since all celestial bodies are effectively at infinite distances from the observer, astronomical telescopes are required to focus only parallel light.

If, to begin with, we set up a mirror whose surface is spherically concave (figure 72), we can make some initial investigations. Let the radius of curvature of the concave surface, which is the distance from the mirror to the center of the imaginary sphere of which its surface forms a part, be R. If a small lamp is set up at the center of curvature, its rays will strike the mirror at right angles (*normally*) to its surface, and will be reflected back along the same path to form an image on the source. If the source is moved away from the mirror, we find that the image moves toward the mirror, since the rays are no longer striking the surface normally. When the source is removed to infinity, the image is formed at a distance of $\frac{1}{2}$R from the mirror, and this distance is, of course, the focal length. It follows from this that a mirror of 48 inches focal length must have a radius of curvature of twice this, or 96 inches.

The image, however, will no longer be sharp, for it will be discovered that different regions of the mirror form their image at different distances from the mirror's center. Careful examination would show that the focal length of the margin of a spherical 6-inch f/8 mirror is .023 inch shorter than that of the center. To make the images formed by the whole of the mirror coincide, it is necessary to shorten the radius of curvature, and

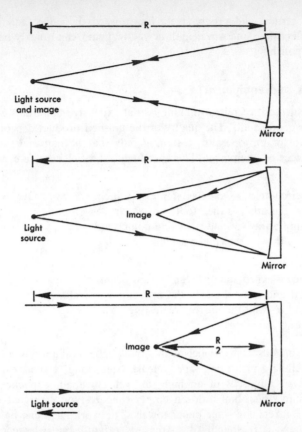

Figure 72. *Image formation by a concave mirror. (a) Light source at radius of curvature. (b) Light source beyond radius of curvature. (c) Light source at infinity.*

hence the focal length, of the central region. The name of the curve which gives the necessary steepening of the center is the *parabola*, and such a mirror that is free from spherical aberration is called parabolic.

Normal optical grinding and polishing procedures produce closely spherical surfaces. The usual technique in mirror-making is, therefore, to grind and polish as spherical a surface as possible, and to reserve the parabolizing, or *figuring*, until the very last stage. To parabolize a spherical 6-inch f/8 mirror requires the removal of only 1/80,000 inch of glass from the center, and progressively less toward the edge; and this small amount (less than one wavelength of yellow light) can quickly be removed by local polishing.

Materials and equipment

A great merit of mirror-making is that expensive apparatus is not necessary for good results. The quality of the finished product depends entirely on the skill of the operator, assuming only that he can make some very simple pieces of equipment. The absolutely essential items are as follows:

To buy
1. Two circular discs of glass about 6⅛ inches across. One will be the mirror, the other is the "tool" on which it is ground.
2. Grinding abrasives and polishing compound.
3. Pitch, with which to make a polisher.

To make
4. A rigid 3-legged stand, or an old oil drum, to which the work can be secured and around which the operator can walk.
5. A stand to hold the mirror for testing.
6. A testing apparatus.

Miscellaneous items, which will probably be available as a matter of course, include a measuring tape at least 8 feet long, a short rule marked in fiftieths or hundreds of an inch, an old saucepan, a pocket torch, a carborundum stone, half a dozen empty squeeze-type detergent bottles, a couple of buckets and some paper towels. The work itself can be done any place where it is reasonably dust-free and private, but a large cellar, because of its constant temperature and freedom from drafts, is best of all. Let us take the items one at a time.

GLASS DISCS.   Ordinary plate glass will serve excellently, but low-expansion glass, of the kind used for ovenware (Pyrex is a common brand), gives less trouble with thermal effects. It is also harder, which means that more work is required to produce the initial curve, but offers the advantage of a less scratch-prone surface at later stages. Overall, then, it is to be preferred for the mirror disc, but plate glass is suitable for the tool. The mirror disc should be an inch thick, for stability; the tool used need not be more than ¾ of an inch thick.

ABRASIVES.   There are a number of different varieties of abrasive. Carborundum powder (silicon carbide) is universally used for roughing out the curve, since it abrades quickly. To fine the surface, ready for polishing, emery or the proprietary Aloxite (both are forms of aluminum oxide) can be used; these powders are preferred to carborundum, since their cutting action is more gentle. It *is* possible to use carborundum powder throughout, but the finished surface may require twice as long to bring to a good polish.

Abrasive powders are commonly graded by mesh size, the number referring to the number of meshes per inch. A recommended series, with the amounts for a 6-inch mirror, would be: 80 carborundum (1 lb.) for grinding out the curve; 120 carborundum (4 oz.); and 2 oz. of grades 220, 320, 400, 600, and 900 carborundum or aluminum oxide. Some workers might omit the 320 grade, preferring to give longer with the 400; but large jumps of grain size should be avoided.

The finer grades of abrasive have various designations. They are frequently referred to in the form of grain size in microns (.001 mm, or 2 wavelengths of yellow light), an equivalent of the 900 mesh size being 17 microns. A powder of this grade will give a reasonable finish, but a very fine powder of about 1500 mesh (10 microns) will save considerable polishing time. A popular finishing powder in the United States is the ultrafine M305 emery, made by the American Optical Company, which is equivalent to a 5 micron grade; but such fine powders must be used with the most rigorous precautions against contamination, if scratches are to be avoided.

POLISHING COMPOUND.   In the old days, jeweler's rouge (iron oxide) was used. Its messiness and slow polishing action have led to its replacement by rare-earth oxide compounds, which are clean, fast-polishing and give excellent results. Barnesite and Cerirouge are two proprietary types. An ounce is plenty with which to polish and figure a 6-inch mirror.

PITCH.   Any optical pitch will do: it may be British (produced from coal) or Swedish, Canadian, or Burgundy (produced from resin). Coal pitch is usually too hard, and will require softening with pure turpentine. A pound of pitch will be more than enough.

WORK STAND.   Its nature does not matter as long as it is the right height for comfortable working, so that the operator does not have to double over in order to exert pressure on the discs. It must be firm enough not to rock, and the base must not be so wide as to prevent the operator from walking slowly round it. Three small wooden blocks are screwed into the top to hold the mirror or tool during working.

394   Principles and Equipment

Mirror stand. A simple stand is shown in figure 73. The mirror must not be able to fall or roll out, and a fine tilting adjustment is most useful.

Testing apparatus. This is needed in the polishing and figuring stage. Some very complex contrivances have been produced, but they are not necessarily more accurate than a simple rig, since it is the judgment of the optician that decides the accuracy of the measurements. The apparatus consists of two parts: an illuminated pinhole, and a knife-edge, and will be described in detail in Chapter 27, when the test itself is discussed.

Obtaining the glass, abrasives, pitch and polishing compound individually may present a problem. If a nearby optical firm is approached, they may sympathize and help out. It must be realized, however, that the time taken to measure out two ounces of powder and keep everything uncontaminated may cost as much as the powder. By far the best solution for the beginner is to buy a complete telescope-making kit from a telescope supplier. The price for a 6-inch kit will be from about $20 (£8), and the purchaser may find that his mirror and tool discs have already been approximately curved to shape, so that he can commence grinding with fine powders. To make this section comprehensive, however, it will be assumed that the mirror-maker is faced with the classical problem of producing the right curve on two pieces of flat glass.

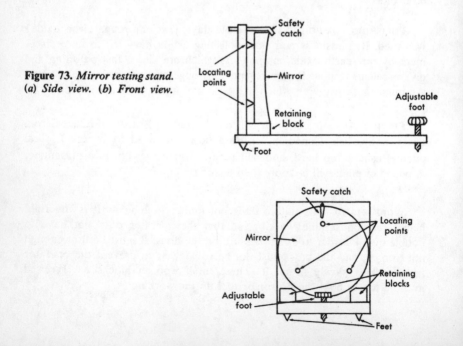

Figure 73. *Mirror testing stand.* (*a*) *Side view.* (*b*) *Front view.*

26

Rough and
Fine Grinding

These are distinctly separate processes. Rough grinding has as its aim the generation of the correct concave curve on the mirror disc, and is performed with the coarsest (usually grade 80) carborundum. Fine grinding leaves the curve practically unchanged, merely removing the pits left by each succeeding grade of abrasive until the surface is in a polishable condition.

Before starting work, bevel the edge of the mirror disc with a fine carborundum stone and water, taking off a chamfer of about $\frac{1}{16}$ of an inch. Aim the grinding stone toward the edge rather than the face, in case a chip of glass is removed from the surface. There is no point in beveling the tool until the rough grinding has ended, since it will be quickly ground away in the initial roughing out.

Secure the tool to the top of the stand inside the three wooden blocks, sprinkle a teaspoonful of 80 carborundum and some drops of water on its face, and put the mirror disc face down on top. Now press as hard as possible on the work, and begin sliding the mirror back and forth across the tool so that its center traverses a distance of about 4 inches. Such a stroke is term a "2-inch" stroke, since this is the amount of overhang produced at its extremity. At the same time, take slow but steady steps around the stand (one revolution per half minute is effective), and revolve the mirror disc by giving it a small twist with the fingers at the beginning or end of each stroke. These two revolutions, which are performed at all stages of grinding and polishing, have the effect of distributing the action evenly around the discs and producing a surface of revolution.

Many textbooks advise pitching a handle to the back of the mirror. However, there is no virtue in this arrangement. The most effective grinding is produced by firm, two-handed pressure low on the mirror. Keeping the back of the disc unobstructed allows the grinding action to be observed. Furthermore, handles can come loose at unfortunate moments, and glass does not bounce well, as the author's first mirror testified!

To begin with, the carborundum will give out an active, crunchy sound. In two or three minutes, however, the grinding mixture will be gray with ground glass and the action will be smooth. This means that the abrasive has been worn down, and a new charge is required. Such a cycle is called a "wet." When using coarse carborundum, the grinding action is accelerated by wiping the surfaces clean of spent grains in between wets, as otherwise the fresh charge is partly absorbed by the sludge.

Judging the amount of water to be added is a matter of experience. Too runny a mixture cuts weakly, the surfaces slipping over each other, while dry abrasive goes patchy. The most convenient way of adding water is from a squeeze bottle.

Examination of the surfaces after just one wet will reveal what is happening. Due to the excess pressure at the end of each stroke, caused by part of each disc being out of contact, the center of the mirror and the edge of the tool both receive extra wear, so that the discs respectively turn concave and convex. The greater the overhang, the more extreme is the effect, and the worker can experiment with different strokes and observe the result. A representation of some different grinding and polishing strokes is given in figure 74, and the "side" stroke is helpful in roughing out, since the mirror disc is overhanging the tool at all times, thereby increasing the curving tendency. On the other hand, gross overhang will produce a markedly irregular curve, and extra time will have to be spent in making

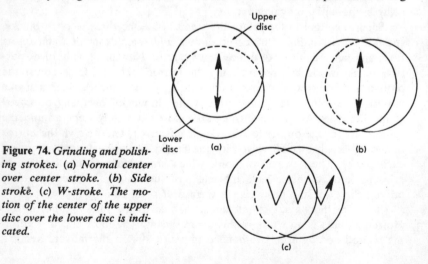

Figure 74. *Grinding and polishing strokes. (a) Normal center over center stroke. (b) Side stroke. (c) W-stroke. The motion of the center of the upper disc over the lower disc is indicated.*

the surface truly spherical. The beginner will be wise to aim at a smooth curve from the beginning, and the 2-inch stroke, with perhaps an inch of side overhang, is safe.

After half an hour's work, a straightedge laid across the mirror will reveal a noticeable concave surface. If the stroke is right, the margins of the disc will be just spotted with carborundum pits. If the glass is quite free of abrasion for perhaps half an inch from the edge, the stroke has been too extreme and should be shortened in the next spell, for we want a smooth curve right up to the edge of the disc.

How deep should the curve be? The depth, or *sagitta*, is given by the formula $r^2/2R$, where r is the radius of the disc, and R is the radius of curvature of the surface. In the present case we have values of 3 and 96 inches respectively, so that the required sagitta is $\frac{9}{192}$, or $\frac{3}{64}$ (.047) of an inch. A depth gauge can be made from a small piece of metal filed down to the right thickness; when it just fits under the straightedge, at the center of the disc, we know that the rough grinding has achieved its objective.

This simple depth method is not particularly accurate, but it will indicate when the time has come to make a more precise check. Sluice the mirror with water to make it reflective, set it up in the stand, and find the reflection of a torch in its surface. Holding the torch near the eye, walk slowly away from the mirror, waving the torch from side to side and watching the reflection. The image will, to begin with, appear to move in the same direction as the torch, but as the center of curvature of the mirror is approached it will grow larger and larger, and its movement less and less, until the mirror appears either dark or else full of light. Beyond the center of curvature, the image reduces in size once more, now moving in the opposite direction to the torch.

The ripply nature of the wet surface will give only an impure image, but it should be possible to establish to within a few inches the position where the image of the torch fills the mirror and shows no detectable side-to-side movement. Torch and eye are now at either side of the center of curvature of the surface, and the radius can be measured with a steel tape. Persevere with the rough grinding until the radius of curvature is within about 5 inches of the required value; if the radius is already too short, reverse the discs for a few wets.

Surface smoothness and "turned edge"

Before washing down and proceeding to the next grade of carborundum, the surface should be checked to see if it is approximately spherical. One good way of doing this is to substitute a much finer powder, say grade

320. There is no need to clean everything thoroughly, but great care must be taken not to contaminate the powder in its container. Using 1-inch strokes, grind one wet with the mirror faceup, and another with it face-down; keep grinding until the abrasive action seems to have disappeared. Then clean and dry the surface. Holding the mirror almost on a level with the eye, catch the oblique reflection of the filament of a clear-glass lamp and lower the mirror (reducing the obliquity) until the image is dull red, and barely visible. If the mirror is now tilted from side to side, so that the image of the filament appears to move across from the right edge to the left edge, its visibility represents the degree of fine grinding received by different parts of the surface. In particular, if it fades out sharply before the edge of the mirror is reached, we know that the margin is relatively flat or "turned down," and has received no fine grinding at all.

A turned edge, whether imparted in the grinding or polishing stage (but clearly much more serious in the former), is the biggest single curse of the mirror-maker, and everything possible must be done to identify and cure the condition at the earliest possible opportunity. In fact, relatively few first-time mirrors are free from this defect, whether slight or gross; not because it is very difficult to avoid, but because their makers have not been told how. This is particularly so in polishing, when, as we shall see, amateurs are repeatedly advised to use methods that are prac-tically certain to produce this insidious defect.

It can be said that fast grinding with long strokes, and particularly if a side stroke is used, is bound to produce a flattened margin, where the original flat glass has not fully conformed to the general curve. A slight turn-down can easily be removed in the subsequent grinding, but a serious effect, such as described above, should be removed with the grade that caused it. The enthusiastic amateur, anxious to see his surface fine-ground and ready for polishing, will want to forge ahead, and it may well be dif-ficult to convince him that spending some more time on the same grade is actually going to help him to reach his goal more quickly. He will not accept this advice until, having reached the finest stage of grinding, he finds the edge of his mirror disc black with untouched 80 pits, and abso-lutely no hope of removing them. Only then does he acknowledge that an extra half hour of work at the basic stage would have saved many, many subsequent hours of labor!

If long strokes have produced a noticeably defective edge, and the radius of curvature is still slightly long, a few wets of short (1-inch) strokes should cure the trouble without shortening the radius too much. If the radius is already too short, turn the mirror faceup with the tool on top, once again using short strokes. This is the most effective way of removing turned edge, the reason being that excess pressure is now brought to bear on the margins of the mirror, so quickly grinding out any anomalous curve. In fact, it is very difficult to produce a turned edge on a faceup mirror, unless the strokes used are so long that the tool teeters on the edge of the glass.

Fine grinding

Coarse grinding completed, which should not take longer than $1\frac{1}{2}$–2 hours of labor, the first of many "sterilizing" processes must take place. No trace of 80 carborundum must remain where it can do any damage, and this involves thoroughly washing down the stand, scrubbing the water bucket, throwing away mopping-up rags (paper towels are preferable), and cleaning arms, hands and fingernails. The sharp edge of the tool should be beveled at this stage, and the bevel on the mirror must be renewed if it has started to wear away; a sharp edge is a sure recipe for fractures and scratches.

The process of cleaning up between grades cannot be done too thoroughly. An old hand, who has made (and probably scratched) several mirrors, can work practically ankle-deep in grit and produce consistently perfect surfaces in conditions where a novice would be overrun with scratches. The difference is one of technique: The experienced worker knows the difference between dangerous and harmless contamination. The beginner, however, would be foolish to take any risks at all, since a particle of grade 80 carborundum dropping on to the mirror at the last stage of fine grinding will require several hours of regrinding if its effects are to be removed. If the worker changes everything changeable and cleans everything else, works with sleeves rolled up and well-scrubbed fingernails, does not allow people to dance on the ceiling overhead, does not wipe his perspiring brow or his overall with carborundum-contaminated hands, and refrains from pausing during the fine grades of emery to do a spot of bricklaying, he has taken all the precautions that could reasonably be expected of him.

The most convenient way of dispensing both 80 and 120 carborundum is from a small screw-top jar, with small holes punched in the lid from the inside (if they are punched from the outside inward, they tend to become blocked). Sprinkle a few drops of water on the surface, and take up the work with short strokes. On the basis of the previous radius measurements, work toward the 96-inch mark by either deepening or lengthening, as required, but never adopt an extreme stroke, as the main purpose of the grinding from now on is to wear the pits down as evenly as possible over the whole surface. The splash test will now enable the radius of curvature to be measured to within a couple of inches. Once it is within this limit, reverse the position of the disc to ensure even smoothing over the whole surface.

An examination of the mirror's surface after half a dozen wets will show how the smoothing is going. Dry the mirror and hold it up to the light; the majority of the surface should now appear of an even texture, but the presence of original deep pits is betrayed by bright speckles. If the smoothing is, as yet, uneven, these speckles will lie thickly either at the center or at the margin. Work must be continued until all these deep pits have vanished, when the surface is examined with a magnifying glass.

The phenomenon of scattered pits tell us something about the abrasive. If the size were uniform, all the pits would be of the same depth; hence, we know that larger particles are present. In the best-graded powders, the difference between the largest and smallest grains is much smaller than the difference in size between successive grades; even so, no matter how long we labor with the 120 grade, the surface will always contain a few pits deeper than the general host, though not as large as those left by 80 grit. Since the carborundum grains are broken down into a finer and finer state during grinding, it follows that the longer a wet is continued, the better the surface that will result. Workers generally grind the last wet of each grade for as long as possible, for this reason.

The beginner will protest at the difficulty of distinguishing between an original deep 80 pit and a new deep 120 pit. The solution is to find an identifiable pit near the margin of the mirror that is much deeper than any of its neighbors. A pencil mark against the side of the disc will act as a guide. Observe the pit's gradual disappearance as the wets proceed, and watch for new ones produced by the 120 abrasive. A little experience will tell when further grinding is unproductive. If there is no steep aspheric curve left from the roughing out, an hour's work, consisting of perhaps twelve wets, will be perfectly safe.

Now proceed down the grades, working faceup and facedown at each stage, and keeping the radius of curvature as close as possible to the desired 96 inches up to the 400 grade, after which further change of depth is inappreciable. At this grade and finer, the splash test gives a well-defined image of the torch bulb, which can be focused on a scrap of fine-ground window glass and the radius of curvature measured to within an inch. Check the center and edge for pits with the magnifying glass, and keep grinding until they have positively gone. Resist the temptation to "leave them for the next grade," which is a first step on the slippery slope to an unpolishable surface. Expect to spend an hour on each grade; if the surface appears perfect after half that time, then perhaps it is—but give another couple of wets for luck.

The grades from 220 onward, and rouge as well, are conveniently dispensed from squeeze-type bottles. Have enough water to keep the mixture sloshy. The 220 and 320 carborundum will have to be shaken up each time, but the fine emeries will stay in suspension for several minutes. Only a few drops of mixture are needed; an excess will, in any case, be expelled when the wet begins.

It is during the grades finer than 600 that the danger of a serious scratch is greater than at any time in the mirror's career. Foreign particles between the surfaces during the coarser grades tend to be held away from the glass by the thickness of the abrasive, while during polishing they stand a chance of sinking into the pitch before doing much harm, or sliding into a facet. But final smoothing brings the surfaces of mirror and tool to within a few microns of each other, with no escape for dirt. It is a good plan, when using these grades, to spread the fresh abrasive over the surface with a finger, feeling for any sharp particle, and then to lower the upper disc into place, first moving it slightly to and fro, under its own weight, to feel for any crunching. The greatest danger of scratching is always at the very beginning and end of the wet. At the end, the danger is that the abrasive has become ground down so fine that the glass surfaces come into physical contact and a seizure or sticking together may occur, particularly if the mixture is watery but rapidly drying.

Seized-up discs are a sore trial to the patient mirror-maker, for, quite apart from the difficulty of separating the discs, the shadow of serious damage in so doing hangs over him. With luck, the discs will come apart freely after being immersed for a few hours in a bucket of water. If they do not, the worker grits his teeth and hits the side of one disc with a mallet while pressing the opposite side of the other disc against a wooden support. This will certainly separate them, but with the risk of scratches or even of a flake of glass being taken out of one of the faces. The moral is clear: If the discs start moving jerkily over each other, separate them at once.

A well-smoothed mirror, finished with fine emery, will have a milky-looking surface and be semitransparent. The general appearance, however, says little of the quality of the work, and the surface should be examined minutely with a magnifying glass for scratches or clusters of deep pits. Dividing the surface up into small squares with strokes of a pencil helps in ensuring that no part has been missed. Fine surface blemishes are most easily seen—in fact, they may otherwise be invisible, at this stage—by using what microscopists call *dark-field* illumination. In other words, instead of the surface being viewed in light transmitted from behind, we arrange an oblique light source to shine on to the surface from one side, and place a piece of black paper at the back of the mirror. Pits and scratches now stand out brilliantly against the dark background. A few isolated pits do not matter, but clusters of them suggest that further fine grinding is required, and the worker will have to use his judgment whether he will save time by backtracking a grade. There should be absolutely no difference in general surface quality between the center and the margins of the mirror.

The mirror having been brought to this condition, the real business of producing an optical surface begins; but it can be concluded successfully only if the preliminary work has been carried out with care.

27

Polishing

It is commonly supposed that abrasives grind and rouge polishes. Yet, if we were to substitute rouge for emery, the mirror would continue to grind, very slowly and finely. The polishing compound itself, then, does not have the property of polishing in its own right. To polish glass, rouge must be applied against a yielding surface. Paper, leather, foam plastic, wax, even human skin will polish glass with rouge, but for fine optical purposes pitch is almost universally employed.

Pitch is a "soft solid." It is grainless and can be made to conform exactly to a molding surface, but it never sets rigidly unless it is frozen. A blow with a hammer will shatter it, but a feather placed on its surface will, with time, leave its own imprint. The dominating factor in the behavior of pitch is temperature, and a rise of just a few degrees will totally alter its mobility. The success of polishing and figuring a mirror is strongly dependent on the operator's ability to cooperate with this strange substance, and to establish how its slow flow changes the figure on the mirror it is polishing. Ways of minimizing the unpredictability of pitch, which can cause so much trouble in the figuring stage, will be described later.

The mirror-maker's first task, having completed the fine grinding, is to clean everything up so that all traces of emery have been banished. This done, the stage is set for the preparation of the polisher or "lap." Experience has shown that at least as many problems are encountered by the beginner at this stage than at any other; and a poor polisher will produce a poor mirror. The fault is, at least, partly that of his instructors, who in some cases seem to want to make things difficult for their readers. Window curtains, netting, saws, rubber molds, and waxed paper strips are just some of the "essential" items to be found. It would be absurd to suggest that these various methods do not, in the right hands, do the job required; but it seems at least equally wrong to complicate the issue. All that is needed to make a polisher is the mirror and tool, a pot of pitch, some rouge and water, and one or two *new* safety razor blades. There is, of course, a secret, which is to have the glass and the pitch at the right temperature.

But first the pitch must be brought to the right degree of hardness or "temper." If it is too soft, it flows quickly under the pressure of polishing and the channels have to be recut too frequently; it may also, in extreme cases, give a poor optical figure to the surface. If too hard, it may be slow to follow the mirror's changing figure and so can give a very bad figure indeed, as well as having a tendency to produce fine, hairlike marks known as "sleeks." If in doubt, then, have the pitch too soft. A useful way of grading the hardness of pitch is to take a sample, *at the working temperature of the room*, and press the edge of a nickel or British penny on it, resting the elbow on the table and just using the weight of the forearm. The weight on the pitch should be about a pound. Now time the number of seconds' pressure required to leave a dent a quarter of an inch long. "Thirty-second" pitch would be too soft for satisfactory working; 60-second pitch is about right, although it should always be remembered that the pitch will warm up during polishing operations, and that after half an hour's work the pitch may be twice as soft as it was when work commenced. One cannot, however, be dogmatic, for the polisher may be subject to unavoidable changes of ambient temperature, which will alter its hardness dramatically. For this reason the constant temperature of a cellar is much to be recommended, and it is the ideal place in which to polish and figure a mirror. If the work has to be done above ground, keep an eye on the thermometer and try to work only when the temperature is within 5°F of an accepted standard.

A sample of the pitch is first tested. Coal pitch is usually about as hard as its namesake, and must be softened. Resinous pitches may be too soft, although they should be about right if they were supplied with a kit. Pitch is hardened by heating it strongly to drive off the more volatile constituents, of which turpentine is the principal one. This is a smelly business, and should be done out-of-doors, preferably down-wind of any neighboring houses; neither should the pitch be boiled, for it may catch fire. Probably at least an hour's heating will be needed to make much difference to a pound of pitch. Cool the sample thoroughly in a bowl of water, which is at the probable working temperature, before testing for hardness.

To soften pitch, turpentine is added. Pure gum turpentine, not white spirit, must be used; the latter does not dissolve properly, and leaves a stringy mess. Add just a teaspoonful, and stir it in vigorously for several minutes before taking a sample.

Making the polisher

This done, prepare to make the polisher. A rouge and water mixture must be made up; it can be either in a flip-top plastic kitchen container, with a clean brush left permanently in it, or else it can be made in a squeeze bottle like the finer abrasives. Brush applications are less wasteful and generally more convenient, since the rouge can be applied thick or thin at will; many people, however, consider a squeeze bottle to be less liable to contamination. When applied, the polishing compound should be well colored, but not so thick as to be lumpy. As to the total amount: If it runs freely down the side of the lower disc, there is a wasteful excess present.

Place the glass tool in a bowl of warm water at about 90°F, and leave it to get thoroughly heated. This is to prevent the cold glass from cooling the warm pitch before it has taken the proper curve. The mirror should be much hotter, yet not too hot to be held comfortably in the hands. Warm it carefully before an electric fire, and, if it is plate glass, be particularly careful not to let any drops of cold water fall on it, or the sudden local contraction could crack the disc. Low expansion glass, not being subject to such severe contraction, can be handled much more freely.

While the mirror and tool are warming, the pitch, having been tempered and warmed to a liquid state, should be cooling. Stir the molten pitch with a clean stick, to keep the more solid material near the sides of the container mixed in with the rest, and periodically let the pitch run from the stick back into the melt with a slow circular motion. Observe how quickly the line of pitch sinks into the molten mass; when it preserves its form for a couple of seconds before vanishing, the pitch is of the right mobility to be poured on to the tool. Some people smear the face of the tool with turpentine, in the belief that this aids adhesion, and it certainly does no harm.

Place the tool on a sheet of old newspaper on a level surface, and pour the pitch on to it with a spiral motion, beginning at the center. If everything is right, the pitch will form a viscous layer about ⅛–³⁄₁₆ of an inch thick. Do not try to hurry the pouring; the tool has been warmed, and the layer will not set hard for some minutes. It is much more important to avoid uneven pouring than to do the job at breakneck speed. Try to make each course of the spiral just touch the preceding one, but do not worry if small gaps are left, since they will probably be filled in during the pressing, and even if not they do no harm. Pour right up to the edge of the tool, so that some pitch spills over the edge on to the paper. Then smear the pitch surface generously with rouge mixture. The water should rise in steam. Before it has dried out, lower the hot mirror very slowly on to the warm pitch, moving it back and forth to minimize the chances of sticking, and observe the contact through the back. If pitch and glass do stick, lift off the mirror at once and scrape off the adhering pitch with a razor blade; sticking means that the pitch is still too hot, and it should be left for half a minute before attempting another contact. If all is well, however, the pitch will flow out under the mirror, the contact region being indicated as an ever-spreading darkness. Keep pressing and moving the mirror until the contact is complete all over, or as near as can be achieved; it is common to find one or two small patches holding back. This does not matter. If a large air bubble forms, slide the mirror back to release the air and press again.

If widespread contact cannot be obtained, the pitch or the mirror, or both, were too cool. Put the half-formed polisher in a bowl of hand-hot water, and reheat the mirror; good contact will probably be obtained at the second attempt.

The mirror is now slid off and laid aside. The embryonic polisher will look anything but beautiful, with molten pitch overflowed down its sides; but if the surface has conformed to the desired curve the optician can congratulate himself, for the worst is over. Put the polisher in a bowl of cold water, to "freeze" the surface of the pitch and make it more amenable to trimming, and cut off the surplus pitch with a new safety razor blade.

As we have seen, pitch slowly flows, particularly when under pressure, and the next task is to cut grooves in the layer, leaving facets. If facets were not present, the pitch would have nowhere to flow to, except over the edge; and the localized buildup of material would destroy the contact with the mirror's surface, producing an irregular curve. Grooves also allow the polishing mixture to remain circulating longer, while there is some evidence that the polishing action is confined largely to the edges of the facets. Generally speaking, the greater the number of facets the better, and it is a good idea, in the case of a 6-inch mirror, to cut major squares about 1 inch across, subsequently dividing them up into smaller ones.

Facets are cut by wetting the razor blade, to prevent the pitch from sticking to it, and attacking the surface with a short shaving action. The major grooves can be cut right down to glass level, and to reach this depth may take three or four cuts; but there is no need to hurry. It is a good idea to cut each groove from both sides, leaving a steep slope of about 60°. The stories one hears of pitch shattering during the grooving process are due either to the use of a blunt blade, or to attempting to cut deeply at each stroke. It should, in any case, take no more than five minutes to cut a set of 1-inch facets. Now take the mirror, smear it with rouge, and press it back into contact. It will probably be found that the center of the polisher has slumped out of contact due to its extra warmth (the center of a disc is always last to cool), and the mirror may need slight warming to effect this re-pressing. The advantage of having a clear back, unencumbered by a handle, is nowhere more apparent than during the pressing and polishing processes.

We now have a warm mirror and polisher in contact, and they must be allowed to cool down to room temperature before polishing begins, a process that will take a couple of hours. If they were set apart to cool, and then brought together, it would be found that the centers were out of contact, due partly to central slumping of the polisher but also because the margins of the warm mirror have expanded less than the hotter center, temporarily flattening the curve. Put the pair of discs aside with the polisher uppermost, so that the warm center tends to sink into contact, and surround them with damp rags or newspaper to prevent the rouge mixture from drying out. An even better precaution is to run a band of waterproof electrical PVC tape around the sandwich. Before doing this, it is a good idea to have a slight chamfer on the edge of the polisher, making the pitch surface very slightly smaller in diameter than the mirror, so as to prevent a slight rim from building up as the discs press, with a possible flaking of the polisher's edge when they are slid apart. Put a 5 lb. weight on the polisher to aid the contact.

The importance of contact

Perfect contact is one of the secrets of polishing. If the contact is not good the polishing cannot be even, and high and low zones will form on the mirror. A gross effect will betray itself to simple examination of the surface, when some zones will have a better polish than others; but even if the polishing appears to be even, it is quite possible for errors of a few millionths of an inch to occur unnoticed until the mirror is tested optically. Removing zones is a slow business, and not always straightforward; and the mirror-making process will be accelerated tremendously by ensuring that zonal errors do not occur during the polishing. When we figure the aim is to transform an approximately spherical shape into a parabolic one; but the process is so delicate that any appreciable initial departure from a smooth surface will mean that the defect must be removed before figuring proper can commence. So the initial aim must be to produce a smooth, spherical surface, and *optical* contact between mirror and polisher is essential. General guides to good contact are smooth movement of mirror over polisher, with no grabbing, and even darkness of the facets when viewed through the back of the mirror disc. Even a slight thickening of the rouge mixture turns the facets gray and indistinct, indicating the need for more cold-pressing to bring them all dark and clear.

Once contact is obtained, the very act of moving the mirror over the polisher tends to deform the outer facets, pushing them out of contact again. The center of the mirror will therefore tend to polish very much more quickly than its edge, producing either a deeper sphere than the one ground in, or, at worst, an aspheric curve which may be a parabola but probably isn't! To overcome this tendency, the polishing should be performed with the same short strokes that were used in fine grinding, and the work should be stopped every half hour or so, so as to allow the mirror and polisher to be cold-pressed, with a 5 lb. weight, for a quarter of an hour. This will also allow the discs to cool somewhat from the effect of the optician's warm hands, as well as letting the optician himself cool down— for polishing should be a strenuous job; the more firmly the operator presses down on the disc, the faster the glass will polish, the limiting weight being defined by the tendency of the pitch to warm up too much and flow too fast, necessitating constant recutting of the channels. Long strokes must be avoided, the requirement being short but firm action. Too rapid strokes can produce a bumpy figure due to local heating, but heavy pressure will do no harm at all.

The polishing process

And so to work. The pressed polisher is examined for filled-in grooves, which are restored. Each facet is then lightly grooved with two or three channels in each direction, the pitch chips are dusted off, and the rouge-coated mirror is set on the polisher. When adding fresh rouge, always press the mirror down firmly for perhaps thirty seconds, so that any hard particles in the mixture will be pushed into the pitch where they can do no harm.

If the stroke is correct, an examination of the mirror after ten minutes' work will show a good semi-polish over the whole surface, the polish at the margins being only slightly inferior to that at the center. If the center or margin is untouched, the polisher is not in contact; but if the extreme edge only is gray a turned-down condition is present, and the fine grinding must be repeated. It is as well to establish this as early as possible, so that pointless polishing is avoided. Even so, it means that the polisher must be destroyed in order to recover the tool's surface, which is why such scrupulous care should be taken to ensure perfect fine grinding.

An hour or an hour and a half of polishing should bring the central 4 inches of the mirror to a perfect polish. Examination of a glass surface for completeness of polish can be done most easily if the back of the disc is polished smooth. Then, if the mirror is held up in front of a clear glass lamp, and the filament itself blocked out from view, residual pits will appear as a hazy grayness. If the back of the mirror is roughly molded, or ground, the surface can still be examined through a magnifying glass (a 1-inch focus eyepiece is ideal), when pits will appear as black specks—as will dust particles, which, however, move when wiped with a finger.

It is important to be rigorous in the examination for polish, since pits which are difficult to see when the glass is still uncoated will stand out depressingly well when the surface has been aluminized. Isolated pits are unimportant, but clusters of them betray regions of bad polishing, or, worse, bad fine grinding, and it can be assumed that the condition extends all around that particular zone of the mirror. If the pits are so large that they appear as holes rather than as dimensionless specks when viewed with a 1-inch magnifier, it is useless trying to polish them out; but, once again, a few deep pits do not matter except to hurt the worker's pride, for the practical effects of dust on the mirror will coat it with thousands of specks within a few minutes of it coming from the aluminizing chamber!

If the central region of the mirror is still very gray after two hours' work, and the magnifier shows clusters of coarse and fine pits mixed together, it is clear that fine grinding has been inadequate, and the work should be returned to the 400 stage.

The center polished, the mirror can now profitably be placed faceup on the stand, and the polisher worked on top. There are two advantages in finishing the polishing of the mirror faceup: firstly, the part most in need of polishing gets the most action; secondly, the tendency for the mirror to get a turned-down edge is reduced. The latter consideration is the more important of the two. Facedown working, particularly if long strokes are used, tends, as we have seen, to produce a deepening effect that does not extend to the very edge of the mirror, leaving a flat curve. On top of this, insufficient cold-pressing at the beginning of the session will leave the center of the polisher low, giving the effect of a ring of high marginal pitch, into which the edge of the mirror plows, so turning down the extreme margin. By observing the common sense precautions of maintaining good contact by cold-pressing, and using only short strokes, the smooth, robust marginal polishing action of a facedown polisher will overcome these subsidiary tendencies and leave a fine clean edge on the mirror.

The objection to faceup working, and the reason why it is not normally recommended, concerns the supposed greater tendency of the mirror to acquire sleeks and even scratches. But scratches, which are due to substantial hard particles finding their way between mirror and polisher, can only occur through carelessness. Sleeks are much harder to avoid, being caused by minute grains of dust in the air or even in the rouge mixture; they cause a smooth-edged shallow mark, commonly not more than half an inch long, which polishes out in a few minutes. Experience has shown that, given "sleeky" conditions, they are as likely to occur with the mirror facedown as faceup. If sleeks occurs, and the ordinary precautions have no beneficial effect, the answer is to use a softer polisher.

If all goes well, the mirror will be completely polished in 3–4 hours, and the stage is set for the most interesting task of all—parabolizing.

28

Testing and Figuring

In order to know what sort of curve the mirror possesses, a testing apparatus is required. The type designed by the French optician Léon Foucault (1859), although far from ideal for the purpose, is almost universally used by amateurs, being simple to make and adequate for testing a 6-inch f/8 paraboloid to the accuracy required for telescopic work.

Foucault's test

Consider a spherical mirror with an illuminated pinhole at its center of curvature. The rays of light from the pinhole everywhere strike the mirror normally to its surface and are reflected back along their original paths, forming a perfect image on the pinhole. If the pinhole is moved slightly to one side, the image moves in the opposite direction by the same amount and can be caught by the eye. Now, positioning the eye slightly behind the image, pass an occulting strip (usually referred to as a "knife-edge," although it can be a razor blade, or a piece of metal with a smooth, sharp edge) across the point where the rays intersect at the image (figure 75). If we suppose the pinhole to be infinitely small, the knife-edge will suddenly cut off the image, the mirror abruptly appearing to darken evenly all over its surface. In practice, of course, a pinhole must have a definite aperture if it is to be able to let any light through, but the accuracy of the test is not affected provided the diameter of the source is not greater than about ⅟₅₀₀th of an inch.

If, now, the knife-edge is passed across the beam a little way *inside* the position of the image, the observer sees a dark band moving across the mirror in the *same* direction as the knife-edge. If, however, it is passed

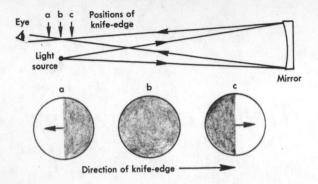

Figure 75. *Appearance of a spherical mirror in the Foucault test. (a), (b), and (c) represent the appearance of the mirror as the knife-edge passes across the beam in the positions indicated.*

across *outside* the image, the dark band appears to move across the mirror in the *opposite* direction. Foucault's test, therefore, enables the optician to know exactly where his knife-edge is in relation to the radius of curvature of his mirror.

But let us suppose that we are testing a mirror that is not spherical but *prolate*, which means that its center has a shorter radius of curvature than its edge. The parabola, of course, is an example of a prolate curve. There is now no unique radius of curvature, so that there is no point where the knife-edge can give an even darkening. If it is placed at the radius of curvature of the central region, it is inside the radius of curvature of the margin (figure 76); if it is set for the margin, it is outside the radius of curvature of the center. It may instead be set for halfway between these two extreme zones, so that it is outside for the center, and inside for the edge. These three positions are the basic testing positions for a parabolic mirror. It should be pointed out that, in practice, a prolate mirror has an infinite number of smoothly merging zones of different radius, so that the shadows are not sharp and black but diffuse gradually into each other, making the establishment of zones of different radius of curvature rather more difficult than may appear at first sight.

We now need a formula giving the required center-to-edge movement of the knife-edge for a parabolic mirror; and this, the *aberration* of the mirror, is given by r^2/R, where r is the radius of the disc and R is the radius of curvature of the central zone of the mirror (for all intents and purposes it can be taken as the radius of curvature of the original sphere before parabolizing commenced, since the difference in the calculated aber-

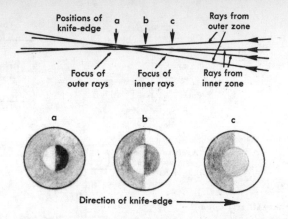

Figure 76. *Appearance of a prolate mirror in the Foucault test. (a), (b), and (c) represent the appearance of the mirror as the knife-edge passes across the beam in the positions indicated. To simplify the explanation, the mirror is represented as having two zones of different radii of curvature.*

ration is infinitesimal). In our case, therefore, we have $\%_6$ or $\%_2$ of an inch as the total displacement of the knife-edge if the mirror is parabolic. Since the position can, with care, be read to about .01 of an inch, this aberration should be measurable to within about 10 percent accuracy, which corresponds to a judgment of the accuracy of the mirror's surface to within better than .1 of a wavelength of yellow light, or 2 millionths of an inch! We could therefore say that the Foucault test is giving a magnification of 5,000 times.

Nothing complicated is required to make a Foucault tester. The knife-edge consists of a razor blade clipped to a little wooden stand (figure 77), with a reference edge cut into the base. A pinhole can be made by setting up a pre-focus bulb (the brighter the better) in a small light-tight can, with a ¼-inch cut in the side opposite the bulb. A piece of cooking foil, pierced with a small hole, is then secured across this aperture in the position of brightest illumination. The pinhole is made by placing the foil on a glass surface and pricking it lightly with a needle, using a twirling motion to produce a round hole. When examined with the 1-inch lens, the hole should be circular and of barely sensible diameter, so that it represents the telescopic image of a star. If it looks more like a planet than a star, the knife-edge shadows will be diluted and the test will lose some of its precision. On the other hand, a very minute pinhole will provide too little illumination for the shadows to be clearly seen, and may also produce undesirable diffraction effects. To overcome these uncertainties, there is no harm in making two or three holes of different diameters close together in the foil so that they are all illuminated simultaneously, the most suitable

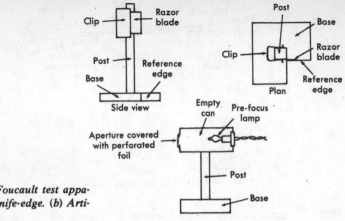

Figure 77. *Foucault test apparatus.* (a) *Knife-edge.* (b) *Artificial star.*

one being selected for actual testing. A large, bright hole in the vicinity of the test hole will also aid in initial location of the image when arranging the test setup.

These requirements may seem in stark contrast with some of the descriptions of Foucault testers to be found in the literature. Some of these devices, with micrometer motions on the knife-edge and even remote-control arrangements for tilting and aligning the mirror, are undoubtedly beautifully made, and presumably do the job they are supposed to do; but it is unlikely that they give *better* results than the simple rig described above, for the reason that the accuracy of the test depends on the judgment of the optician; there is little point in reading a scale to a thousandth of an inch if the knife-edge cannot be placed, with confidence, closer than a hundredth!

Of equal importance to the judgment of the optician is the steadiness of the air in the testing room; once again, if a cellar can be found it will prove ideal, since air movement will be at a minimum. A knife-edge test of a mirror shows up not only the nature of the optical surface, but also the thermal currents in the air, just as an astronomical telescope reveals turbulence on the atmosphere. What is required, therefore, is a uniform temperature with no swirling heat currents to ripple across the view. If the optician has been walking around setting everything up for testing, he will probably find a seething mass of currents roaring across his mirror; after a couple of minutes they will have subsided, and in ten minutes the view may be perfectly clear. Since the mirror, too, needs time to cool down after each figuring session, the best practice is to set it up on its stand and check its alignment before leaving it and the air to stabilize, so that testing can be carried out with the minimum of disturbance.

The work of testing is greatly facilitated if the mirror can be replaced on the stand at exactly the same position each time, so that the business of hunting for the pinhole image does not have to be repeated again and again. Repeatability of setting is helped by making an orientation mark somewhere on the edge or back of the disc, so that the same spots on the back are always located against the three points on the stand.

It does not matter how the pinhole and knife-edge are arranged on the table or test stand, provided that the optician can sit comfortably and view the mirror with no muscular strain. Relaxation and comfort are just as important as when making an astronomical observation. Normally, if the right eye is used, the pinhole is placed to the right of the knife-edge, which is moved in a left-to-right direction as it cuts across the beam; but the arrangement can be suited to the individual. The pinhole and knife-edge should be within an inch or so of the same distance from the mirror, and they should be as closely on line as possible to avoid the astigmatic effects that arise from a separation of more than three or four inches.

Judging the surface

Having brought the image to the knife-edge, find the position of most even darkening and examine the surface as the shadow forms. If the darkening is even, with no residual bright zones, the mirror must be generally spherical. If there is unevenness, however, the shadows must be interpreted to establish the high and low regions. Let us suppose that the mirror is prolate. The knife-edge must therefore be moved toward the mirror to make the center darken evenly, which means that the shadow at the margin will be moving across the mirror in the same direction as the knife-edge, but not having accomplished its passage, when the center is completely dark. Conversely, setting the knife-edge to give even darkening at the edge means that the shadow at the center has moved in from the opposite side from the knife-edge, but has not completed its passage, by the time all the margin has darkened. An *oblate* mirror, whose central region has a longer radius of curvature than the edge, will give the opposite effect. Representations of these and other appearances are given in figure 78.

The general form having been established, it is time to measure the aberration. A sheet of white paper is taped to the surface on which the knife-edge slides, and a pencil mark is made against the locating face.

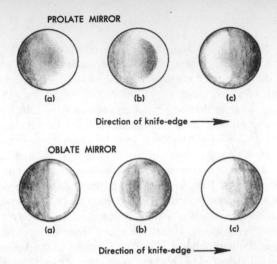

PROLATE MIRROR

(a) (b) (c)

Direction of knife-edge ⟶

OBLATE MIRROR

(a) (b) (c)

Direction of knife-edge ⟶

Figure 78. *Foucault shadows. Appearance of prolate and oblate mirrors at radius of curvature of (a) inner zone, (b) intermediate zone, (c) outer zone.*

When the central and marginal positions have been marked, the aberration can be directly measured. This record is a permanent one, and it is most useful to paste the marks, with a general description of the surface and the figuring action taken, into a work log.

If something was wrong with the stroke or the contact, distinct high or low zones may have developed across what is basically a smooth spherical or aspheric surface. Once again, interpretation is simplified by thinking of a high zone as convex (longer radius of curvature) and a depressed zone as concave (shorter radius of curvature). In reality, of course, there can be no convex curve anywhere on the mirror; the expressions are used solely in relation to the basic sphere, which looks flat under the Foucault test, so that high and low zones *appear* convex and concave respectively. Difficulties of interpretation and action begin to appear when zones appear on a steeply aspheric surface, so that the optician has no "flat" surface to refer to; and this is why the Foucault test is not suitable for making short-focus, highly aspheric mirrors.

The general aspect of the surface has now been assessed, but the edge must be the subject of a special examination. A broad turned-off margin will certainly show up as a dark line at the side where the knife-edge moves in when it is placed at mean radius of curvature (figure 79), but a sharp, narrow turn-down may be barely detectable except by an experienced eye.

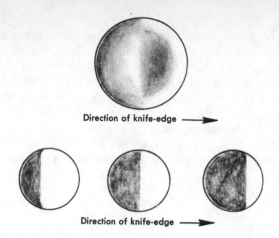

Direction of knife-edge ——→

Direction of knife-edge ——→

Figure 79. *Turned edge. (a) A severe case. (b) Test for slight turned edge.*

It can, however, be demonstrated by moving the knife-edge about half an inch inside the radius of curvature. If the mirror is generally spherical, the dark band crossing the disc will have a straight edge right up to the margin of the mirror, but a turned-down edge will result in the top and bottom of the band having a sharp curve. This curve straightens as the band reaches the mirror's meridian, and turns the other way as it passes towards the opposite margin.

The supreme test of a mirror's edge is *diffraction*. When a light ray encounters a sharp boundary it is scattered or diffracted; to observe this effect, lean a rule against the mirror on test and pass the knife-edge across in the normal way, when the outline of the ruler will be fringed with a narrow, bright band of diffracted light. Correspondingly, if the edge of the mirror cuts off sharply at the bevel, with no turn-off, a diffraction line will form all round the mirror as the surface darkens. If the edge is not quite perfect, the line will be seen only on the side opposite the knife-edge; however, a mirror showing even this quality is effectively perfect as regards the margin, and very few mirrors indeed show a diffraction edge around their entire circumference.

The only other initial imperfection likely to worry the optician is astigmatism. An astigmatic mirror performs as if a truly revolutionary disc had been bent slightly across a diameter, so that the radius of curvature varies in different directions. Normal care will ensure that astigmatism is neither ground nor polished in (when working the mirror faceup, it should be placed on a resiliant backing, such as a thin layer of foam plastic, to support it evenly); but some glass, particularly plate of doubtful origin, may have been cast with strains that were not relieved by annealing so that the tremendous tension in the disc twists the surface. Such discs are, fortunately, very rare. In the case of large mirrors, which are often relatively thin and subject to flexure, astigmatism can be a most serious problem. To test for the effect, view the pinhole image with an eyepiece as described on page 54.

Initial figuring

The optician's first task in figuring, and one which can be commenced, in a general way, as soon as polishing has made the surface of the mirror reflective, is to achieve a smooth surface with a good edge, from which subsequent parabolizing operations can commence; if, in the course of removing zones or trueing the edge, the figure acquires a general prolateness or oblateness, it does not matter. Primary attention must always be given to the edge and the marginal regions generally, for three-quarters of the area of a mirror lies outside a circle of half its diameter (known as the 50 percent zone), while half lies outside a circle of .707 of its diameter (the 71 percent zone). The outer $1\frac{1}{2}$-inch annulus of a 6-inch mirror is, therefore, of paramount importance, the contribution of the inner 3-inch circle being diminished still further by the permanent obstruction of a part of it by the telescope's diagonal mirror.

It is, initially, most important to get the edge right. The most direct way of achieving this is to polish the surface down to the level of the turned-off margin, and this can normally be done by working the mirror faceup, with generous strokes and light pressure to avoid too rapid slumping of the pitch. The effect of this is to lengthen the general radius of curvature somewhat, but this goes unnoticed because the knife-edge is automatically brought to the new cutoff position. A test after 15 minutes should show some change, and firm cold-pressing after this interval is recommended. As the turned-off zone narrows, it may in fact appear sharper and more serious until it is finally polished out. If the edge does not respond to this treatment, suspect the contact of the polisher and make sure that the marginal grooves have not filled up, presenting a raised circle of pitch. The short-stroke, facedown method often advocated is a far less efficient way of removing a turned edge, since the polishing action is slower and not so directly aimed at the offending region. The desire on the part of many mirror-makers to work their mirrors facedown at all stages, regardless of the shape of the surface, is probably a hangover from the early days of widespread amateur mirror-making when soft plate glass and scratchy rouge were universal. The excellent rare-earth rouges available today, if used carefully, are quite trustworthy, and the advent of hard low-expansion glasses of the Pyrex type has produced a strongly blemish-resistant surface.

Admirable though the desire for a superb, sleek-free surface may appear, it must never be forgotten that the optical accuracy of the mirror is the main consideration, and to obtain a good parabolic shape it may well be worth risking a few marks. After all, a mirror becomes coated with minute dust particles within seconds of being unwrapped, and nothing can be done about it; so there is no point in fussing about a few sleeks appearing during the figuring process. Infinitely more important is a good general polish, particularly at the edge.

Zones present the mirror-maker with another problem; it is not easy to polish them out if they are at all severe, since they indent their form on the polisher during cold-pressing, so perpetuating their existence unless very long strokes are used, which may themselves bring other evils. The most direct way of treating zones is by local polishing to smooth them out somewhat, and then to use the full-size polisher. A raised zone (figure 80) is the easier to handle. Having located the crest of the zone, run a circle round the back of the mirror with a waterproof pen, dot the zone with a few drops of rouge, and set to work with the thumb, principally on the crest but also allowing brief excursions into the surrounding area. Keep walking slowly but also regularly around the stand, and do not apply more force to the polishing than is already in the weight of the arm. Not more than five minutes of this should be attempted at first, for the action is rapid and very local and can easily produce its own deep zone until experience has been gained. The aim in all small-polisher work, whether the thumb or a small soft pitch polisher is used, must be to blend in the surface to prevent fresh zones from being formed.

Figure 80. *A raised zone.* (a) *Appearance on test.* (b) *Correct path of small polisher.*

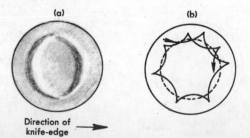

(a)   (b)

Direction of
knife-edge ⟶

A depressed zone (figure 81), clearly, cannot be raised up; so the surrounding surface needs to be blended down to its level. First of all, the high regions on either side of the zone are worked down in the manner shown. This will, effectively, raise the center, making the mirror more oblate, since we have shortened the radius of curvature of the margin and lengthened that of the center. This oblateness will, however, be amenable to full-size polisher treatment on the faceup mirror, which will tend to lengthen the margins and make the mirror more prolate.

Without a considerable amount of skill, which can be acquired only by practice, this local treatment will probably have the initial effect of making the mirror look worse than it was before, since several narrow zones will replace the single major one. However, we are not concerned with the number of the zones as much as with their depth. A dozen very fine zones will look ugly, but will quickly polish out, since the surface is closely spherical and the polisher can easily be brought to general conformity. It is in matters such as these that experience tells. The aim of local polishing is to produce a generally spherical surface; it is the task of the full-size polisher to render this surface optically smooth, with no short-period errors. Once the surface is smooth, be it slightly oblate or prolate, with a crisp edge, the way to a parabola should be a direct one. It is absolutely useless trying to parabolize a surface unless it is zone-free.

Parabolizing

So far, the knife-edge test has been used mainly qualitatively, to judge the general nature of the mirror's surface, but once parabolizing starts it is time to make actual measurements. Let us suppose, then, that we have a smooth but aspheric surface. We must now measure the radii of curvature of the inner and outer zones, to see what the aberration is, so that the rate of change of the aberration as figuring proceeds can be judged.

Figure 81. *A depressed zone.* (a) *Appearance on test.* (b) *Correct paths of small polisher to reduce outer and inner crests.*

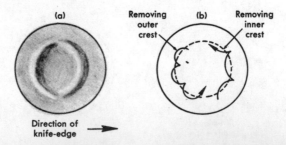

It is easy enough to find the radius of curvature of the center. Set the knife-edge some way toward the mirror and observe the way the dark shadow crosses. As the knife-edge is brought back toward the eye, we observe that the edge of the shadow is no longer straight, but that some parts jump forward before the rest. This means that the knife-edge is close to the radius of curvature of this particular zone. If the mirror is prolate, the center of the shadow will bulge, until, when the radius of curvature of the center is reached, the inner 30 percent or so of the mirror darkens evenly while the margin opposite the margin is still brightly illuminated. Make a mark on the paper against the reference edge at this point. Then move the knife-edge back still more, and observe how the central shadow now forms more strongly on the side away from the knife-edge and moves back, as it were, against the tide, producing a darker crescent, when the margin darkens evenly.

The radius of curvature of the margin is not so easy to judge, since the contrast of the marginal shadow on the side of the knife-edge with the lighter central crescent makes it appear deceptively dark; there is, therefore, a tendency to make the radius of curvature too long. A good way of minimizing this effect is to watch the top quadrant of the mirror and to note the point at which the upper margin, from about half-past ten to half-past one o'clock, darkens evenly. Make another mark, and measure the distance between them. The mean of at least two tests should always be taken.

If the mirror is prolate, then the edge-darkening position will be reached first as the knife-edge is brought back, but the technique of making the measurements is the same.

One reason the paper-and-pencil method of recording the aberration is better is that the readings are taken down "blind," and there is no possibility of adjusting the setting on the knife-edge to give the wanted result! If the testing room is not completely dark, and it is possible to see the lines as they are drawn on the paper, a further aid to subconscious honesty is to take the readings so that the knife-edge stand always covers the previous mark; that is, working backward toward the observer, as described.

For practical purposes, an "approximately spherical" mirror is one with an aberration of not more than $\frac{1}{8}$ of an inch at this stage; but let us admit at once that an oblate surface is going to be much further from a parabola than the equivalent prolate one, and will need more work to be done on it. If the mirror is prolate, an aberration of as much as $\frac{3}{16}$ of an inch is within the capabilities of a small polisher, although so large a difference suggests violent strokes or, possibly, poor contact, and the surface could probably be improved by using short strokes with the mirror facedown. Such a mirror is hyperbolic or "over-corrected," and the treatment of such a mirror will be discussed later. For the moment let us suppose that the mirror is under-corrected, whether slightly prolate, spherical, or oblate. The problem now is to shorten the radius of curvature of the central zone relative to that of the margin.

Several ways have been used to achieve this action, for it is a problem that has faced opticians for three centuries. In the case of small mirrors, the required asphericity can be produced by a long stroke alone, since the amount of glass to be removed is only slight (this represents the current case). More severe action can be obtained by trimming the outer facets of the polisher, so reducing their polishing action; the great objection to this method is that the polisher is spoiled for other requirements, and much experience is needed if the process is to be anything more than hit-and-miss. Finally, the unwanted glass can be polished locally with sub-diameter polishers. This is the most economical way, since glass is directly removed under the operator's intimate control; and steeply aspheric surfaces are usually worked in this way.

It is advisable, where possible, to use the full-size polisher, since the tendency to remove material unevenly, producing zones, is at a minimum. A spherical 6-inch mirror can be given a superficially parabolic form in just a few minutes' polishing with a long (2–2½ inch) stroke and side or W-stroke, but there is a severe danger of the edge being left behind in the process, giving a serious turn-down, unless figuring spells are limited to a minute or two with considerably longer pauses for cold-pressing. A generally better, though slower, technique is that of using a long stroke with the mirror faceup. In this case the center is being left largely alone, and the outer regions are being flattened. As we have seen, such a method should keep the edge crisp, and a 15-minute session of long W-stroke should produce a distinctly prolate tendency. If no change of figure is noticed, the stroke and overhang can be increased. A somewhat soft polisher is recommended, both to encourage the aspherizing tendency and to reduce the likelihood of sleeks.

Once an aspherizing stroke has been found, keep to it and see if the aberration builds up toward the required value. Frequent pauses and thorough cold-pressing are required if control over the figure is to be maintained. Before testing, allow the disc to cool for at least 30 minutes.

The drawback of the large polisher technique lies in the fact that the operator has only limited control over what is happening to his mirror. It is largely a matter of trial and error, being considerably affected by the temperature of the pitch—which itself is very sensitive to temperature—so that a good deal of experience, and much attention to the conditions, are essential for predictable results. Nevertheless, this is the method that should be tried first, for it is unlikely to produce a very bad shape and the chances are good that something approaching a parabola will be achieved. Since, however, small polishers may be needed to touch up the general curve, this method certainly deserves mention.

The right size for a small polisher depends on the task it has to perform. As we have seen, irregular zones on a 6-inch mirror can be treated with the thumb. For general aspherizing, something of the order of 2 inches in diameter is required. A disc can be turned out of wood and varnished with cellulose lacquer, before pouring pitch on to one side and pressing it on a splash of rouge mixture on the mirror. Channel it into very small facets. The pitch for a sub-diameter polisher can be very soft, since we are not concerned with maintaining contact over a large surface, and a small tin of pitch, softened with a generous amount of turpentine, can be prepared for the purpose.

Small-polisher work is very efficient, and the speed with which it removes glass from small areas of the mirror makes it all the more important for the operator to test frequently. A stroke of the kind shown in figure 82 will rapidly deepen the center, but care must be taken to ensure that the work is not too localized, producing a central "hole" rather than a smooth asphericity. Therefore, the polisher does not always pass over the very center of the mirror, but over the intermediate zones as well, besides touching the very edge of the mirror at the extremity of each sweep to keep the curve blended in. It will at once be obvious that here is a very sensitive way indeed of controlling the surface; extra rubbing can be applied to any zone at will, in a way that is not possible with a full-size polisher.

A test after 5 minutes of small-polisher working will show a detectable difference in the aberration, and spells should always be kept short because of the rapidity with which the figure changes. Most important of all is to keep the stroke regular across all diameters of the mirror, which it will be automatically, provided the operator keeps walking steadily around his stand.

Figure 82. *Aspherizing stroke using small polisher.*

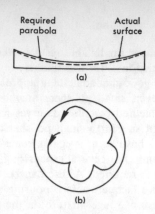

Figure 83. *Reducing a hyperbola. (a) Correction required. (b) Correct path of small polisher.*

The small-polisher method is particularly useful if the mirror, after polishing, is found to be very prolate. The "dreaded hyperbola" figures prominently in much of the older literature. It can be seen from figure 83 that the way to reduce the aberration of a hyperbola is to shorten the radius of curvature of the outer region by depressing the intermediate zone; this also has the effect of lengthening the center, and, with care, the two regions can be brought to the correct relative state. This work is best done with a polisher about 1½ inches across, since a relatively narrow zone of the mirror is being worked. If the edge acquires a turn-down, it means that the polisher is not reaching the edge of the mirror at the outward pass of the stroke, so that the original curve is not being replaced.

Small-polisher work is quick, but the writer would certainly not claim it to be easy. It is, however, a technique well worth mastering if the amateur intends progressing to short-focus mirrors, such as are to be found in Cassegrain telescopes, which require the removal of a relatively large amount of glass in the figuring process. Object-glasses, too, are always figured by working on the high zones with small polishers.

The 71 percent zone

Let us now suppose that the mirror, when tested, gives an aberration immeasurably different from that desired. This tells us much, but it does not reveal everything. Although the center and edge are relatively correct, how about the intermediate zone? It could be either above or below the correct parabolic shape, and in either case it will not be focusing the light where it should go.

We must, therefore, have some means of measuring the radius of curvature of the intermediate zone, and the one that concerns us is the 71 percent zone, for in terms of area this marks the division between the inner and outer parts of the mirror. In terms of aberration, too, the radius of curvature of this zone comes exactly midway between the center and edge positions. Therefore, the knife-edge should have to move back $\frac{3}{64}$ of an inch from center darkening to 71 percent zone darkening, and another $\frac{3}{64}$ of an inch to the edge-darkening position. When testing this zone, it can be indicated by making two marks on the horizontal diameter of the mirror with a felt-tip pen, or a light piece of wood, with two pins sticking up in the right position, can be placed across the mirror so as to rest on two projections from the stand. The distance between the marks is $4\frac{1}{4}$ inches.

The midway position of the knife-edge gives us the classical prolate shadows, as already shown in figure 59. If, at this position, the inner edge of the outer shadow and the outer edge of the inner shadow meet at the 71 percent radius (figure 84a), we know that the overall contour of the mirror is correct (small departures may occur in the regions between the measured zones, but they are unlikely to be significant in the case of a 6-inch f/8 mirror). If the knife-edge, when placed at the midway position, shows the shadows touching too far out (figure 84b), it means that the center of the mirror is flattened, like a bowl; if they touch too far in (figure 84c), there is a central hole. The latter condition is a frequent result of ill-controlled small-polisher work. In either case, the remedy is to take a small polisher and work on the high regions, as shown. If a full-size polisher is being used, the stroke should be lengthened for the second case, and shortened for the third.

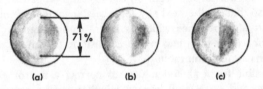

(a)   (b)   (c)

Figure 84. *Appearance of mirror at intermediate knife-edge position. (a) Correct 71 percent crest. (b) Flat center (crest too far out). (c) Deep center (crest too far in).*

It may now be clearer why there are some reservations about the testing method described. The shadows on a 6-inch f/8 mirror being faint and diffuse, the optician requires accurate judgment to decide just where "the" shadow starts and ends, which is why the Foucault test, though easy to set up, is not always simple to interpret.

Making a star test

The ultimate test of a telescope mirror is its astronomical defining power, and it is very advisable to try the mirror out on a star before going to the expense of having it aluminized. If the tube is not yet finished (no mounting being necessary), the test can be made by setting the mirror and aluminized flat at their correct respective positions in a rough wooden jig, and aiming the arrangement at Sirius or some other bright star, which will, using an uncoated 6-inch mirror, appear about twice as bright as it does to the unaided eye. Using an orthoscopic or other well-corrected eyepiece of about ½-inch focal length, locate the focus and examine the expanded image of the star about ⅛ of an inch inside and outside focus. The ideal appearance of these images is a small disc, more brightly illuminated toward the margin, with a crisp, not flared, edge and a small black spot (the outline of the flat) in the center. The more equal they appear, the better the mirror. If the black spot appears larger inside the focus, the center of the mirror is too high and the surface is under-corrected; if the reverse, it is over-corrected. If the margin of the inside disc is fuzzy, while that of the outside disc is sharp, a turned-down edge is present. Combined with a large central spot outside focus, this appearance indicates a generally hyperbolic figure.

Do not be disheartened if your evaluation of the mirror on the Foucault test fails to equate with the sky's verdict; the latter is the one to be trusted and, knowing now the true state of the mirror, it is possible to understand a little better the vagaries of the Foucault test. Very likely, the error will be obvious at first sight on the re-test, and can be corrected with confidence, so that on the next sky test the star image is immensely better. It is this sort of experience which is learnt rather than taught, and then often in an uncertain way, making its acquisition that much more precious, and its lessons more memorable.

It is possible for a skilled maker to correct a mirror entirely on the basis of intra- and extra-focal images, whether using a real star or an artificial one in the testing room; but the amateur will be better advised to seek first the advice of the Foucault test—which is excellent for revealing zonal errors—and then to appeal to a real star to discover the overall degree of correction of his mirror. An intelligent combination of the two tests will allow the maker to produce an excellent optical surface.

General comments

A few words should be addressed to the choice of a diagonal mirror, which should be in the form of an ellipse with a minor axis of about 1½ inches. Having gone to the trouble of producing a fine mirror, it would be foolish to buy anything but the best. Some unscrupulous makers merely cut diagonals out of unworked plate glass, coat them, and sell them; and the amateur is unlikely to have the equipment necessary to test them in advance. The best course of action is to seek the advice of an experienced telescope-maker at the local club who may—who knows?—happen to have just the thing tucked away on a shelf, available at a bargain price. Typical prices of new aluminized 1½-inch diagonals are $15 (£8), and the price is not always commensurate with quality.

Much the same advice applies to the choice of an optical firm to whom to send the mirror for aluminizing. Coating charges vary considerably, but about $15 (£8) for a 6-inch mirror is representative. A transparent coating of silicon dioxide is usually evaporated over the aluminum to protect it from tarnish, and the small extra cost is well worthwhile. A well-coated mirror should retain most of its reflective efficiency for two or three years, provided it is covered with a dust-proof cap when not in use and is washed with detergent every few months; a badly-coated mirror suffers from poor adhesion of the coat due to inadequate cleaning of the glass surface, and the aluminum breaks up from the edge inward in a few weeks or months. Even reputable firms sometimes produce a bad coat, and to protect their name they may well offer to replace a defective coat free of charge.

Do not hesitate to read widely, and to consult others, on the subject of mirror-making. A plethora of conflicting advice will be obtained, but this is all to the good, showing as it does the wide choice of techniques available to the amateur. The writer has frequently come across sincere and radical differences of opinion between opticians, both amateur and professional, which sometimes appear almost mutually exclusive. The explanation is that the worker exerts his own unconscious technique on the molding of his mirror's surface; and these differences may produce a completely different result from what, apparently, is the same procedure. The final dictum can only be to "try it and see," based on the general advice that these chapters have aimed at giving.

To sum up:

1. Use, if possible, Pyrex or similar low-expansion glass, because of its rapid recovery from thermal effects and resistance to sleeks.

2. Maintain a good bevel at all times.

3. Make sure that pits from the preceding grade are all removed before going on to the next grade of abrasive.

4. Be scrupulous over washing-up between each grade.

5. Ensure a crisp edge by fine-grinding and polishing the mirror faceup for part of the time.

6. Maintain good contact and well-channeled facets during polishing.

7. Obtain a smooth surface before attempting to parabolize.

8. If possible, star-test the mirror before it is aluminized to confirm that the images are good.

29

Advanced Optical Work

Few people are content to stop at one mirror. They may give up in despair, and never want to look at a disc of glass again; but, having completed a telescope entirely with their own hands, the thought of making a more powerful one soon becomes an obsession.

In some ways this is a pity, for an astronomer turned optician is probably an astronomer lost from the eyepiece; and the science needs more observers rather than more telescopes. It is, in truth, thoroughly depressing to consider how many fine and elaborate instruments have been made and then abandoned in favor of other telescope-making projects, while many keen but ham-fisted observers have to make do with inadequate equipment. However, the individual is free to choose his own hobby, and optical work is undoubtedly a fascinating and rewarding pursuit.

Making an object glass

Before proceeding along the classic sequence of ever-larger Newtonian mirrors, the amateur would do well to consider the alternative optical systems that his increasing skill opens up to him. In particular, refracting telescopes offer a number of advantages over reflectors but are almost unavailable commercially except in small sizes; and a 6-inch object glass alone will cost at least $500 (£200). The myth that an object glass is too difficult for the average amateur to attempt is totally unfounded; indeed, from the optical viewpoint, it is easier to make an object glass than a paraboloid mirror of the same aperture, because the accuracy of the lens surfaces need be only one quarter that of the mirror. Again, an object glass of f/15 or f/20 gives the finest definition that can be obtained from a telescope of a given aperture, and the planetary images given by a 6-inch refractor may prove a revelation to the user of a typical 6-inch reflector. The slight bluish halo, or secondary spectrum, of a refractor is of little practical importance in apertures up to 6 or 8 inches. The disadvantage of a long tube—perhaps 8 feet in the case of a 6-inch refractor—can be over-come by "folding" the light beam with two flat mirrors (figure 85), the resultant tube being shorter than that of a 6-inch f/8 reflector!

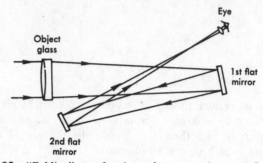

Figure 85. *"Folding" a refracting telescope, using two flat mirrors.*

Contrary to popular opinion, little in the way of special equipment is necessary for the manufacture of an object glass. A lathe is certainly not essential these days, either for making a cell or for edging the lenses, since the glass supplier can provide discs in whatever diameter is required; while the cell can consist of nothing more elaborate than a short piece of brass or aluminum alloy tubing of suitable wall thickness, with three outer flanges carrying the squaring-on screws. The lenses themselves can be retained inside their cell with cutout rings screwed to the cell wall. Of course, a more elegant cell can be produced if a lathe is available, but the point to be made is that it is not *essential*.

Object glass designs

There are a great many designs of object glass, all being variations on the convex crown/concave flint arrangement described on page 9. Un-doubtedly the most suitable pattern for the amateur to attempt is the *Littrow* design (figure 86), in which the crown lens is equi-convex and the flint lens is plano-concave, all three curved surfaces having the same

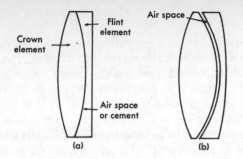

Figure 86. *Different types of object glass.* (*a*) *Littrow design with three equal curves.* (*b*) *Fraunhofer* (*coma-free*) *type.*

radius of curvature. Both sides of the crown lens can, therefore, be ground convex by working them underneath the flint, which itself becomes concave. The plane side of the flint is worked on a disc of plate glass. It is clear that a good deal more labor is involved in the grinding and polishing of an object glass than in a mirror, but the relaxation of figuring tolerances means that the amateur should have relatively little difficulty in producing a telescope giving superb definition.

Let us look more closely at the choice of optical glass. A sample of glass is defined by two parameters: its *refractive index* (n) and its *dispersive power* (V). The refractive index is a measure of the amount by which light is deviated in its passage through the glass; if two identically curved lenses are made of glass of different refractive index, the one with the higher value will refract the light more strongly, and have a shorter focal length. Typical value of n for crown and flint glass are about 1.5 and 1.6 respectively (where water is 1.33 and air is 1.00). Dispersion is indicative of the amount by which white light is separated into prismatic colors by a wedge of the glass; here, the lower value indicates the stronger dispersion, typical values for crown and flint glass being 60 and 36. These constants are accurately determined by the manufacturer for each sample of glass he sends out, and a crown glass which is widely used in telescope objectives has n nominally 1.519 and V nominally 60.4. This would be written in the catalogue as "519604," the "1" and the decimal point being omitted.

The tremendous range of optical glass means that an object glass design can be satisfactory only for a given selection of crown and flint glass. The essential feature of an achromatic combination is that the focal lengths of the two components are inversely proportional to their dispersive powers. Given this, manipulating the actual curves of the four surfaces will vary the other characteristics of the combination, such as spherical aberration, coma, distortion, and so on, and the various designs of object glass are characterized by their own particular merits. For example, the *Fraunhöfer* design, which is difficult to make because the curves on all four surfaces are of different radii, is free from the off-axis effect of coma, and has very slight residual spherical aberration even down to f/10. The simple Littrow design given here will provide, at f/15 or longer, an excellent achromatic system with only a small amount of spherical aberration, which can, in any case, be figured out in a short time. This assumes, of course, that the surfaces are closely spherical at the end of polishing. It will be found, however, that the relatively steep curves of an object glass hold their sphericity better than the shallow curve of the typical Newtonian mirror.

The recommended glasses for this design are the hard crown (519604) and dense flint (620361) made by Chance-Pilkington (England), but the American firm of Bausch & Lomb would be able to offer near equivalents. In this case, the radii of curvature of the first three surfaces would be 0.419 of the intended focal length, while the fourth surface (the rear of the flint) is, of course, plane. For example, if an object glass of 100 inches focal length was being made, the required radii would be 41.9 inches. This design originated with H. E. Dall, a well-known amateur telescope maker of Luton, England, and it is probably the best form of simple object glass that has ever been devised. A source of optical glass for the amateur is the Coulter Optical Co., 8217 Lankershim Blvd., North Hollywood, Calif. 91605. For further details of these and other designs see works listed in the Bibliography.

When ordering glass for an objective, it is best to state what the glass is for and to ask for the appropriate quality. For lenses of up to 6 inches in aperture, normal fine annealing is adequate; but different manufacturers have their own way of expressing what they mean by this term. Only when a great thickness of glass is used is superfine quality required; such glass is tested most stringently for flaws, and because of this special selection it costs perhaps three times as much. The crown and flint elements for an f/15 object glass should have a diameter/thickness ratio of about 12:1 and 15:1 respectively, but the manufacturer will normally supply the discs in the closest standard thickness above that requested. It is possible to buy some common crown/flint combinations in the form of molded blanks at a rather lower price than that asked for special edging to a required diameter, but the worker will then need some means of his own of producing accurately matched circular discs from the rough castings.

The spherometer

A very useful gadget for all optical work, but particularly so when making object glasses and other components requiring fairly accurate curves, is a *spherometer*. This consists of a robust circular metal disc with three ball-ended projections set equally around the circumference, and some form of depth gauge, whether a micrometer or a dial gauge, at the center. By setting the gauge to zero on a truly plane surface, and knowing the radius of the circle marked out by the three feet, the radius of curvature of any surface, convex or concave, can be found by setting the depth gauge to contact. Then, if R is the radius of curvature of the unknown surface, S the change of setting from zero, and r the radius of the spherometer, we have

$$R = \frac{r^2}{2S}$$

By calculating S in advance, the surfaces of the lens can be ground to truth without having to make regular splash tests, so that the work proceeds much more readily. The best way of obtaining a value for r is to set the spherometer on a curve of known radius (such as a telescope mirror), and to evaluate r^2 directly from the formula.

The amateur may well inquire how he is to obtain a flat surface for setting the spherometer to zero, and for checking the plane surface of the flint component. The answer is to take his flint disc, and a plate glass disc of the same diameter, and to grind them together, with regular reversals, until they are in contact all over. Then take spherometer readings on both. If the reading, whatever it is, is the same, they must both be flat; otherwise, one must be as convex as the other is concave, and the spherometer will show different depths. By working the convex disc on top, they can both be brought to flatness. If the spherometer reads to 1/10,000 of an inch, and the feet approximate to the edge of the disc, a curvature of only 5 wavelengths of light can be measured (representing an error of only 2½ wavelengths on each disc, if they match exactly). This is quite accurate enough for the purpose, an error of 5 wavelengths in the flatness of the rear surface being unimportant, provided the surface is free from zones. The plate glass disc should be polished when it has served its immediate purpose, for it will be useful as a reference flat and for testing the finished object glass.

Eliminating "wedge"

Another requirement, when making a lens of any sort, is to ensure that the two surfaces are accurately square-on to each other, which is another way of saying that the edge thickness must be the same all the way round. Error in edge thickness is known as "wedge," and serious departure from truth will produce a comatic image and even colored fringes due to the prismatic effect. A simple wedge-testing rig is shown in figure 87. The lens

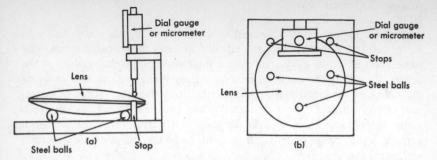

Figure 87. *Wedge-tester.* (*a*) *Side view.* (*b*) *Plan.*

rests on three points, which can be ball-bearings cemented to a metal plate, and is pushed laterally against two stops. A depth gauge is set ᴛᴏ contact near the edge of the lens, which is rotated so that readings can be taken around the circumference. The error in edge thickness in each component of the finished object glass should not exceed .0005 of an inch, the excess glass being removed by extra pressure on the thick side in the normal course of grinding. Gross errors, which will occur only if the discs were not truly parallel sided to start with, should be removed at the 80 stage, while the last .005 of an inch of wedge can be taken out with the 220 or 320 powders. Particular care must be taken to ensure that any irregular flattening effect cause by this eccentric grinding is removed before the next grade of abrasive.

Procedure

The choice of abrasive grades and general grinding and polishing technique differs in no way from the method of making a mirror. The three curves are first roughed out with 80 carborundum by working the flint alternately for two or three wets on each side of the crown disc. Once the curve approaches the right steepness, wedge-removing operations can begin on the crown by marking the thickest part of the edge and pausing at this point of the revolution around the stand to exert extra wear. Work can begin on the crown by marking the thickest part of the edge and pausing at this point of the revolution around the stand to exert extra wear. Work can stop a few percent short of the true curve, which is then arrived at with the 120 grade, once again working alternately on both sides. The grades from 220 to 400 should be worked both with the flint on top and underneath, and particular care must be taken to ensure that the edges of all three surfaces get adequate fine grinding.

The flint and one surface of the crown are now taken through to the finest emery; once it is established that the crown surface is completely free from any defects, the other side is ground down through the fine grades, checking with the spherometer against the completed convex surface to ensure that the two surfaces of the crown are as equally curved as possible. This method is preferable to alternating right down through the grades, for it will usually be found, no matter how much care is taken, that one side of the crown fits the flint better than the other, so that a part of the other surface fails to receive adequate fine grinding.

Any noticeable wedge in the flint component, assuming that the plane side has already been fine-ground for testing the spherometer, is removed at the same time as that in the crown by pressing extra hard on the thick edge as the disc is revolved in the fingers.

Polishers can be poured on the glass, but well-lacquered plywood discs, with their faces turned to the approximate convex and concave curves, are a preferable substitute. It is not advisable to pour curved pitch polishers on to a flat backing, since the unequal thickness of the pitch will cause uneven slumping and loss of contact. It will be found that the crown glass polishes faster than the flint, since the latter is softer and is ground more coarsely by the abrasives.

Testing and figuring

The maker of an object glass enjoys not only the advantage of relaxed figuring tolerances; he has, at his command, a very sensitive test of figure. Let us suppose that he polishes the flat glass tool used for grinding the rear surface of the flint lens, and sets up the test shown in figure 88. If the illuminated pinhole is at the focus of the object glass, the light emerges as parallel rays from the front of the lens and is reflected back by the polished flat surface, returning to form an image at the focal point. This so-called *autocollimation* test, in which the object glass produces its own parallel or "collimated" rays, is doubly sensitive because the light passes through the lens twice, so amplifying any errors present. A perfect lens will show the surface darkening evenly all over, like a spherical mirror in

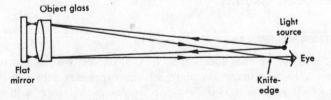

Figure 88. *Autocollimation test for object glass.*

the Foucault test, as the knife-edge passes across. If the surface shows residual aberration, judgment of the zones can be made in the ordinary way, with the important proviso that the errors appear opposite to those visible on a mirror (a bulge appearing as a depression, and vice versa). This excellent test, and the relative insensitivity of refracting surfaces to errors of figure, justify the claim that it is easier to make a first-class object glass than a mirror of the same imaging quality.

It is not important to know on which surface of the lens the remaining errors occur. Figuring can be carried out on any of the four faces, but the front of the crown or the rear of the flint is the most convenient, for the two components can be taped together round their edges, with three foil spacers placed around the circumference, during the whole of the figuring process. These foil spacers, which can be tiny squares of baking foil cemented in place to one of the components to prevent them from coming loose, are retained when the object glass is permanently mounted in its cell. Cementing should never be used if the lens is a 3-inch or larger, for it may, when hard, strain the components out of true and ruin the telescope's defining power.

Much has been made of the need for the plane reflector to be accurately flat. An error of a few wavelengths of light in the flatness (to the limit, say, of measurement with a good spherometer) will have a negligible effect on the result. Freedom from zones is much more important, since the zones may incorrectly be imputed to the lens. By testing with the flat and object glass slightly displaced, however, it can be seen to which circumference the zone conforms. Neither is it necessary for the surface to be coated. Should the back of the disc be polished, it will be necessary to smear it with thick oil or grease in order to dull confusing reflections.

An object glass figured to an even cutoff on the autocollimation test will perform brilliantly when turned to the sky. At f/15, stars will show their spurious discs and diffraction rings when the air is calm, and the lack of diffraction from any diagonal mirror and its supports will enhance the contrast in planetary markings. The initial cost of the optical glass for a 6-inch lens is high, in the region of $50 (£20), but the results will undoubtedly justify the outlay.

The Maksutov telescope

The amateur who has successfully tackled one or two paraboloids, and an object glass, can undertake more advanced projects with considerable confidence. If portability is desirable, a *Maksutov* telescope will need no extra recommendation. The Maksutov is a *catadioptric* (lens and mirror)

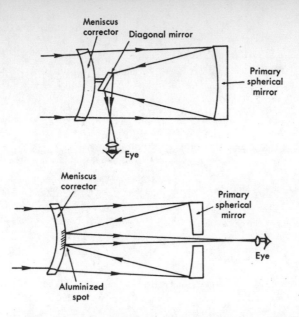

Figure 89. *Maksutov telescope.* (*a*) *Newtonian form.* (*b*) *Cassegrain form.*

system, which overcomes the difficulty of figuring a parabolic mirror of short focal ratio by passing the light first through a steeply-curved negative meniscus lens (figure 89). This lens is only very weakly divergent, but it has sufficient spherical aberration to compensate that of the spherical primary mirror with which it is matched. Although the curves of the meniscus have to be ground very accurately, with a permissible error of only about 0.1 percent on their radii, it is arguably easier to do this than to parabolize a mirror of f/3 or less. The Maksutov can be used as a short-focus, wide-field Newtonian telescope, or as a Cassegrain with, in a special case often employed, the convex secondary mirror formed by aluminizing a central spot on the rear (convex) surface of the meniscus. It is possible to produce a Cassegrain system of about f/14 effective focal ratio with a primary as short as f/2.5, having a very short and convenient tube.

Maksutov telescopes have become popular because of their compactness and permanence, the reflective optical surfaces being protected from dirt and tarnish by the completely closed tube. A number of articles on the construction of this fascinating instrument have appeared, and a particularly useful publication produced by *Sky & Telescope* is detailed in the Bibliography. Catadioptric telescopes and wide-field cameras offer a whole new world of interest for the amateur optician, and one that has, as yet, been by no means fully explored.

Appendixes

Forthcoming Lunar Eclipses
1987–2000

DATE		TYPE	DATE		TYPE
1987	October 7	*Partial*	1994	May 25	*Partial*
1988	August 27	*Partial*	1995	April 15	*Partial*
1989	February 20	*Total*	1996	April 4	*Total*
	August 17	*Total*		September 27	*Total*
1990	February 9	*Total**	1997	March 24	*Partial*
	August 6	*Partial**		September 16	*Total**
1991	December 21	*Partial*	1999	July 28	*Partial*
1992	June 15	*Partial*	2000	January 21	*Total*
	December 10	*Total*		July 16	*Total**
1993	June 4	*Total*			
	November 29	*Total*			

*An asterisk indicates that the moon at mid-eclipse cannot be seen from anywhere in the United States.

Forthcoming Solar Eclipses
1987–2000

DATE		TYPE	REGION OF VISIBILITY
1987	March 29	*Total*	Mid-Atlantic Ocean
	September 23	*Annular*	U.S.S.R., China, Pacific Ocean
1988	March 18	*Total*	Indian Ocean, East Indies, Pacific Ocean
	September 11	*Annular*	Indian Ocean, Antarctic
1990	January 26	*Annular*	Antarctic
	July 22	*Total*	Finland, U.S.S.R., Pacific Ocean
1991	January 15/16	*Annular*	Australia, New Zealand, Pacific Ocean
	July 11	*Total*	Pacific Ocean, Brazil
1992	January 4/5	*Annular*	Pacific Ocean
	June 30	*Total*	South Atlantic Ocean
1994	May 10	*Annular*	Pacific Ocean, Mexico, U.S.A., Canada
	November 3	*Total*	Peru, Brazil, Atlantic Ocean
1995	April 29	*Annular*	Pacific Ocean, Peru, Brazil
	October 24	*Total*	Iran, India, East Indies
1997	March 9	*Total*	U.S.S.R.
1998	February 26	*Total*	Pacific Ocean, Panama, Atlantic Ocean
	August 22	*Annular*	Indian Ocean, East Indies, Pacific Ocean
1999	February 16	*Annular*	Indian Ocean, Australia, Pacific Ocean
	August 11	*Total*	Atlantic Ocean, Europe, Turkey, India

Elongations of the Inferior Planets
1987–2000

MERCURY

Evening			**Morning**		
DATE		ELONGATION (°)	DATE		ELONGATION (°)
1987	February 12	18	1987	March 26	28
	June 7	24		July 25	20
	October 4	26		November 13	19
1988	January 26	18	1988	March 8	27
	May 19	22		July 6	21
	September 15	27		October 26	18
1989	January 9	19	1989	February 18	26
	May 1	21		June 18	23
	August 29	27		October 10	18
	December 23	20			
1990	April 13	20	1990	February 11	25
	August 11	27		May 31	25
	December 6	21		September 24	18
1991	March 27	19	1991	January 14	24
	July 25	27		May 12	26
	November 19	22		September 7	18
				December 27	22
1992	March 9	18	1992	April 23	27
	July 6	26		August 21	18
	October 31	24		December 9	21
1993	February 21	18	1993	April 5	28
	June 17	25		August 4	19
	October 14	25		November 22	20
1994	February 4	18	1994	March 19	28
	May 30	23		July 17	21
	September 26	26		November 6	19
1995	January 19	19	1995	March 1	27
	May 12	22		June 29	22
	September 9	27		October 20	18

MERCURY *(cont'd)*

	Evening			**Morning**	
DATE		**ELONGATION (°)**	**DATE**		**ELONGATION (°)**
1996	January 2	20	1996	February 11	26
	April 23	20		June 10	24
	August 21	27		October 3	18
	December 15	20			
1997	April 6	19	1997	January 24	24
	August 4	27		May 22	25
	November 28	22		September 16	18
1998	March 20	18	1998	January 6	23
	July 17	27		May 4	27
	November 11	23		August 31	18
				December 20	22
1999	March 3	18	1999	April 16	28
	June 28	26		August 14	19
	October 24	24		December 2	20
2000	February 15	18	2000	March 28	28
	June 9	24		July 27	20
	October 6	26		November 15	19

VENUS

	EVENING ELONGATION	INFERIOR CONJUNCTION	MORNING ELONGATION	SUPERIOR CONJUNCTION
1987			January 15	August 23
1988	April 3	June 12	August 22	
1989	November 8			April 5
1990		January 18	March 30	November 2
1991	June 13	August 22	November 2	
1992				June 13
1993	January 19	April 1	June 10	
1994	August 25	November 2		January 17
1995			January 13	August 21
1996	April 1	June 10	August 19	
1997	November 6			April 2
1998		January 16	March 27	October 30
1999	June 11	August 20	October 31	
2000				June 10

Because of the almost perfect circularity of the orbit, elongations of Venus are always between 46° and 48°.

APPENDIX IV
Oppositions of Mars, Jupiter, and Saturn 1987–2000

MARS

Date		Diameter	Magnitude
1988	September 28	23·8	−2·6
1990	November 27	17·8	−2·0
1993	January 7	14·0	−0·9
1995	February 12	13.9	-0.8
1997	March 17	14.2	-1.0
1999	April 24	16.2	-1.4

JUPITER

Date		Diameter	Magnitude
1987	October 18	49·8	−2·5
1988	November 23	48·7	−2·4
1989	December 27	47·5	−2·3
1991	January 28	45·7	−2·2
1992	February 28	44·7	−2·1
1993	March 30	44.2	-2.0
1994	April 30	44.5	-2.0
1995	June 1	45.7	-2.1
1996	July 4	47.0	-2.2
1997	August 9	48.7	-2.3
1998	September 16	49.6	-2.4
1999	October 23	49.6	-2.5
2000	November 28	48.4	-2.4

SATURN

Date		Magnitude	
1987	June 9	+0·2	
1988	June 20	+0·2	Ring system at its widest,
1989	July 2	+0·2	north face presented
1990	July 14	+0·2	
1991	July 26	+0·3	
1992	August 7	+0·4	
1993	August 19	+0.5	
1994	September 1	+0.7	
1995	September 14	+0.8	Earth passes through ring plane
1996	September 26	+0.7	
1997	October 10	+0.4	
1998	October 23	+0.2	
1999	November 6	0.0	
2000	November 19	−0.1	

Minor Planet Data

No. & Name	Diam- eter (miles)	Max. Mag.	Orbital Period (years)	Inclina- tion (°)	Ampli- tude (mag.)	Fluctuation Period
						h m
(1) Ceres	480	7·0	4·6	10·6	0·04	9 05
(2) Pallas	304	6·3	4·61	34·8	0·13	5^h—6^h or 10^h— 12^h
(3) Juno	118	6·9	4·36	13·0	0·15	7 12·6
(4) Vesta	236	5·5	3·63	7·1	0·13	5 20·5 or 10 41·0
(6) Hebe	70	7·1	3·77	11·7	0·16	7 17
(7) Iris	78	6·7	3·69	5·5	0·29	7 05
(8) Flora	56	7·8	3·27	5·9	0·04	13 36
(9) Metis	78	8·1	3·69	5·6	0·26	5 04·6
(15) Eunomia	?	7·4	4·30	11·8	0·53	6 05·0
(20) Massalia	66	8·2	3·74	0·7	0·20	8 05·9

These are among other minor planets reaching magnitude 9·0 or brighter at a favorable opposition:

NO. & NAME	MAGNITUDE	NO. & NAME	MAGNITUDE
(5) Astraea	8·7	(42) Isis	8·8
(10) Hygeia	8·8	(43) Ariadne	8·8
(11) Parthenope	8·7	(44) Nysa	8·8
(12) Victoria	8·1	(97) Clotho	8·9
(14) Irene	8·7	(129) Antigone	8·9
(16) Psyche	8·8	(132) Aethra	8·5
(18) Melpomene	7·7	(192) Nausicaa	7·5
(19) Fortuna	8·7	(194) Procne	8·9
(23) Thalia	8·9	(216) Cleopatra	8·4
(25) Phocaea	9·0	(313) Chaldaea	9·0
(27) Euterpe	8·5	(324) Bamberga	7·4
(29) Amphitrite	8·6	(372) Palma	8·8
(39) Laetitia	8·8	(387) Aquitania	8·2
(40) Harmonia	8·9	(393) Lampetia	8·6
(41) Daphne	8·7	(433) Eros	6·5

Tables of Precession
for Ten Years

1. Precession in Right Ascension

R.A.—If N., read top; if S., read lower

Dec.	0,12ʰ	1,11ʰ	2,10ʰ	3,9ʰ	4,8ʰ	5,7ʰ	6ʰ
	m	m	m	m	m	m	m
80°	+0·51	+0·84	+1·14	+1·40	+1·60	+1·73	+1·77
70°	0·51	0·67	0·82	0·94	1·04	1·10	1·12
60°	0·51	0·61	0·70	0·78	0·84	0·88	0·90
50°	0·51	0·58	0·64	0·70	0·74	0·77	0·78
40°	0·51	0·56	0·61	0·64	0·67	0·69	0·70
30°	0·51	0·54	0·58	0·60	0·62	0·64	0·64
20°	0·51	0·53	0·55	0·57	0·58	0·59	0·59
10°	0·51	0·52	0·53	0·54	0·55	0·55	0·55
0°	0·51	0·51	0·51	0·51	0·51	0·51	0·51
	0,12ʰ	23,13ʰ	22,14ʰ	21,15ʰ	20,16ʰ	19,17ʰ	18ʰ

R.A.—If N., read top; if S., read lower

Dec.	18ʰ	19,17ʰ	20,16ʰ	21,15ʰ	22,14ʰ	23,13ʰ
	m	m	m	m	m	m
80°	−0·75	−0·70	−0·58	−0·38	−0·12	+0·19
70°	−0·10	−0·08	−0·02	+0·08	+0·21	0·35
60°	+0·13	+0·14	+0·18	0·24	0·32	0·41
50°	0·25	0·26	0·28	0·32	0·38	0·44
40°	0·33	0·33	0·35	0·38	0·42	0·46
30°	0·38	0·39	0·40	0·42	0·45	0·48
20°	0·43	0·43	0·44	0·45	0·47	0·49
10°	0·47	0·47	0·48	0·48	0·49	0·50
0°	0·51	0·51	0·51	0·51	0·51	0·51
	6ʰ	5,7ʰ	4,8ʰ	3,9ʰ	2,10ʰ	1,11ʰ

If positions are being converted to an earlier epoch, the signs must be reversed.

2. Precession in Declination

R.A. h	PREC. ′	R.A. h	PREC. ′
0,24	+3·3	7,17	−0·9
1,23	3·2	8,16	1·7
2,22	2·9	9,15	2·4
3,21	2·4	10,14	2·9
4,20	1·7	11,13	3·2
5,19	0·9	12	3·3
6,18	0		

If positions are being converted to an earlier epoch, the signs must be reversed.

APPENDIX VII

The Latitude of the Center of the Sun's Disk (B_0) Throughout the Year

DATE		B_0	DATE		B_0	DATE		B_0
Jan.	4	$-3 \cdot 5°$	May	4	$-3 \cdot 8°$	Sept.	1	$+7 \cdot 2°$
	9	4·0		9	3·2		6	7·3
	14	4·5		14	2·7		11	7·2
	19	5·0		19	2·1		16	7·1
	24	5·5		24	1·5		21	7·0
	29	5·9		29	0·9		26	6·9
Feb.	3	$-6 \cdot 2$	June	3	$-0 \cdot 4$	Oct.	1	$+6 \cdot 7$
	8	6·5		8	$+0 \cdot 3$		6	6·4
	13	6·8		13	0·9		11	6·1
	18	7·0		18	1·5		16	5·7
	23	7·1		23	2·0		21	5·3
	28	7·2		28	2·6		26	4·9
							31	4·4
Mar.	5	$-7 \cdot 3$	July	3	$+3 \cdot 2$			
	10	7·2		8	3·7	Nov.	5	$+3 \cdot 9$
	15	7·1		13	4·2		10	3·3
	20	7·1		18	4·7		15	2·8
	25	6·8		23	5·1		20	2·1
	30	6·6		28	5·5		25	1·5
							30	0·9
Apr.	4	$-6 \cdot 4$	Aug.	2	$+5 \cdot 9$	Dec.	5	$+0 \cdot 2$
	9	6·0		7	6·3		10	$-0 \cdot 4$
	14	5·6		12	6·6		15	1·0
	19	5·2		17	6·8		20	1·6
	24	4·8		22	7·0		25	2·2
	29	4·3		27	7·2		30	2·9

The positive and negative prefixes refer to north and south latitudes, respectively.

Sidereal Time at 00ʰ U.T.
Throughout the Year

The conversion is given for 8-day intervals. It can be calculated for intervening days by adding 3·9 minutes for every complete day. By remembering that a sidereal day consists of 23 hours 56 minutes, the sidereal time for any hour during the day can be calculated with sufficient accuracy for most purposes. The conversion is subject to an error of ± 2 minutes due to leap-year adjustments.

Date		Sidereal Time h m	Date		Sidereal Time h m	Date		Sidereal Time h m
Jan.	1	6 40	May	1	14 33	Sept.	6	22 58
	9	7 11		9	15 04		14	23 29
	17	7 43		17	15 36		22	0 01
	25	8 14		25	16 07		30	0 32
Feb.	2	8 46	June	2	16 39	Oct.	8	1 04
	10	9 17		10	17 11		16	1 35
	18	9 49		18	17 42		24	2 07
	26	10 21		26	18 14	Nov.	1	2 38
Mar.	6	10 52	July	4	18 45		9	3 10
	14	11 24		12	19 17		17	3 41
	22	11 55		20	19 48		25	4 13
	30	12 27		28	20 20	Dec.	3	4 44
Apr.	7	12 58	Aug.	5	20 51		11	5 16
	15	13 30		13	21 23		19	5 48
	23	14 01		21	21 54		27	6 19
				29	22 26			

Glossary

Aberration. The apparent annual displacement of a star caused by the "bending" of its light, due to the earth's orbital motion. Its effect is to make every star appear to revolve around a fixed point, its maximum distance from this point being about 20"·5. *Chromatic aberration:* The formation of a colored fringe around the image produced by a simple lens. *Spherical aberration:* Imperfect image caused by the lens or mirror not bringing all rays to a point.

Absolute magnitude. The brightness a star would appear to have if viewed from a standard distance (10 parsecs, or 32.6 light-years). It is therefore a measure of a star's actual luminosity.

Albedo. The ratio of light reflected to that received. Approximate albedos for the planets are: Mercury .06; Venus .65; Earth .39; Moon .07; Mars .15; Jupiter .42; Saturn .45; Uranus .46; Neptune .53. Some authorities consider the four last-named planets to be as reflective as Venus.

Almanac. A yearly publication containing relevant astronomical data. The principle one is *The American Ephemeris and Nautical Almanac;* in Great Britain this is published as the *Astronomical Ephemeris.* Other almanacs are the *Connaissance de Temps* (France) and the *Astronomisch-Geodatisches Jahrbuch* (Germany).

Altazimuth mount. A telescope mounted with axes in the horizontal and vertical planes.

Annular eclipse. A solar eclipse occurring with the moon near apogee, so that it appears smaller than the sun and cannot block it out completely.

Ansae. Term used to describe the eastern and western extremities of Saturn's rings as viewed from the earth.

Aphelion. The point on a planet's or comet's orbit at the greatest distance from the sun.

Apogee. The point on a satellite's orbit at the greatest distance from its primary.

Apparent magnitude. The brightness of a star as seen from the earth, a value dependent on both its absolute magnitude and its distance.

Appulse. A close approach of one celestial body to another, as seen from the earth. It is a line-of-sight effect, and does not imply physical proximity.

Asteroid. An alternative name for a minor planet.

449

Astronomical Unit. The mean distance from the earth to the sun, now taken as 92,900,000 miles.

Autocollimation. Optical test for a telescope mirror or object glass in which the light passes through the system twice, making it doubly sensitive.

Bailey's beads. A phenomenon occurring at the beginning and/or end of a total solar eclipse, when fragments of the photosphere shine out brilliantly through deep rifts in the moon's limb.

Barlow lens. A concave (negative) achromatic lens placed a short distance inside the focal point of the objective. This has the effect of increasing the image scale.

Barycenter. The point of the imaginary line joining two bodies under mutual gravitational influence, around which they revolve.

Binary system. Two or more stars revolving around each other.

Bode's law. An empirical guide to the relative distances from the sun of all the planets except Mercury, Neptune, and Pluto. It is obtained by taking the series 3, 6, 12, and so on, and adding 4 to each.

Cassegrain telescope. A reflecting telescope using a convex as well as a concave mirror. This increases the effective focal length and gives a large image scale.

Catadioptric telescope. A telescope which uses both reflection and refraction in the formation of its primary image.

Celestial sphere. An imaginary sphere surrounding Earth and carrying all the celestial objects. It rotates in 23 hours 56 minutes and is inscribed with the celestial equivalents of the terrestrial poles, equator, latitude, and longitude.

Chronograph. A device for recording the instant at which an observation is made.

Circumpolar. An object so close to the celestial pole that it remains permanently above the horizon.

Conjunction. Strictly speaking, the condition of two celestial bodies when their R.A. or Dec. become the same. In practice, it is equivalent to an appulse. For instance, a superior planet is in conjunction when it appears near the sun in the sky. An inferior planet is said to be in *inferior* or *superior* conjunction when it appears nearest to the sun on the near side and far side, respectively, of its orbit.

Constellation. One of the 88 defined regions of the celestial sphere.

Culmination. The condition of a celestial object when at its greatest possible altitude above the horizon. Unless very near the pole, this occurs when it is due south (to an observer in the northern hemisphere).

Cusp. A horn of the moon, Mercury, or Venus when in the crescent phase.

Declination (*Dec.*). The angular distance of a celestial body north (+) or south (—) of the celestial equator.

Diffraction. The scattering effect of light passing through a small aperture, or encountering a sharp boundary.

Dispersive power. A measure of the degree to which a sample transparent material will spread white light into its component colors.

Doppler effect. The shift of a source's spectral lines due to its motion toward or away from the observer.

Earthshine. Illumination of the moon's dark side due to sunlight reflected back from the earth.

Eclipse.   Passage of the moon wholly or partly across the sun, or the passage of a satellite wholly or partly through its primary's shadow.

Eclipsing variable.   A binary system at so great a distance that its individual components cannot be seen from the earth. The light of the "single" star fluctuates as one component is occulted by the other.

Ecliptic.   The apparent path of the sun around the celestial sphere, approximately marking the plane of the solar system.

Elongation.   The condition of an inferior planet when at its greatest angular distance from the sun.

Emersion.   The reappearance of an object after occultation.

Ephemeris.   A table giving the calculated future positions of a celestial body.

Ephemeris time.   A time system used in astronomical computing that ignores the slight irregularities in the earth's rotation upon its axis. During this century, the difference between "observed" and "predicted" rotation has so far amounted to more than 30 seconds.

Epoch.   Generally speaking, the date for which star positions (as on a chart) are correct.

Equation of time.   The discrepancy between the sun's southing and the instant of noon. This varies, because the sun does not move along the ecliptic in a regular manner. The sun is 14 minutes slow in February and 16 minutes fast in October-November.

Equatorial mount.   A telescope stand with one axis parallel to that of the earth, making it easy to follow the diurnal motion of a celestial body.

Equinoxes.   The two points at which the ecliptic crosses the celestial equator. The sun reaches these points in March and September.

Foucault test.   A test which reveals the optical contour of a concave mirror by examining the image of an illuminated pinhole at its center of curvature.

Gibbous.   Phase intermediate between half and full.

Greenwich Mean Time (G.M.T.).   Standard world time system, with 0 and 24 hours occurring at midnight on 0° longitude. This line passed through the old Greenwich Observatory in England. Usually referred to as U.T.(q.v.).

Gregorian telescope.   Mainly obsolete form of reflecting telescope, using a second concave mirror. Most "compound" telescopes use the Cassegrain system.

Halation.   The appearance of a star as a disk in a long-exposure photograph, due to its light spreading into the emulsion.

Hour angle.   The interval, measured in sidereal time, since a given celestial object was last on the meridian.

Hyperbola.   The path followed by many comets. Unlike an ellipse, it is not continuous; this means that a body traveling along such an orbit can never return to the vicinity of the sun.

Immersion.   The disappearance visually of a body when occulted.

Inclination.   The angle formed by the plane of one orbit with respect to another plane.

Inferior planet.   A planet whose orbit is smaller than that of the earth.

International Astronomical Union (I.A.U.).   A body coordinating the work of astronomers throughout the world. A number of committees have been established to specialize in various important departments of astronomy; these hold discussions at symposia held in various countries. I.A.U. conferences are held every three years.

Irradiation.   The apparent augmentation in size of a celestial body due to its brightness against the dark sky.

Julian date.   The number of days that have elapsed between the day in question and 1 January 4713 B.C. It is expressed in days and decimals of a day, and is used in much computing work. The selection of the original date is arbitrary.

Libration.   The axial swing of the moon with respect to the earth, due to its varying orbital velocity.

Light-year.   An arbitrary measure of distance, taken as the distance traveled by light in one terrestrial year. It is equal to 5,880,000,000,000 miles.

Limb.   The "edge" of the sun, moon, or a planet, as seen from the earth.

Lunation.   The period elapsing between successive similar phases of the moon, also known as the *synodic period.* It is roughly 29¾ days. The term also refers to the phase progress from new to new.

Magnitude.   The classification of a star's real (*absolute*) or *apparent* brightness.

Maksutov telescope.   A reflecting telescope using a spheroidal instead of paraboloidal main mirror. Before reaching this, the light passes through a concave lens almost as large as the mirror itself to remove the sperical aberration.

Meridian.   The great circle passing through the zenith and touching the horizon at the north and south points. The meridian of the sun, moon, or a planet is the straight line joining the north and south poles and passing across the center of the disk.

Meridian circle.   A telescope constrained so that it can be pointed only at the meridian. It is used to time the instant at which stars pass across the meridian and so to keep a check on the rotation of the earth.

Micrometer.   A device in the eyepiece of a telescope for measuring the angular dimensions of a celestial object.

Nadir.   The point on the celestial sphere directly beneath the observer.

Newtonian telescope.   A reflecting telescope using a concave paraboloidal mirror, with a small plane mirror to reflect the converging rays out of the tube.

New General Catalogue (N.G.C.).   Standard list of the brighter star clusters, nebulae, and galaxies, published in 1888 and listing more than 7,000 objects.

Node.   The apparent crossing of two paths, or orbits. An example is the intersection of the ecliptic and the celestial equator at the equinoxes.

Nutation.   A minute oscillation of the earth's axis, due to lunar perturbations, superimposed on the much more marked precession.

Objective. The focusing and light-gathering agent of the astronomical telescope, whether mirror or lens.

Occultation. The passage of a nearby celestial body in front of a more remote one.

Opposition. The condition of the moon or a planet when opposite the sun in the sky as seen from the earth.

Orbit. The path followed through space by a celestial body.

Parallax. The apparent displacement of a body against its background when seen from different stations.

Parsec. A distance of about 3˙26 light-years, at which the earth's orbit with a diameter of 186,000,000 miles, would subtend an angle of 1″.

Penumbra. The partly illuminated outer region of the shadow cast by a solid body from a light source of appreciable diameter.

Periastron. The point in their orbit at which the two components of a binary pair are closest to each other.

Perigee. The point on a satellite's orbit which is closest to its primary.

Perihelion. The point on a planet's or comet's orbit which is closest to the sun.

Period. The time taken for a planet, comet, or satellite to achieve one circuit of its orbit.

Personal equation. The fractional discrepancy between the observation of an instantaneous phenomenon (such as an occultation) and the recording of it.

Perturbation. The gravitational influence of a nearby mass, causing a body to deviate from its true path.

Phase. The percentage area of a body seen illuminated.

Phase angle. The angular distance between the sun and the earth as seen from the moon or a particular planet.

Position angle. The bearing of the fainter member of a double star measured from its primary. It is reckoned in degrees, starting at the north point and working counterclockwise.

Precession. A slow oscillation of the earth's axis which takes 25,900 years to complete. This has the effect of changing the celestial coordinates, though to a very small extent.

Quadrature. The condition of the moon or a superior planet when it subtends a right angle with the sun as seen from the earth.

Radial velocity. The speed at which a celestial body, particularly a star, appears to be moving toward or away from the earth.

Radiant. The point on the celestial sphere from which meteors appear to radiate during a shower.

Red shift. The displacement toward the red end of the lines in the spectrum of a distant galaxy. This is taken to imply recession of the galaxy.

Refractive index. A measure of the degree to which a sample of transparent material will bend a light ray entering the material from its original path.

Resolving power. The ability of a telescope to resolve, or separate, fine detail, such as the individual components of a close double star. The limiting angular distance is a direct function of aperture.

Retrograde motion.   Real or apparent motion of a planet, comet, or satellite in the opposite sense to that usual in the solar system. It can also be applied to binary stars.

Right ascension (R.A.).   The celestial equivalent of longitude, measured eastward from the spring or vernal equinox.

Saros.   An interval of roughly 18 years 10¼ days, after which the sun and moon are in almost exactly the same relative positions in the sky. It was an ancient method of predicting eclipses.

Sagitta.   Term given to the depth of curve of a convex or concave surface.

Scintillation.   More usually known as "twinkling," this is the flickering of a star when viewed with the naked eye, caused by heat currents in the atmosphere.

Secondary spectrum.   The residual color, usually blue, that cannot be brought to a common focus in an achromatic lens.

Seeing.   The steadiness of the telescope image, which is affected by atmospheric currents. Some observers rate it from 1 to 10, 1 being hopelessly bad and 10 unattainably good.

Shadow transit.   The passage of a satellite's shadow across the disk of Jupiter or Saturn.

Sidereal period.   The time it takes for a planet or satellite to achieve one circuit of its orbit, relative to a fixed point.

Sidereal time.   Time system based on the true rotation of the earth (relative to a fixed point and not to the sun), which takes only 23 hours 56 minutes.

Sleeks.   Fine smooth scratches formed on a glass surface during optical polishing, due to dirt particles.

Solar Time.   Time system based on the earth's rotation relative to the sun, the basis of all civil time reckoning.

Solstices.   The two points on the ecliptic farthest removed from the celestial equator; the sun is at these points at midsummer and midwinter.

Southing.   A celestial object's crossing of the meridian.

Spectroscope.   A device for dispersing the light received from a source into its component wave lengths. Examination of the light distribution in the spectrum can give information about the composition of the source.

Spherometer.   Device for indicating the radius of curvature of an optical surface by measuring the sagitta (q.v.).

Superior planet.   A planet whose orbit is larger than that of the earth.

Synodic period.   The time it takes for a planet or satellite to achieve one circuit of its orbit as seen from the earth.

Terminator.   The division between the illuminated and dark hemispheres of the moon or a planet.

Time zone.   A division on the earth's surface in which civil time is taken as a whole number of hours earlier or later than G.M.T.

Transit.   There are three meanings of the word. A star or planet transits when it Crosses the meridian; a detail on a planet's disk transits when it is carried across the planet's meridian by its rotation; and a satellite (or its shadow) transits when it crosses in front of its primary's disk.

Umbra. The region of the shadow, cast by a solid body, in which all direct light from the source is cut off.

Universal Time (U.T.). The 24-hour time system used by astronomers all over the world. The reckoning is the same as for G.M.T.

Vertex. The point on the limb of the sun, moon, or a planet that is highest above the horizon.

Wedge. Uneven edge thickness around a lens, due to the two surfaces not facing each other squarely.

Zenith. The point on the celestial sphere directly above the observer.

Zodiac. A zone 18° wide, centered along the ecliptic, inside which the major planets (except Pluto) and many minor planets are always to be found. The twelve constellations through which it passes are called the zodiacal constellations.

Bibliography

Perhaps the best way of obtaining a comprehensive grounding in all departments of astronomy is through the 8-volume series of extracts from articles published over the last thirty years in *Sky & Telescope*. The work is published by Macmillan (New York), and is known as the *Sky & Telescope Library of Astronomy*.

CHAPTER 1

Barlow, Boris V. *The Astronomical Telescope*. New York: Springer-Verlag, 1975. The emphasis is on the mechanical and optical design of large professional instruments.

Bell, Louis. *The Telescope*. New York: Dover Publications, 1981. A classic work on the development of the astronomical telescope, first published in 1922.

King, Henry C. *History of the Telescope*. New York: Dover Publications, 1980. A very detailed study, with extensive references. The standard work.

Land, Barbara. *The Telescope Makers*. New York: Thomas Y. Crowell, 1968. A historical account, written in a popular style.

Warner, Deborah J. *Alvan Clark & Sons, Artists in Optics*. Washington: Smithsonian Institution Press, 1968. An excellent account of the art of telescope-making during the great days of this famous family firm, 1844–1897.

Woodbury, David O. *The Glass Giant of Palomar*. London: Heinemann, 1940. Also: Wright, Helen. *Explorer of the Universe*. New York: Dutton, 1966. Two biographies of George Ellery Hale, the famous solar astronomer whose enthusiasm led to the building of a number of great telescopes, including the 200-inch reflector at Mount Palomar. Hale was always observing and constructing at the limits of current knowledge, and his pioneering achievements make fascinating reading.

CHAPTER 3

Howard, Neale E. *Standard Handbook for Telescope Making*. New York: Thomas Y. Crowell, 1959. An excellent introduction to mirror-making and the construction of a reflecting telescope.

Ingalls, Albert, ed. *Amateur Telescope Making* (3 vols.). This work, whose volumes first appeared between 1935 and 1953, form a classic assembly of features by amateur telescope-makers, and probably contains more useful information (some of it now rather dated) than does any other work.

457

Moore, P. A., ed. *Practical Amateur Astronomy.* London: Lutterworth, 1963. A useful survey of instrumentation and observing technique.

Muirden, James. *Beginner's Guide to Astronomical Telescope Making.* London: Pelham Books, 1978. An introduction to the manufacture of mirrors and lenses, with a description of a simple telescope mounting.

Paul, Henry E. *Telescopes for Skygazing.* Cambridge, Mass.: Sky Publishing Corp., 1966. A popular book written by a well-known amateur.

Sidgwick, J. B. *Amateur Astronomer's Handbook.* London: Pelham Books, 4th edition 1979. A standard work on the theory and function of optical instruments, together with treatment of photography, spectroscopy, physiology of the eye, discussion of errors, etc.

Texereau, J. *How to Make a Telescope.* New York: Interscience Publishers, 1963. The well-known French optician gives much useful advice both on making and mounting a mirror.

Thompson, Allyn J. *Making Your Own Telescope,* revised ed. Cambridge, Mass.: Sky Publishing Corp., 1973. A book which has influenced a whole generation of telescope-makers.

CHAPTER 4

Becvár, A. *Atlas of the Heavens* (formerly *Atlas Coeli*). Cambridge, Mass.: Sky Publishing Corp., 1977. This standby of most advanced amateurs covers the whole sky down to about mag. 7.7 on a scale of 7.5 mm/°. Nebulae, clusters, and galaxies are included, while doubles and variables are indicated. Also by Becvár are three other atlases on a scale of 2 cm/°, showing stars to about mag. 9.5, with their spectral class indicated by a color code: *Atlas Borealis* (Dec. +90° to +30°); *Atlas Eclipticalis* (Dec. + 30° to −30°); and *Atlas Australis* (Dec. −30° to −90°).

Lampkin, Richard H. *Naked Eye Stars.* Edinburgh: Gall & Inglis, 1972. A handy listing of the magnitude and position of 3,047 naked-eye stars, covering the whole sky down to about mag. 5.5.

Norton, Arthur P. *Norton's Star Atlas & Reference Handbook.* Edinburgh: Gall & Inglis, 17th edition 1978. A classic work, showing the whole sky down to the 6th magnitude, with tables and much other useful information. The best atlas for the newcomer to astronomy.

Papadopoulos, Christos. *True Visual Magnitude Photographic Star Atlas.* New York: Pergamon Press, 1979. A two-volume work covering the whole sky down to about mag. 13.5, on a scale of 3 cm/°, and containing 456 separate charts. It has the advantage over other photographic atlases (and hand-drawn charts derived from photographic magnitudes) that the star magnitudes correspond very closely to those sensed by the eye—normally, blue stars are shown too bright, and red stars too faint.

Roth, Gunter D., ed. *Astronomy: A Handbook.* Cambridge, Mass.: Sky Publishing Corp., 1975. A comprehensive survey of observational fields open to the amateur.

Sidgwick, J. B. *Observational Astronomy for Amateurs.* London: Pelham Books, 4th edition 1982. This long-established work discusses observational techniques in depth, and is aimed at the serious amateur with some experience behind him.

Tirion, Wil. *Sky Atlas 2000.0.* Cambridge, Mass.: Sky Publishing Corp., 1981. On a slightly larger scale than *Atlas of the Heavens* (see above), this is essentially an improved and updated version of the earlier work, showing stars down to mag. 8.0. The star positions are correct for the year 2000, unlike those in the other atlases listed here, which are for the epoch 1950.

Vehrenberg, Hans. *Photographic Star Atlas* and *Atlas Stellarum.* Cambridge,

Mass.: Sky Publishing Corp. These atlases were compiled in the 1960s by a well-known German amateur. The first shows stars down to about mag. 13, on a scale of 15 mm/°; the second has twice the scale, and descends to about mag. 15.

Periodicals

The Astronomical Almanac (formerly *Nautical Almanac, American Ephemeris, etc.*). Washington: U.S. Government Printing Office, annually. The standard work, containing ephemerides for the sun and moon, the major planets and their satellites, sunrise and sunset tables, and much other information.

Handbook (annually) and *Circulars* of the British Astronomical Association, Burlington House, Piccadilly, London W1. The *Handbook* gives ephemerides for the sun, planets, satellites, and comets, as well as listing current meteor showers, binary stars, etc. *Circulars* give notification of important discoveries.

IAU Circulars are issued by the Smithsonian Astrophysical Observatory, 60 Garden Street, Cambridge, Mass. 02138, which also transmits telegrams and telex messages worldwide.

Observer's Handbook (annually), issued by the Royal Astronomical Society of Canada, 252 College Street, Toronto 130, Ontario, is similar in scope to the BAA *Handbook* (above).

Sky & Telescope. Sky Publishing Corp., 49 Bay State Road, Cambridge, Mass., 02238, monthly. The international magazine of astronomy, with contributions from both amateurs and professionals, news notes, reviews, sky diary, etc. Highly recommended.

Yearbook of Astronomy. P. A. Moore, ed. London: Eyre & Spottiswoode, annually. Contains sky maps and observing notes for each month, as well as articles on recent developments in astronomy.

CHAPTER 6

Alter, Dinsmore. *Pictorial Guide to the Moon.* New York: Thomas Y. Crowell, 1973. Although rather dated, this book contains excellent photographs of the surface taken by various space missions.

Atlas & Gazetteer of the Near Side of the Moon. Washington: U.S. Government Printing Office, 1971. This NASA publication contains about 400 *Orbiter* photographs, and indicates every feature that had then received an official I.A.U. designation.

Cherrington, Ernest H. *Exploring the Moon Through Binoculars.* New York: McGraw-Hill, 1969. A description of the formations observable on a night-by-night bases throughout a complete lunation.

Kopal, Z. *A New Photographic Atlas of the Moon.* New York: Taplinger Publishing Co., 1971. A collection of about 200 photographs taken from both earth and space.

Lunar Orbiter Photographic Atlas of the Moon. Washington: U.S. Government Printing Office, 1971. A NASA publication, with 675 close-up photographs.

Meeus, J. & Mucke, M. *Canon of Lunar Eclipses, −2002 to +2526.* Vienna: Astronomisches Büro Wien, 1979. The standard work.

Moore, P. A. *A Survey of the Moon.* New York: Norton, 1974. Written by a well-known amateur, who emphasizes the work still open to the lunar observer.

Povenmire, Harold R. *Graze Observer's Handbook.* New York: Vantage Press, 1975. This book describes how to observe grazing lunar occultations, and of what use the results can be.

Wilkins, H. P. and Moore, P. A. *The Moon.* London: Faber & Faber, 2nd edition

1961. An excellent verbal description of the lunar features, accompanied by a 100-inch diameter map—probably the last great amateur map that will ever be produced.

CHAPTER 7

Baxter, W. M. *The Sun and Amateur Astronomy.* New York: Norton, 2nd edition. A useful but rather elementary introduction to solar observation, written by an amateur specialist.

Bray, R. J. and Loughhead, R. F. *Sunspots.* New York: Dover Books, 1979. Based on the authors' high-resolution photography, this book contains a great deal that will interest the serious amateur.

Eddy, John A. *A New Sun.* Washington: U.S. Government Printing Office, 1979. Although based primarily on data received by the orbiting *Skylab* satellite, this book should be read by all observers who wish for an up-to-date account of our knowledge of the sun.

Maag, R. C. *et al,* eds. *Observe and Understand the Sun.* Washington: Astronomical League, 1977. A small book, but full of practical advice for amateur projects.

von Oppolzer, T. *Canon der Finsternisse.* New York: Dover, 1963. This re-issue of an old classic gives information on both solar and lunar eclipses from -1207 to $+2163$.

CHAPTER 8

Baum, R. M. *The Planets: Some Myths and Realities.* Newton Abbot, England: David & Charles, 1973. Some interesting observing puzzles, such as a satellite of Venus and a bright ring around Uranus—both of which were once seriously suggested—are examined and explained. A book for the practical amateur as well as for the historian.

Chapman, Clark R. *The Inner Planets.* New York: Charles Scribner's Sons, 1977. Written by a planetologist, this book describes the methods and the results of space-age planetary study.

Moore, P. A. *The New Guide to the Planets.* New York; Norton, 1972. The emphasis in the book is on practical observation.

Roth, Gunter D. *Handbook for Planet Observers.* New York: Van Nostrand Reinhold Co., 1970. Although rather dated, the amateur will find its discussion of instruments and accessories particularly useful.

CHAPTER 16

Moore, P. A. *Guide to Mars.* New York: Norton, 1977. An up-to-date survey of the planet, with hints for the observer.

CHAPTER 12

Peek, B. M. *The Planet Jupiter.* London: Faber & Faber, 2nd edition 1980. A detailed description of each region of the planet, based on the observational work of the past century.

CHAPTER 13

Alexander, A. F. O'D. *The Planet Saturn.* New York: Dover Books, 1980. A reissue of this definitive survey of practically every published observation of the planet and its satellites made up to 1960.

CHAPTER 14

Alexander, A. F. O'D. *The Planet Uranus.* London: Faber & Faber, 1965. A definitive survey of practically every published observation of the planet and its satellites, including those made before its planetary nature was known.

Grosser, Morton. *The Discovery of Neptune*. Cambridge, Mass.: Harvard University Press, 1962. This book recounts the detective work that led to the discovery of this new outer planet, and the recriminations that followed.

Tombaugh, Clyde W. and P. A. Moore. *Out of the Darkness: The Planet Pluto*. Harrisburg, Pa.: Stackpole Books, 1980. A description, by the discoverer, of the arduous search that led to the detection of Pluto in 1930.

CHAPTER 16

Brown, P. L. *Comets, Meteorites and Men*. New York: Taplinger Publishing Co., 1974. Comet-hunters, in particular, will be interested in the historical anecdotes related by this English amateur.

Delsemme, A. H., ed. *Comets, Asteroids, Meteorites*. University of Toledo, 1977. Papers from an IAU colloquium present recent findings and discoveries.

Dubyago, A. D. *The Determination of Orbits*. New York: Macmillan, 1961. A well-known work on orbit calculation, though partly dated now by the widespread introduction of computers.

Ley, Willy. *Visitors from Afar*. New York: McGraw-Hill Book Co., 1969. A popular account of some famous comets and comet-hunters.

Richter, N. B. *et al*. *The Nature of Comets*. New York: Dover Books, 1963. An authoritative work, although now somewhat dated.

CHAPTER 17

Kresak, L., ed. *Physics and Dynamics of Meteors*. Reidel, 1968. A text for the serious student.

Povenmire, Harold R. *Fireballs, Meteors and Meteorites*. Florida: JSB Enterprises, 1980. Written by an enthusiastic observer for amateurs who want to undertake meteor observation.

Sears, D. W. *The Nature and Origin of Meteorites*. New York: Oxford University Press, 1978. A detailed discussion of both the orbits and the physical nature of meteorites.

CHAPTER 18

Aitken, R. G. *The Binary Stars*. New York: Dover Books, 1963. Aitken was one of the most assiduous of all double-star observers, and this book, first published in 1935, is a classic.

Couteau, Paul. *Observing Visual Double Stars*. Cambridge, Mass.: MIT Press, 1981. Information on the measurement of double stars and the calculation of their orbits, together with extensive lists of objects, written by an experienced professional astronomer.

Glyn Jones, K. *Messier's Nebulae and Star Clusters*. London: Faber & Faber, 1968. A scholarly compendium, discussing the discovery and significance of each object, and including with each one a pair of maps to aid location and identification.

Glyn Jones, K., ed. *Webb Society Deep-Sky Observer's Handbook*. Hillside, N.J.: Enslow Publishers. A four-volume work for the practical amateur. Vol. 1: *Double Stars;* Vol. 2: *Planetary and Gaseous Nebulae;* Vol. 3: *Open and Globular Clusters;* Vol. 4: *Galaxies*. All contain extensive lists, charts, drawings, and other useful information.

Hartung, E. J. *Astronomical Objects for Southern Telescopes*. Cambridge, Mass.: Cambridge University Press, 1968. An Australian amateur provides lists of the often neglected southern celestial objects, based on his own observations.

Mallas, J. H. and Kreimer, E. *The Messier Album*. Cambridge, Mass.: Sky Publishing Corp., 1978. A description of each Messier object, accompanied by a photograph and a drawing.

Peltier, Leslie C. *Guideposts to the Stars*. New York: Macmillan, 1972. One of America's greatest amateur observers describes his method of finding a way around the sky.

Vehrenberg, Hans. *Atlas of Deep-sky Splendours*. Cambridge, Mass.: Sky Publishing Corp., 3rd edition 1978. Both N.G.C. and Messier objects are shown in small-scale (3 cm/°) and close-up photographs.

Webb, T. W. *Celestial Objects for Common Telescopes*. New York: Dover Books, 1962. A reissue of the famous classic, in two volumes. Vol. 2 (The Stars) is particularly valuable for its long lists of stellar objects. It is in need of extensive revision, however, since the last edition was in 1917.

CHAPTER 19

Kukarkin, B. V. *et al. General Catalogue of Variable Stars*. Moscow, 3rd edition, 1970. The international variable-star catalogue, containing information on 20,437 stars, extended and updated in subsequent supplements.

Scovil, Charles E. *The AAVSO Variable Star Atlas*. Cambridge, Mass.: Sky Publishing Corp., 1980. Covering the whole sky to about mag. 9, this atlas shows the location of numerous variable-star fields, with comparison stars identified, as well as indicating the positions of deep-sky objects. It is therefore useful as a general sky map.

CHAPTER 20

Burnham, Robert, *Burnham's Celestial Handbook* (3 vols.). New York: Dover Publications, 1978. A detailed survey of each constellation in the sky, containing lists of double and variable stars, as well as deep-sky objects and general notes. One of the most comprehensive sky-guides now available.

Muirden, James. *Astronomy with Binoculars*. New York: Thomas Y. Crowell, 1979. This book includes a constellation by constellation description of the interesting objects that are observable with binoculars or a very small telescope.

CHAPTER 21

Paul, Henry E. *Outer Space Photography*. New York: Amphoto, 1976. A well-known introduction to the subject.

Rackham, Thomas W. *Astronomical Photography at the Telescope*. London: Faber & Faber, 1972. A useful handbook of photography for the amateur.

Index